SPRINGER
LAB MANUAL

Springer-Verlag Berlin Heidelberg GmbH

Michael T. Rogan (Ed.)

Analytical Parasitology

With 24 Figures

Springer

Dr. MICHAEL T. ROGAN
Department of Biological Sciences
University of Salford
Salford, M5 4WT
UK

ISBN 978-3-642-47810-9   ISBN 978-3-642-60345-7 (eBook)

DOI 10.1007/978-3-642-60345-7

Library of Congress Cataloging-in-Publication Data. Analytical parasitology/Michael T. Rogan (ed.). p. cm. – (Springer lab manual) Includes bibliographical references and index. ISBN 3-540-58919-8 (alk. paper) 1. Diagnostic parasitology – Laboratory manuals. I. Rogan, Michael T., 1960– . II. Series. QR255, A43 1997 616.9'6 – dc20 96-33385

Originally published by Springer-Verlag Berlin Heidelberg New York in 1997

Cover Design: Design & Production GmbH, Heidelberg

Typesetting: Scientific Publishing Services (P) Ltd, Madras

SPIN: 10469573   39/3137/SPS – 5 4 3 2 1 0 – Printed on acid-free paper

In memory of Sophie North, Dunblane, 13 March 1996

Preface

Parasitic diseases still affect millions of people every year, especially in the tropics, causing considerable morbidity or death. Such infections within livestock are probably an even bigger problem, leading to poorer productivity, condemnation of infected meat and considerable economic loss. Parasitological research has, however, helped the situation in some cases and the development of novel drugs, vaccines and diagnostics has improved our chances of controlling these diseases.

Research into parasitic infections is, therefore, often goal-orientated. However, the study of parasites and host/parasite relationships still remains one of the most exciting and interesting aspects of biology. Scientists, from undergraduate students to research professors, frequently ponder over how endoparasitic organisms can survive within the most alien of environments – inside another organism. The nutritional, reproductive and survival strategies which have evolved within each group of parasites have allowed the development of highly specific host-parasite relationships and allow the successful transmission of the parasite from one host to another. A considerable amount of research is therefore directed at improving our understanding of various aspects of parasite biology.

This manual is designed to provide an experimental approach to analytical parasitology and is focused on obtaining information on the nature, function and location of significant parasite components. It is largely aimed at post-graduate and post-doctoral scientists and will provide protocols for some basic techniques in immunological, microscopical and molecular analysis along with more specialized approaches such as analysis of proteinases and other excretory/secretory substances, membrane biology and neurobiology. Clearly, there are good, more extensive, laboratory manuals covering each of these approaches individually, but the production of a multidiciplinary manual will give researchers a good grounding in experimental parasitology and allow the development of more than one line of investigation within a project.

Salford, June 1996   MICHAEL T. ROGAN

Contents

Chapter 1 Biochemical Pathways in Parasites
JOHN BARRETT 1

1.1   Introduction 1
1.2   Preparation of Enzyme Fractions 1
1.3   Enzyme Assays 6
1.4   Metabolite Measurements 15
1.5   Use of Inhibitors 27
      References 29

Chapter 2 Electrophoresis of Parasite Proteins
EILEEN DEVANEY 32

2.1   Introduction 32
2.2   Lysis of Parasites and Extraction
       of Proteins 32
2.2.1 Solubilisation in SDS Sample Cocktail 33
2.2.2 Detergent Solubilisation 33
2.3   SDS-PAGE 36
2.4   Modifications of Standard Method 47
2.4.1 Gradient Gels 47
2.4.2 Tricine Gels for the Analysis
       of Very Low Molecular Weight Polypeptides . . 48
2.5   Applications of SDS-PAGE
       to the Characterisation
       of Individual Polypeptides 50
2.5.1 Glycoprotein Analysis
       Using Specific Enzymes 51
2.5.2 Peptide Mapping 53
2.5.3 Preparative SDS-PAGE and Electroelution . . . 55
2.6   2D Gel Electrophoresis 57
2.6.1 Non-Equilibrium pH Gradient Electrophoresis
       (NEPHGE) 59

2.6.2   Isoelectric Focusing (IEF) 63
        References 65

**Chapter 3   Molecular Techniques
in Analytical Parasitology**
STEVEN HEATH 66

3.1   Introduction 66
3.2   Isolation of Parasite Nucleic Acid 66
3.2.1   Preparation of DNA 67
3.2.2   Preparation of RNA 68
3.2.3   Quantification of Nucleic Acid 69
3.3   Gel Electrophoresis and Blotting
        of Nucleic Acid 69
3.3.1   Gel Electrophoresis 69
3.3.2   Blotting of Nucleic Acid 72
3.4   Labelling and Hybridisation
        of DNA Probes 73
3.4.1   Radiolabelling of DNA Probes 73
3.4.2   Non-Isotopic Labelling of DNA Probes 75
3.4.3   Hybridisation and Signal Detection 76
3.5   Polymerase Chain Reaction (PCR) 78
3.5.1   Reverse Transcriptase – PCR (RT-PCR) 79
3.5.2   Touchdown PCR 81
        References 82

**Chapter 4   The Production and Analysis of Helminth
Excretory-Secretory (ES) Products**
ALAN BROWN, GARY GRIFFITHS,
PETER MICHAEL BROPHY,
BARBARA ANNE FURMIDGE,
and DAVID IDRIS PRITCHARD 83

4.1   Introduction 83
4.2   Collection of Excretory-Secretory (ES) Products 84
4.3   Acetylcholinesterase 91
4.3.1   Quantitation of Acetylcholinesterase Activity.. 92
4.3.2   Detection of AChE Activity
        Following Polyacrylamide Gel Electrophoresis . 95
4.3.3   Purification of AChE 95
4.3.4   Identification and Separation
        of AChE Molecular Forms 98
4.3.5   Kinetic Properties of AChE 99

4.4   Proteolytic Enzymes 99
4.5   Anticoagulants, Blood Coagulation Assays ... 100
4.5.1   The Collection of Plasma 101
4.5.2   Recalcification Time Test (Intrinsic Pathway) . 103
4.5.3   Activated Partial Thromboplastin Test
         (Intrinsic Pathway) 104
4.5.4   Prothrombin Time Test (Extrinsic Pathway) .. 105
4.5.5   Stypven Clotting Time Test
         (Common Pathway) 105
4.5.6   Platelet Aggregation 106
4.6   Potential Immunomodulators 110
4.6.1   Glutathione s-Transferases 110
4.6.2   Superoxide Dismutase (SODs) 115
4.6.3   Detection of Immunomodulation
         of Antibody Production 117
4.6.4   Neutrophil Inhibitory Factor (NIF) 120
4.7   Concluding Remarks 124
         References 125

Chapter 5 Parasite Proteinases

         MICHAEL J. NORTH 133

5.1   Introduction 133
5.2   Sample Preparation 138
5.3   Substrates and Proteinase Assays 139
5.3.1   Optimal Conditions for Proteinase Activity .. 139
5.3.2   Protein and Peptide Substrates 140
5.3.3   Synthetic Fluorogenic
         and Chromogenic Substrates 148
5.3.4   Exopeptidase Substrates 154
5.4   Proteinase Inhibitors 158
5.4.1   Standard Procedures for Enzyme Inhibition .. 158
5.4.2   The Selection and Design
         of Specific Inhibitors 160
5.5   Electrophoretic Methods 161
5.5.1   Substrate-SDS-PAGE – General Procedure ... 161
5.5.2   Substrate-SDS-PAGE – Additional Techniques 167
5.5.3   Affinity Labelling 168
5.6   Purification Methods 170
5.7   Endogenous Inhibitors 177
5.8   Final Comments 178
         References 180

Chapter 6 Neurobiology of Helminth Parasites
AARON G. MAULE, DAVID W. HALTON,
TIMOTHY A. DAY, RALPH A. PAX,
and CHRIS SHAW 187

6.1   Introduction 187
6.2   Immunocytochemical Localisation
        of Neuroactive Substances
        in Helminth Parasites 187
6.2.1   Production of Antisera 188
6.2.2   Immunocytochemistry Protocol 191
6.2.3   Confocal Scanning Laser Microscopy (CSLM) . 197
6.2.4   Electron Microscope Immunogold Labelling . . 198
6.3   Characterisation and Purification
        of Neuropeptides 200
6.3.1   Tissue Extraction 201
6.3.2   Peptide Detection 205
6.3.3   Primary Purification Steps 207
6.3.4   Reverse-Phase HPLC 208
6.3.5   Endoproteinase Digestion 213
6.4   Physiological Action of Transmitters
        and Drug Assessment In Vitro 215
6.4.1   Recording of Motor Activity
        in Whole Animals or in Muscle Strips 215
6.4.2   Action of Transmitters and Drugs
        at the Cellular Level 219
        References 224

Chapter 7 Electron Microscopy in Parasitology
ANDREW HEMPHILL and SIMON L. CROFT 227

7.1   Introduction 227
7.2   Preparation for Transmission Electron
        Microscopy (TEM) 228
7.2.1   Fixation: General Background 229
7.2.2   Standard Fixation Methods 233
7.2.3   Standard Fixation Protocols 234
7.3   Cytochemistry and Staining Protocol in TEM . 239
7.3.1   Introduction 239
7.3.2   Standard Stains 240
7.3.3   Special Cytochemical Stains 241
7.3.4   Enzyme Stains 242

7.4   Preparation for Scanning Electron
        Microscopy (SEM) 243
7.5   Immunogold-Labelling Techniques 245
7.5.1   Basic Information 245
7.5.2   Approaches for Immunogold-Labelling 251
7.5.3   Applications of Immunogold EM Techniques . 262
        References 263

Chapter 8 **Studies of the Surface Properties,
           Lipophilic Proteins and Metabolism
           of Parasites by the Use of Fluorescent
           and "Caged" Compounds**
           J. MODHA, C.A. REDMAN, S. LIMA,
           M.W. KENNEDY, and J. KUSEL 269

8.1   Introduction 269
8.2   Incorporation of Fluorescent Compounds
        into Membranes 270
8.2.1   Surface Membrane Incorporation
        of Fluorescent Compounds 270
8.2.2   Examination of Location of Fluorescent
        Probe in Membrane 276
8.2.3   Uptake of Fluorescent Fatty Acids
        and Cholesterol 281
8.2.4   Labelling Parasite Surfaces
        with a "Stable" Cell-Labelling Reagent
        (Sigma, PKH2 and PKH26) 282
8.3   Using Fluorescent Techniques
        to Measure Membrane Fluidity 284
8.4   Distribution of Actin 287
8.5   Assessment of Parasite Membrane Damage .. 288
8.6   Localisation of Probes in Organelles 290
8.7   Measurement of pH at the Surface
        of Parasites and Intracellular pH 291
8.8   The Use of Caged Compounds 292
8.9   Use of Fluorescent Probes in the Analysis
        of Parasite Proteins 295
        References 301

**Chapter 9 Detection and Identification
of Parasite Surface Carbohydrates
by Lectins**
GEORGE A. INGRAM 304

9.1   Introduction 304
9.2   Lectin-Mediated Trypanosomatid
         Parasite Agglutination 305
9.3   Fluorescein-Isothiocyanate
         (FITC)-Conjugated Lectins 307
9.4   Peroxidase-Labelled Lectins 310
9.5   Biotinylated Lectins 310
9.6   Fixatives and Enzymatic Treatments 311
9.7   Histological Sections 312
9.8   Controls 313
         References 317

**Chapter 10 Immunological Analysis
of Parasite Molecules**
MICHAEL T. ROGAN 320

10.1   Introduction 320
10.2   Parasite Extracts 321
10.3   Production of Antibodies
         to Parasite Molecules 323
10.3.1   Production of Polyclonal Antibodies 324
10.3.2   Production of Monoclonal Antibodies 328
10.3.3   Labelling of Antibodies 336
10.4   Enzyme Linked Immunosorbent Assays
         (ELISA) 338
10.4.1   The Basic Technique 340
10.4.2   Optimizing the Assay 341
10.4.3   Capture ELISA 345
10.5   Immunoblotting 347
10.5.1   The Basic Technique 348
10.5.2   Probing the Blot with Antibody 349
10.6   Affinity Chromatography 353
10.6.1   Antibody Purification 353
10.6.2   Antigen Purification 355
         References 358

Subject Index 361

List of Contributors

JOHN BARRETT
Institute of Biological
Sciences
The University of Wales
Aberystwyth
Dyfed, Wales SY23 3DA
UK

PETER MICHEL BROPHY
Department of
Pharmaceutical Sciences
De Montfort University
Leicester
LE1 9BH
UK

ALAN BROWN
Department of Life Sciences
University of Nottingham
Nottingham NG7 2RD
UK

SIMON CROFT
Department of Medical
Parasitology
Electron Microscopy
and Histopathology Unit
London School of Hygine
and Tropical Medicine
Keppel Street
London WC1E 7HT
UK

TIMOTHY A. DAY
Department of Zoology
Michigan State University
East Lansing, MI 48824
USA

EILEEN DEVANEY
Department of Veterinary
Parasitology
Veterinary School
University of Glasgow
Bearsden Road
Glasgow G61 1QH
UK

BARBARA ANNE FURMIDGE
Department of Life Sciences
University of Nottingham
Nottingham NG7 2RD
UK

GARY GRIFFITHS
Department of Life Sciences
University of Nottingham
Nottingham NG7 2RD
UK

DAVID W. HALTON
Comparative
Neuroendocrinology
Research Group
Schools of Clinical Medicine,
Biology and Biochemistry
The Queen's University
of Belfast
Belfast BT7 1NN
UK

STEVEN HEATH
Department of Biological
Sciences
University of Salford
Salford M5 4WT
UK

ANDREW HEMPHILL
Institute of Parasitology
Faculty of Veterinary
Medicine and Faculty
of Medicine
University of Bern
Langgasse-Strasse 122
3012 Bern
Switzerland

GEORGE A. INGRAM
Department of Biological
Sciences
University of Salford
Salford M5 4WT
UK

M.W. KENNEDY
Department of Infection
and Immunity
I.B.L.S
University of Glasgow
Glasgow
UK

J. KUSEL
Division of Biochemistry
and Molecular Biology
I.B.L.S
University of Glasgow
Glasgow
UK

S. LIMA
UFMG
Belo Horizonte
Brazil

AARON MAULE
The Queen's University
of Belfast
School of Biology
and Biochemistry
Belfast BT7 1NN
UK

J. MODHA
Division of Biochemistry
and Molecular Biology
University of Glasgow
Glasgow G12 8QQ
UK

MICHAEL J. NORTH
Department of Biological
and Molecular Sciences
University of Stirling
Stirling FK9 4LA
UK

RALPH A. PAX
Department of Zoology
Michigan State University
East Lansing, MI 48824
USA

DAVID IDRIS PRITCHARD
Department of Life Sciences
University of Nottingham
University Park
Nottingham NG7 2RD
UK

C.A. REDMAN
Division of Biochemistry
and Molecular Biology
I.B.L.S
University of Glasgow
Glasgow
UK

MICHAEL T. ROGAN
Department of Biological
Sciences
University of Salford
Salford M5 4WT
UK

CHRIS SHAW
Comparative
Neuroendocrinology
Research Group
Schools of Clinical Medicine,
Biology and Biochemistry
The Queen's University
of Belfast
Belfast BT7 1NN
UK

Chapter 1

Biochemical Pathways in Parasites

John Barrett

1.1
Introduction

The unambiguous demonstration of a metabolic pathway requires a number of criteria to be satisfied. All of the proposed enzymes in the pathway must be present in the tissue, in the correct cellular compartment and with sufficient activity to account for the observed rates of flux. When labelled substrates are added, the intermediary metabolites should become labelled in the correct sequence and in the expected positions. Finally, specific inhibition of one or more of the intermediate enzymes should result in a decrease in the overall flux rate and the accumulation of upstream metabolites. These conditions are rarely, if ever, met, and all too often the demonstration of one or two key enzymes is taken as being sufficient to indicate the presence of a complete pathway.

In general, more attention has been paid in parasites to the enzymes involved in energy metabolism than synthetic reactions. Catabolic enzymes usually have a high specific activity, making assay and purification easier, whilst synthetic enzymes are often under tight metabolic control. This chapter will deal with enzyme assays, metabolite measurements and the use of inhibitors.

1.2
Preparation of Enzyme Fractions

The preparation of enzyme fractions from helminths presents a number of problems, the main ones being the small size of many parasites and the problem of contamination with host material. With helminths, except for some of the larger nematodes, it is not usually possible to isolate different tissues. So one is dealing with whole-organism homogenates, with enzymes originating from different tissues, and different tissues can have very different enzyme profiles. In adult female parasites, for example, there may be a major biochemical contribution from eggs or larvae in the uterus. Taking only specific parts of the parasite can enhance the contribution of specific tissues, e.g. anterior end for nervous tissue, central section of *Fasciola* for Mehlis' gland. Host contamination can be minimized by repeated washing in extraction media prior to homogenization. In digeneans, incubation in Tyrode's solution for 1–2 h at 37 °C can be used to induce them to void their gut contents. In nematodes, gut bacteria have been identified as a problem in studying certain synthetic pathways (Shahkolahi and Donahue 1993). High levels of antibiotics and antimycotics (up to 1%) are often included in incubation media to suppress bacterial and fungal growth. However, care must be taken to ensure that the antibiotics or antimycotics do not interfere with the pathway under study. In vitro culture is one way of overcoming the difficulty of obtaining large numbers of parasites; however, care must be taken to ensure that the metabolism of the cultured forms does not differ substantially from that of the parasitic forms. Where host contamination is impossible to avoid, it may be necessary to compare isoenzyme patterns between parasite and host tissues in order to determine the extent of contamination. Some helminths contain large amounts of lipid which can interfere with enzyme assays, and this can be effectively removed by filtering homogenates through glass wool.

When surveying a range of enzymes, it is usual to use whole homogenates, rather than attempting to purify each individual enzyme. Using whole homogenates allows a more direct comparison between enzyme activities, and in some pathways the enzymes show characteristic constant activity ratios (Crabtree and Newsholme 1972). Purification can also lead to the loss of important cellular regulators. However, in vitro, enzymes are assayed under substrate-saturating conditions and, at their optimum pH, this does not necessarily reflect their in vivo activities. In cells, the protein concentration is 100–300 mg/ml, in in vitro assays the protein level is usually 0.1–1 mg/ml. Also in crude homogenates the presence of other enzymes can interfere with specific assays or result in unacceptably high control values.

Where possible, enzyme fractions should be prepared from fresh material. If material has to be stockpiled before use, storage in liquid N_2 is recommended. Enzyme activity is preserved better in frozen tissue than in frozen homogenates. Repeated cycles of freezing and thawing must be avoided, so it is better to freeze material in a series of small aliquots.

Tissues are normally homogenized in an approximately neutral (pH 7) buffer, of relatively low ionic strength, containing protecting agents such as EDTA to chelate heavy metal ions and sulphydryl protecting agents such as mercaptoethanol or dithiothreitol. A useful general homogenizing medium is: **Extraction buffers**

10 mM Triethanolamine-HCl, pH 7.6
 1 mM EDTA
 2 mM Dithiothreitol

Volume of tissue to extraction buffer 1:10 or 1:5. Sodium or potassium phosphate buffers give good general protection for most enzyme systems and are extensively used in extraction media. Tris buffers can be inhibitory (for example in glycosidase reactions) particularly at pH values above 8, and are best avoided. Some enzyme systems (for example, enolase and many of the enzymes involved in CoA metabolism) are inhibited by sodium ions; for these, use organic buffers, or add 1 M KCl to the medium to overcome the inhibition.

If proteases are a problem, 1 mM PMSF (phenylmethylsulphonate) can be added to the extraction buffer, phosphatases in tissue extracts can be inhibited by 20 mM NaF. The extraction of membrane-bound enzymes can be improved by including 0.1% Triton X-100 in the extraction buffer.

A variety of specialized extraction media has also been developed for particular enzyme assays, in order to optimize activity; however, these have usually been optimized for mammalian enzymes, not parasite enzymes.

Large pieces of tissue should be minced before homogenization. The homogenizer and extraction buffer must be **Homogenization** kept cold on ice and, if the homogenizer is hand-held, wrapped in a towel to prevent warming and to protect the hand in case of breakage. Glass homogenizers with a glass pestle will homogenize most tissues, including nematodes. Glass/PTFE homogenizers are suitable for softer tissues and the preparation of sub-cellular fractions. A Waring blender can be used for really large samples.

Sonication of the homogenate can help solubilize membrane-bound proteins. Sonication on its own can also be used to break up small organisms. When using sonication or power-driven homogenizers, care must be taken not to let the homogenate overheat.

Some parasitic stages, such as eggs or cysts, can be extremely difficult to disrupt, and it may be preferable to hatch or excyst the stage first. A French pressure cell has been found very effective in disrupting *Ascaris* eggs; grinding with carborundum powder has been used for protozoan cysts. Other techniques include freeze thawing and membrane disruption with organic solvents. Finally, freeze clamping, followed by a percussion mortar, will reduce even the most recalcitrant tissues to a fine powder (see Sect. 1.4).

Before use, crude homogenates should be centrifuged at 1000 *g* for 10 min to remove debris and undamaged parasites.

Acetone powders ("dry homogenates") are used for some enzymes (e.g. arylamine acyltransferase). The tissue is homogenized in 10 vol of acetone, precooled to −20 °C. The homogenate is filtered by suction in a Buchner funnel and the material washed with a further 10 vol of cold acetone. The "cake" is broken up by hand and allowed to air dry. Freeze-clamped material (see Sect. 1.4) is a good starting point for preparing acetone powders.

Note: Acetone is extremely inflammable and great care must be taken at all stages.

Subcellular fractions The basic technique is homogenization in an isotonic medium followed by differential centrifugation. A basic protocol for helminths is described.

1. Homogenize in
   0.22 M mannitol
   70 mM sucrose

2 mM HEPES buffer, pH 7.4
1 mM EDTA
0.2 mM Dithiothreitol
0.5 g/l bovine serum albumin

2. Centrifuge at 1000 g for 10 min at 4 °C to remove nuclei and large cell debris. Wash pellet by resuspending in extraction buffer (use an homogenizer to disperse pellet) and recentrifuge. Take resulting supernatant and add to the original 1000 g supernatant fraction.

3. Centrifuge supernatant at 14 000 g for 15 min at 4 °C to give mitochondrial pellet. Pellet washed as before and washings added to the 14 000 g supernatant.

4. Supernatant recentrifuged at 100 000 g for 1 h at 4 °C to give a microsomal pellet and a cytosolic fraction. Pellet washed as before and washings added to the cytosolic fraction.

Helminth mitochondria require slightly higher g forces and slightly longer centrifugation times for sedimentation than mammalian mitochondria, and often show a greater degree of size heterogeneity. A simple homogenizing medium for the preparation of *Ascaris* mitochondria is:

0.24 M sucrose
0.5 mM EDTA
0.15% bovine serum albumin,
adjusted to pH 7

If it is important to avoid organic compounds in the extraction medium, the tissue can be homogenized instead in isotonic (0.15 M) KCl. Details of methods to prepare glycosome microbodies from trypanosomatids can be found in Hort and Opperdoes (1984), and for the preparation of tegumental membranes from *Hymenolepis diminuta* in Knowles and Oaks (1979).

The purity of the cellular fractions should be checked by electron microscopy or by marker enzymes. If possible, two marker enzymes should be used for each fraction but the distribution of enzymes in helminths may not, of course, be the same as in mammals. For example, NAD-linked malic enzyme is a good marker for *Ascaris* mitochondria, but in mammals it is a cytoplasmic enzyme. Some useful marker enzymes are lactate dehydrogenase for cytosol, succinic dehydrogenase and cytochrome oxidase for mitochondria, and glucose-6-phosphatase for microsomes. Glucose-6-phosphatase activity, however, is often extremely low in helminths, but there is no other suitable marker. Alkaline phosphatase and phosphodiesterase I are good markers for platyhelminth tegumental membranes.

1.3
Enzyme Assays

Enzyme activity is determined by measuring either the rate of disappearance of substrate or the rate of appearance of product. Both methods are susceptible to error in crude homogenates because of the activities of contaminating enzymes removing the substrate or product. Measurement of the rate of appearance of product is probably the least prone to error. Enzyme assays can be followed either continuously (kinetic) or discontinuously by incubating for a fixed time, stopping the reaction, and then analyzing the reaction mixture. Kinetic assays are to be preferred, as they provide more information.

It is important, in all enzyme assays, to have adequate controls. One control should be minus enzyme, the other minus substrate; if more than one substrate is involved, there should be separate controls for each component. This will eliminate spurious results due to non-enzymatic reactions, and interfering enzyme reactions. For example, tissues frequently contain quite active NADH oxidases, which will interfere with any NAD-linked enzyme assay (these oxidases can be inhibited by the addition of 1 mM KCN or 1 μg/ml oligomycin to the assay mixture). Other controls may be based on plus/minus a specific activator or inhibitor. To check that there are no non-enzymatic components in the enzyme extract which are effecting the assay, a boiled enzyme control can be used. Haem compounds, for example, can show pseudoperoxidase activity and also act as superoxide dismutases. However, boiling causes the protein to flocculate, and this can interfere with spectrophotometric assays. In these cases, chemical inactivation, e.g. with acid or hydrogen peroxide, is better. In many recording spectrophotometers, the blank (cuvette 1) is automatically subtracted from the others. If the control cuvette is put in this position an automatic correction is made. However, it can be useful to know what is actually going on in the control – is there, for example, a high non-enzymatic rate or a significant reaction in the absence of substrate? For this reason it is a good idea to have just buffer in the blank position and have the chemical and enzyme controls in the experimental positions.

There are a number of different ways of expressing enzyme activity. The most frequently used are:

- Specific activity = μmoles of substrate utilized min^{-1} mg enzyme protein^{-1}

- International unit (IU) = amount of substrate which will catalyze the utilization of 1 μmol of substrate min^{-1}

- SI unit (katal) = amount of enzyme which will catalyse the utilization of 1 mol s^{-1}. Therefore 1 μmol min^{-1} = 1 IU = 16.67 nkat

- Molecular activity = moles of substrate utilized min^{-1} mole of enzyme^{-1}

- Catalytic centre activity = moles of substrate utilized min^{-1} catalytic centre^{-1}

Specific activity is the most widely used method (also sometimes expressed as μ moles of substrate min^{-1} g fresh tissue^{-1}), but the values depend on the degree of purity of the enzyme. The other methods of expressing activity really require purified enzyme and knowledge of the molecular weight and number of active centres.

In protein determination, it is usually wise to run a standard curve at the same time as the unnown. BSA is the usual "universal" standard, but dilute BSA standard solutions are prone to degradation by micro-organisms and must be made up regularly. Protein precipitates can be dissolved in 0.5 M or 1 M NaOH at room temperature or by gently heating to 100 °C. Frozen material must be allowed to thaw completely before withdrawing the sample.

Protein determination

A number of different methods are available for protein determination, all of which have some limitations.

- UV Absorption at 280/260 nm (use quartz cuvettes). The amount of protein is calculated from the formula, protein (mg/ml) = 1.55 OD_{280}–0.76 OD_{260}. This method is sensitive down to 10 μg/ml, but it is a relative method, i.e. the values depend on the amino acid composition of the protein, so for highly accurate work a standard curve should be prepared with a pure sample of the protein under study.

- Lowry, sensitive down to 20 μg/ml (Lowry et al. 1951). Again, a relative method and rather sensitive to interference by buffers and polyhydroxy compounds.
  - Solution A: 2% Na_2CO_3 in 0.1 M NaOH
  - Solution B: 2% potassium sodium tartrate (or sodium citrate) + 1% $CuSO_4$ in water
  - Folin reagent: commercial Folin-Ciocalteau reagent diluted 5:12 with water
  - Copper reagent: (prepare fresh before use), 1 ml of B added to 100 ml of A

A set of dilutions is set up of standards and unknowns including a water blank (range 10–100 μg of protein) in a final volume of 0.8 ml. Add 4 ml of copper reagent, mix, wait 10 min.

Add 0.4 ml of Folin reagent (mix each tube immediately after addition). After 30 min, read OD at 750 nm.

The range of the assay can be extended to 250 μg by reading at 500 nm.

- Biuret is an absolute method; it measures the number of amide bonds, but is much less sensitive than the other methods (1 mg/ml). Standards, unknown and control are made up in 1 ml, 4 ml of Biuret reagent are added, mixed and the OD read at 550 nm. The Biuret reagent is made by dissolving 6 g of potassium sodium tartrate and 1.5 g $CuSO_4.5H_2O$ in 500 ml water; to this is slowly added 300 ml of 2.5 M NaOH while constantly stirring. The mixture is then made up to 1 l; the addition of 5 g/l KI improves stability.

- Coomassie brilliant blue, dye binding assay, is sensitive down to 20 μg/ml, but again a relative method (Sedmak and Grossberg 1977).
- Coomassie reagent: 0.1% Coomassie G250 in 3% perchloric acid (dissolve dye in 1 ml absolute ethanol then add the perchloric acid, stir for 4–5 h and filter). Store in dark, absorbance at 465 nm of stock reagent should be 1.3–1.5 OD, if necessary dilute with 3% perchloric acid.
- Standards and unknowns are made up in 0.5 ml and 0.5 ml of Coomassie reagent added. The OD is read at 465 and 620 nm and the 620/465 ratio calculated for each sample.
- A calibration curve is prepared of sample OD_{620}/OD_{465}-blank OD_{620}/OD_{465} against protein concentration.

Enzymes are assayed in the presence of excess substrate to give zero-order kinetics (i.e. the rate is independent of the substrate concentration) and the rate of reaction is calculated from the progress curve, which should be linear (one of the criteria for determining whether or not a reaction is enzymatic is that enzyme reactions should be linear with time and linear with protein concentration). However, progress curves are frequently non-linear and the initial rate (v_o) has to be estimated by drawing a tangent to the curve at time 0. Non-linear progress curves can sometimes be linearized by re-plotting as 1/v against 1/t. An alternative approach to non-linear progress curves is to use the "flying start" method. Here the time for change between two set optical densities is taken. For this, the reaction is started and when the OD reaches, say, 0.1, the timer is started, and stopped when the OD reaches 0.2. Using this method, for all reactions carried out under the same conditions, enzyme concentration will be proportional to reaction rate. It is, however, a relative method. The conversion of enzyme "units" determined from non-linear progress curves into specific activities can only be approximate.

The general formula for calculating enzyme activity is

$$\frac{\Delta OD}{\epsilon \times t} \times V \times L\frac{1}{pr} \times \frac{1}{S} \times 10^9 = \text{nmol min}^{-1} \text{ mg protein}^{-1},$$

where ΔOD is the change in optical density over t min, ϵ is the extinction coefficient, V is the volume of the assay mixture in the cuvette (ml), L is the path length of the cuvette (usually 1 cm), pr is the protein concentration of the enzyme extract (mg/ml) and S is the volume of enzyme extract added (in ml).

Some confusion is caused by two different extinction coefficients being in common use. The molar coefficient is the optical density of a 1 M solution in a 1-cm cell, and has units of M^{-1} cm^{-1} and is usually in the range 10^3–10^5. The other extinction coefficient is the optical density of 1 mol in 1 cm^3 in a 1-ml cell, it has units of cm^2 mol^{-1} and is usually in the range 10^6–10^8. The second coefficient is 10^3 times the first. The above formula is for the second type of coefficient; if the molar coefficient is used, the 10^9 is replaced by 10^6. Other extinction coefficients are sometimes used based on millimoles rather than moles or a 1% solution, and the formula has to be modified accordingly.

Not all enzymes follow zero-order kinetics; a few, like catalase, demonstrate first-order kinetics (the rate is dependent on the substrate concentration). In this case, the straight line calculation is not appropriate; instead, the velocity constant (k) must be calculated from the formula

$$ k = \frac{2.3}{t_1 + t_2} \times \log_{10} \frac{x_1}{x_2} , $$

where x_1 and x_2 are the OD at t_1 and t_2, respectively and k has the units s^{-1}. The specific activity, K'_1, is given by k/enzyme molarity and has the units M^{-1} s^{-1}.

Another important enzyme whose activity is difficult to express is superoxidase dismutase. Here, one unit is the amount of enzyme causing 50% inhibition of an O_2^- dependent reaction under the conditions specified, e.g. SOD inhibition of NBT reduction. The unit is, therefore, heavily dependent on the degree of purification of the enzyme.

Coupled assays If an enzyme assay cannot be followed directly, it may be possible to couple the reaction with a second enzyme which can be followed. Coupled assays usually show a lag phase following initiation. The bigger the V_{max}/K_m ratio of the coupling enzyme is, compared with the V_{max}/K_m ratio of the first enzyme, the shorter will be the lag phase. As an approximate guide, the specific activity of the coupling enzyme must be at least ten times that of the first enzyme. If a chain of coupling enzymes are involved, the second coupling enzyme should have an activity 100 times that of the first enzyme, and the third 1000 times.

Isotopic methods are not extensively used in enzyme assays; their success depends on being able to easily separate substrates and products and measuring either the transferred (product) or residual (substrate) radioactivity. In general, it is preferable to measure the production of radioactive product rather than disappearance of substrate. A number of different methods have been employed to separate substrates and products.

Isotopic assays

- Volatile products, the most widely used is $^{14}CO_2$, from, for example, decarboxylase reactions. The labelled CO_2 is driven out of solution at the end of the assay by the addition of HCl and absorbed by alkali in a well or on a suspended filter paper.

- Solvent extraction, with or without derivatization, for example inorganic phosphate can be extracted as its molybdate complex into isobutanol.

- Precipitation is used for nucleic acids and polysaccharides.

- Ion exchange chromatography.

- Thin-layer chromatography or HPLC.

Care must be taken in radiochemical assays to ensure that the compounds are pure, since even very small amounts of a high specific activity contaminant can result in erroneous results. If the radioactive substrate is volatile (e.g. carboxylic acid), this can again give rise to spurious results, so in both cases careful controls are required. Because of the high sensitivity of isotopic methods, often very small quantities of substrate are used and microorganism decomposition can be a problem if sterile precautions are not taken.

In vitro enzymes are assayed under what is hoped are optimal conditions to give the maximum V_{max}. Often, with parasite material the assays used are based on those developed for mammalian enzymes, and the conditions have not been optimized for the parasite enzyme.

Factors influencing enzyme activity

pH

Intracellular pH is thought usually to be around 6.9, although there may be local variations. Enzymes often show a fairly narrow pH optimum; the optimum may vary with different substrates and with temperature, and can also be influenced by the nature and ionic strength of the buffer. When constructing a pH curve, a series of different buffers with overlapping pHs should be used to eliminate buffer effects.

Care should also be taken to ensure that substrates and any other additions to the reaction mixture are neutralized, if necessary, before use. The adenine nucleotides (ATP, ADP, AMP), for example, are very acidic, as are glutathione and EDTA; KCN is strongly alkaline, and failure to neutralize these compounds will cause inhibition. Extremes of pH in the assay mixture can also affect substrate stability.

Temperature

The temperate coefficient for enzyme activity is approximately 10% per °C. There is a convincing argument that enzymes should be assayed at their normal physiological temperature, i.e. 37 °C for a mammalian parasite. At 37 °C, time must be allowed for the cuvette contents to equilibrate; however, if the solutions are prewarmed, equilibration times are reduced. Pipettes are usually designed to work at 20 °C; if the liquid being dispensed is warmer or colder than ambient, considerable errors can occur, especially with air-displacement pipettes.

Solutions from the refrigerator must be allowed to warm up to room temperature before pipetting. If cold solutions are put into cuvettes to warm, small bubbles may form on the faces of the cell as air comes out of solution. These can be removed by "scavenging" with large air bubbles blown in from a Pasteur pipette, or the problem can be avoided by degassing the solutions beforehand.

Assaying at 37 °C may, in some cases, run into problems with protein stability or substrate stability. For this reason, many workers prefer to carry out their assays at 30 °C as a compromise temperature.

Substrate Concentration

Activity increases with substrate concentration until the rate reaches V_{max}. In order to achieve V_{max}, the K_m should be negligible in comparison to substrate concentration. In practice, the substrate concentration should be at least ten times the K_m. If substrate inhibition is suspected the assay should be carried out at a series of increasing substrate concentrations.

It is important that the substrate should be free from contaminants which might act as substrates for other enzymes. For example, in the assay of malate dehydrogenase with oxalao-acetate as substrate and following NADH oxidation, contamination of the oxaloacetate with pyruvate will result in a contribution from lactate dehydrogenase.

Protein Concentration

Enzyme activity should be linear with protein concentration. However, some enzymes are much less stable at low protein concentrations, a problem that can usually be overcome by adding 0.1% BSA to the assay mixture. If the enzyme extract contains an endogenous activator or inhibitor, dilution can sometimes result in a spurious decrease or increase in apparent enzyme activity.

When reporting enzyme activities, it is important to give **Statistics** some idea of the variation. One way is to quote the mean value and range, or the standard deviation or standard error can be calculated. If the replicates are homogeneous (i.e. all done on the same enzyme extract), there is no problem in calculating standard deviations or standard errors. However, if the replicates have been accumulated from several different enzyme preparations, properly an analysis of variance should be carried out to distinguish between sample variation from between assay variation.

When there are only a small number of observations (less than ten), as is often the case with enzyme assays, special statistical methods are needed to estimate standard deviations. A very useful procedure is given by Dean and Dixon (1951); based on the range of the data, the method can also be used to exclude doubtful values. The data is arranged in ascending order of magnitude and the range (w) calculated.

Table 1. Deviation factors and rejection quotients for two to ten observations

No. of observations n	Deviation factor Kw	Rejection quotient Q
2	0.89	–
3	0.59	0.94
4	0.49	0.76
5	0.43	0.64
6	0.40	0.56
7	0.37	0.51
8	0.35	0.47
9	0.34	0.44
10	0.33	0.41

The standard deviation can be estimated by multiplying the range by the deviation factor K_w for the number of observations (Table 1). To exclude doubtful values at the extremes, the distance of the doubtful value from its nearest neighbour (x_2-x_1) is divided by the range to give the Q value.

$$Q = \frac{x_2 - x_1}{w} .$$

If Q exceeds the tabulated value, the observation may be rejected with 90% confidence (i.e. a deviation this great or greater would occur by chance only 10% of the time at either end of a set of observations from a normal distribution). Example:

a set of observations: 126.5, 115.7, 124.4, 124.8. The doubtful value is 115.7

$$w = 126.5 - 115.7 = 10.8 \ (x_2-x_1) = 124.4-115.7 = 8.7$$

$$Q = \frac{8.7}{10.8} = 0.805.$$

For $n = 4$, $Q = 0.76$; thus the value may be rejected and the remaining three values used in subsequent calculations. Kinetic constants (K_m, V_{max}) are usually determined from double reciprocal plots of $1/v$ against $1/s$. Fitting the line to this type of data, where the deviations become larger at lower substrate concentrations, requires a weighted linear regression. A suitable method has been described by Wilkinson (1961), which also allows the errors on the K_m and V_{max} to be estimated. Statistically, an s/v against s plot is preferable to a

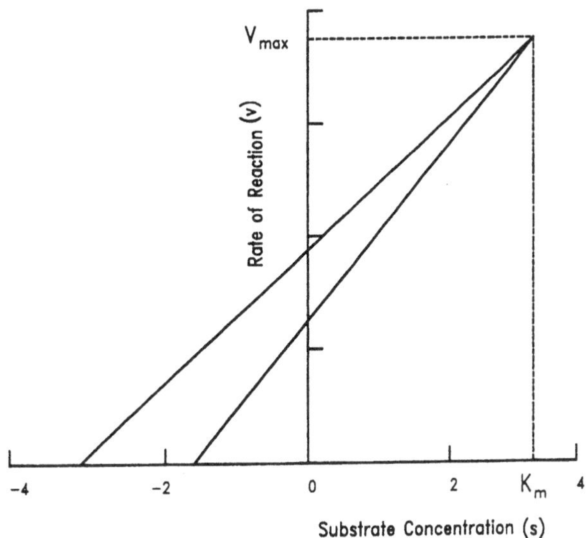

Fig. 1. Determination of V_{max} and K_m by direct linear plot. For each assay a *line* is drawn from *s* (substrate concentration) on the abcissa through *v* (*rate*) on the ordinate, the point of intersection gives the coordinates of the best fit for K_m and V_{max}. (Eisenthal and Cornish-Bowden 1974)

double reciprocal plot. The problems inherent in the double reciprocal transformation can be avoided by fitting the rectangular hyperbola directly using the method of Raaijmakers (1987). An alternative and very straightforward method of estimating K_m and V_{max} values and their errors is the direct linear plot of Eisenthal and Cornish-Bowden (1974). An example of a direct linear plot is shown in Fig. 1.

1.4
Metabolite Measurements

The concentration of intermediary metabolites in cells is usually of the order of 1–100 micromolar, with a few metabolites such as glucose-6-phosphate and ATP reaching millimolar levels. Metabolites are normally determined in cell-free extracts using enzymatic methods or, more recently, NMR or HPLC. An alternative approach is to use internal probes or reporter molecules for specific compounds.

Preparation of extract In making extracts from whole organisms, tissue differences are lost, as is any differential distribution between intracellular compartments. The two key problems in preparing extracts for metabolite measurements are first halting metabolism sufficiently rapidly to avoid post-mortem changes and, secondly, preventing any enzyme activity which might alter the relative metabolite concentrations during subsequent steps. This can be achieved by freeze clamping followed by low temperature denaturation of the enzymes with 6 M perchloric acid.

1. Tissue is clamped between two aluminium blocks (provided with insulated handles) which have been precooled in liquid nitrogen. The tissue is compressed into a thin layer; any incompletely frozen tissue protruding over the edge of the blocks is broken off and discarded. If necessary, frozen tissue can be wrapped in aluminium foil and stored in liquid nitrogen until needed.

2. A stainless steel precussion mortar and pestle are precooled in liquid N_2; the tissue is placed in the mortar and reduced to a fine powder by striking the pestle with a wooden mallet. The tissue must not be allowed to thaw at any stage.

3. A porcelain mortar is cooled with solid CO_2 and a quantity (5–10 ml) of 6 M perchloric acid is run into the mortar so that it freezes as a thin film over the surface. A quantity of the frozen powder is quickly weighed and added to the frozen perchloric acid in the mortar (1 g powder to 5 ml perchloric acid). The tissue powder and perchloric acid are ground together whilst still frozen, and the mixture is then allowed to melt slowly. Care must be taken when weighing out the frozen powder to avoid condensation from the atmosphere.

4. The tissue/perchloric acid mixture is transferred to a glass vessel and kept on ice for 30 min. The surface of the glass catalyzes the destruction of resistant enzymes such as adenylate kinase, which will not be completely denatured if plastic tubes are used

5. The perchloric extract is then neutralized with 3.75 M Na_2CO_3 using an internal or external indicator. The

neutralized mixture is allowed to stand for a further 30 min on ice whilst the potassium perchlorate precipitates. The mixture is centrifuged at 16 000 g for 10 min and the supernatant fraction taken for metabolite measurements.

Modifications

A modification of this technique suitable for small quantities of tissues is:

1. An all-glass homogenizer is cooled to −20 °C. The fresh tissue is dropped in, quickly followed by 6% perchloric acid in 15% ethanol, also precooled to −20 °C.

2. The tissue is rapidly homogenized and the extract treated as above.

Because of their stability in alkali, a specific extraction technique can be used for NADH and NADPH, which destroys NAD and NADP and other metabolites.

1. As above, only instead of perchloric acid, 0.5 M KOH in 50% ethanol, precooled to −20 °C, is added and the tissue rapidly homogenized.

2. The mixture is neutralized using 0.5 M Triethanolamine HCl. The neutralized mixture is left at room temperature for 15 min to allow proteins to flocculate, then centrifuged at 16 000 g for 10 min and the supernatant taken for assay.

Metabolite concentrations are usually expressed as nmole/ gram fresh weight of tissue or, more rarely, as per mg of protein. One gram of tissue is approximately equivalent to 1 ml, but in attempting to estimate intracellular metabolite concentrations, account must be taken of the extracellular water, which is usually somewhere around 40%. The extracellular space in tissue can be estimated using [carboxyl $^{-14}$C] inulin (Levy 1969).

Extracellular volume

1. The tissue is cut into thick slices, quickly washed with physiological saline, then incubated at 37 °C in a suitable medium containing labelled inulin and unlabelled carrier (0.4 mg/ml).

2. At the end of the incubation (60 min), the tissue is removed from the incubation, quickly rinsed to remove surface label and then samples (100–200 mg) placed in scintillation vials.

3. The tissue sample is digested by adding 0.2 ml of 60% perchloric acid followed by 0.4 ml of 100 vol (30%) H_2O_2. The mixture is heated to 80 °C for 1 h (do not heat any higher than this), then allowed to cool before scintillant is added. Vials can be stored in the dark for 24 h to allow chemiluminescence to decay before counting. If the tissue sample will not dissolve with perchlorate oxidation, try 12 M NaOH to dissolve the tissue (leave for several days), neutralize with 12 M HCl before adding scintillant.

Colorimetric methods for determining inulin are unsuitable in tissue samples due to interfering compounds.

Enzymatic analysis If a metabolite acts as a specific substrate for an enzyme, the enzyme can be used to quantify the metabolite. There are two ways in which this can be done, end-point methods and kinetic assays.

End-point Methods

Providing the enzyme converts the substrate almost completely to product, and either the substrate or the product has a characteristic property which can be measured, this is the basis for an end-point assay.

Enzyme
$S \rightarrow P$.

If either S or P has a clear absorption spectrum, the reaction can be followed spectrophotometrically, or the compound may have a characteristic fluorescent spectrum, or P could be an acid whose production can be followed by pH change. End-point methods require the equilibrium constant for the reaction to be small, so that the equilibrium position lies far over to the right, and for practical purposes S is completely converted to P. The enzyme must also have a sufficiently high activity to bring about the conversion in a reasonable time. Manipulation of the assay conditions, such as altering the pH, can be used to shift the equilibrium to the right, or it is sometimes possible to effectively remove the product P by adding a trapping agent. However, care must be taken to ensure that the trapping agent (e.g. 0.02 M semicarbazide-HCl or 0.4 M hydrazine sulphate to remove aldehydes) does not inhibit the enzyme. If at equilibrium there is still an appreciable amount of substrate remaining, this has to be taken into account in the final calculation and can be estimated from the equilibrium constant.

Many reactions involve a co-factor and the co-factor may provide the means of following the reaction. In this context, NAD/NADH and NADP/NADPH with their characteristic OD difference at 340 nm are particularly useful ($E = 6.22 \times 10^6$ $cm^2\ mol^{-1}$).

$$S + NAD \rightarrow P + NADH.$$

Many metabolite assays are coupled to the oxidation or reduction of the adenine nucleotides, and these assays are sensitive down to 10^{-8} mol/ml. Changes in NAD/NADH or NADP/NADPH can also be followed fluorimetrically (excitation wavelength 340 nm, emission 464 nm); this is about 100 times more sensitive than UV absorbtion and allows determination down to 10^{-10} mol/ml. However, unlike extinction coefficients, fluorescence is measured in arbitrary units and there is no linear relationship between fluorescence and concentration, so a calibration curve has to be constructed each time.

If the change in substrate or product cannot be followed directly, it may be possible to couple product formation with a second indicator reaction.

$$
\begin{array}{ll}
\text{Primary} & \text{Indicator} \\
\text{enzyme} & \text{enzyme} \\
S \rightarrow P_1 & \rightarrow P_2
\end{array} \quad ,
$$

where the formation of P_2 can be followed. Again, the activity of the indicator enzyme must be sufficiently high to bring about the conversion in a reasonable time. The indicator enzyme can also be used to increase the specificity of the assay. For example, hexokinase will phosphorylate both glucose and fructose; by coupling this reaction with glucose-6-phosphate dehydrogenase, which is specific for glucose-6-phosphate, the assay can be made specific for glucose.

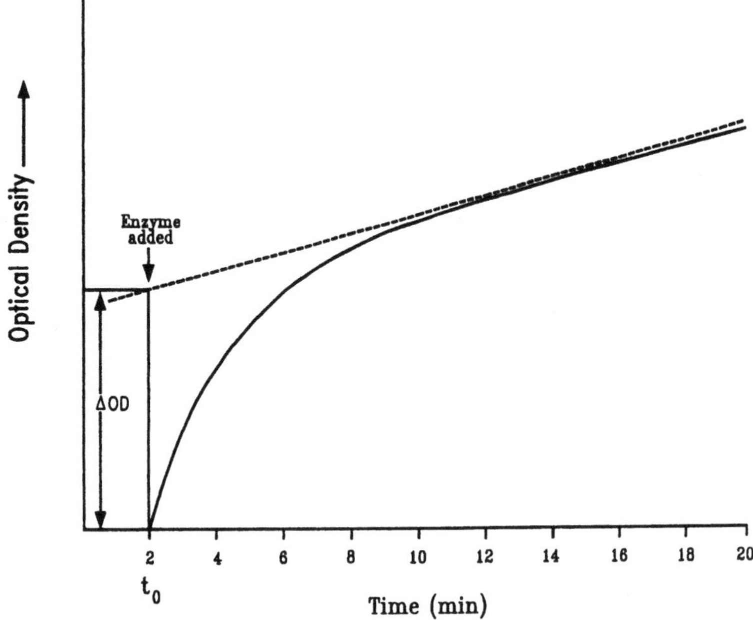

Fig. 2. Metabolite assay with drifting baseline, extrapolation to time zero

Ideally, end point assays should give a clear end point; however, in practice there may be a creeping baseline which makes the precise end point difficult to decide (Fig. 2). This is particularly a problem if the reaction mixture contains NAD or NADP in an alkaline buffer. Under these circumstances, the drifting baseline must be extrapolated back to zero time. (i.e. the time when the coupling enzyme was added). This can often conveniently be done by replotting the data as a double reciprocal plot.

By sequential addition of enzymes and co-factors, it is possible to determine several metabolites in the same assay. Tables 2–5 list sequential assays for determining all of the glycolytic and tricarboxylic acid cycle intermediates (based on Barrett and Butterworth 1982). Known amounts of metabolites are added, as internal standards, after each determination during the course of the assay. The changes in NADH or NADPH levels are followed at 340 nm or fluorimetrically. In the latter case, the coupling enzymes must be checked for inherent fluorescence and great care needs to be

Table 2. Sequential assay for ten metabolites

Addition (final concentration)	Metabolite determined
1 Malate dehydrogenase (10 μg/ml)	Oxaloacetate
2 Glycerol-3-phosphate dehydrogenase (5 μg/ml)	Dihydroxyacetone phosphate
3 Triosephosphate isomerase (5 μg/ml)	Glyceraldehyde-3-phosphate
4 Aldolase (50 μg/ml)	Fructose-1,6-bisphosphate
5 NADH 0.03 mM Aspartate 0.4 mM Glutamate-oxaloacetate transaminase (5 μg/ml)	2-Oxoglutarate
6 Lactate dehydrogenase	Pyruvate
7 ADP 0.2 mM Pyruvate kinase (5 μg/ml)	Phosphoenolpyruvate
8 Enolase	2-Phosphoglycerate
9 2,3-Diphosphoglycerate 0.001 mM Phosphoglyceromutase (5 μg/ml)	3-Phosphoglycerate
10 Pyruvate 0.02 mM; wait until all of the NADH has been oxidized, then add: MnCl$_2$ 0.1 mM NADP 0.05 mM Isocitrate dehydrogenase (25 μg/ml)	Isocitrate

Buffer: triethanolamine, 0.1 M pH 7.6; NADH, 0.02 mM; MgCl$_2$, 2mM; KCl, 40 mM; neutral perchlorate extract 0.1–0.2 ml, control without neutral perchlorate extract. The OD is measured before and after each addition.

taken to exclude small particles of dust from the cuvette; routine filtration of the reaction medium through a 0.45-μ Amicon filter is recommended.

The high specific activity of the purified enzymes used in metabolite determinations means that great care must be taken to avoid cross-contamination. It is a good idea to take an aliquot of the stock enzyme, use this for the assays and discard it at the end of the experiment. Never pipette directly from the stock enzyme into the assay cuvette unless the pipette tip is discarded each time.

Kinetic Assays

If enzyme rate is a linear function of substrate concentration, then the rate of reaction can be used to estimate the concentration of substrate. However, very low concentrations of substrate cannot be measured kinetically because they are consumed too quickly to give a linear rate.

Table 3. Sequential assay for seven metabolites

Addition (final concentration)	Metabolite determined
1 Malate dehydrogenase (10 μg/ml)	Oxaloacetate
2 Citrate lyase (20 μg/ml)	Citrate
3 Lactate dehydrogenase (10 μg/ml)	Pyruvate

4 Pyruvate 0.02 mM
  NADP 0.05 mM
  Glucose-6-phosphate dehyrogenase (2 μg/ml)
	Glucose-6-phosphate

5 Phosphoglucose isomerase (5 μg/ml)	Fructose-6-phosphate
6 Glucose-1,6-bisphosphate 0.001 mM Phosphoglucomutase (5 μg/ml)	Glucose-1-phosphate
7 Glucose (10 mM) Hexokinase	ATP

Buffer: Tris-HCl, 0.1 M pH 7.6; NADH, 0.02 mM; $ZnCl_2$, 0.04mM; $MgCl_2$, 2mM; neutral perchlorate extract 0.1–0.2 ml, control without neutral perchlorate extract. The OD is measured before and after each addition.

Table 4. Sequential assay for three metabolites

Addition (final concentration)	Metabolite determined
1 Lactate dehydrogenase (10 μg/ml)	Pyruvate
2 Phosphoenolpyruvate 1mM Pyruvate kinase (5 μg/ml)	ADP
3 ATP 0.1 mM Myokinase (5 μg/ml)	AMP

Buffer: Triethanolamine, 0.1 M, pH 7.6; $MgSO_4$, 1 mM; KCl, 10 mM; EDTA, 0.001mM; NADH, 0.05 mM; neutral perchlorate extract 0.05–0.15 ml, control without neutral perchlorate extract. The OD is measured before and after each addition.

Table 5. Sequential assay for four metabolites

Addition (final concentration)	Metabolite determined
1 Glycerol-3-phosphate dehydrogenase (5 μg/ml)	Glycerol-3-phosphate
2 Malate dehydrogenase (15 μg/ml)	Malate
3 Fumarase (25 μg/ml)	Fumarate
4 Lactate dehydrogenase (15 μg/ml)	Lactate

Buffer: 0.4 M hydrazine, 1 M glycine, pH 9.5; EDTA, 5 mM; NAD, 0.15 mM; neutral perchlorate extract 0.025–0.05 ml, control without neutral perchlorate extract. The OD is measured before and after each addition.

In a catalytic assay, the metabolite concentration remains unchanged, but the increase in rate of an indicator enzyme is measured and the result compared with a previously prepared standard curve. An example of this is the measurement of flavin mononucleotide (FMN) with lactate oxidase (Friedman 1974).

$$\text{L-lactate} + O_2 \xrightarrow{\text{L-lactate oxidase}} \text{Acetate} + CO_2 + H_2O.$$

The cofactor of lactate oxidase is FMN and, within limits, the enzyme activity, is proportional to FMN concentration. In the assay, the uptake of oxygen is followed manometrically or with an oxygen electrode.

Enzyme cycling can be used to greatly increase the sensitivity of catalytic assays. In this technique, the metabolite of interest is constantly regenerated, and the accumulation of a second metabolite is used as an indicator. For example, NADPH can be measured in a cycling assay with glutamate dehydrogenase.

$$\text{NADPH+2-oxoglutarate+NH}_4^+ \xrightarrow{\text{glutamate dehydrogenase}} \text{NADP+glutamate.}$$

The NADP is recycled back to NADPH by glucose-6-phosphate dehydrogenase.

$$\text{NADP+glucose-6-phosphate} \rightarrow \text{NADPH+6-phosphogluconate.}$$

The system is allowed to cycle for a set time (30 min) and then the reaction stopped by destroying the enzyme (by placing in 100 °C water bath). The accumulated glutamate is measured in a separate assay and the results compared with a standard curve prepared under the same conditions. With enzyme cycling, the sensitivity of the assay can be increased to 10^{-11} to 10^{-13} mol/ml. In all kinetic assays, rigorous adherence to the experimental conditions is essential in order to obtain reproducible results.

Finally, a metabolite which is not a substrate or cofactor, but which is an activator or inhibitor, can be measured kinetically by its effect on a standard assay. Examples include the activation of isocitrate dehydrogenase by magnesium, the inhibition of pyruvate kinase by heparin or the inhibition of acetylcholine esterase by several organophosphorus compounds. Again, a calibration curve is required and these often turn out to be non-linear, but can usually be linearized by a semi-log plot.

Luminometry Firefly luciferin/luciferase can be used to measure ATP levels in tissue extracts or permeabilized cells by luminometry (DeLuca 1978). The method is extremely sensitive (10^{-13} mol/ml) and can be used to measure any metabolite which can be coupled with ATP production (phosphoenolpyruvate, ADP, AMP, arginine phosphate, creatine phosphate). In the presence of excess enzyme, one molecule of ATP produces one photon of light, which can be measured in a luminometer. The extreme sensitivity of the method means that care must be taken not to contaminate any of the reagents with ATP.

The luciferin/luciferase preparation is normally supplied lyophilized and reconstituted in 0.02 M Tris/MgSO$_4$ buffer, pH 7.7. The samples and ATP standards are added by an automatic injection system.

A standard ATP curve is constructed (range 10^{-6} to 10^{-10} M), then the unknown sample measured. It is also necessary to use an internal standard and, for this, a known quantity of ATP is added to the sample. This corrects for quenching and interfering effects in the sample, since both colour and turbidity affect light detection.

Nuclear magnetic resonance and high performance liquid chromatography Enzymatic analysis is extremely sensitive and highly specific, but you only detect what you are looking for. NMR and HPLC detect everything which is present and this can lead to new and hitherto unsuspected results. For example, phosphorus NMR has led to the discovery that approximately 40% of the free phosphorus compounds in parasites is glycerolphosphoryl-choline, something which would never have been found by enzymatic analysis. The challenge with NMR and HPLC is to identify the different components in the often extremely complex spectra obtained.

NMR spectroscopy is becoming established as an important technique in the non-destructive analysis of biological samples and can be used for tissue extracts and for looking at whole organisms. Several workers have investigated the NMR spectra of whole parasites using incubation chambers with internal restraints, usually in the form of a mesh, to keep the parasites within the sensitive volume of the radiofrequency receiver coil (Mathews et al. 1985; Behm et al. 1987). The ratio of parasite to medium is kept high in these studies, typically 1 g of parasite to 5 ml of medium. It is, therefore, important to perfuse the chamber to prevent a buildup of end products and maintain pO_2 levels. Perfusion also allows the incubation conditions to be modified during the course of the experiment, e.g. with and without glucose; it also enables pharmacological agents such as serotonin to be added. An alternative, if the parasite is small enough, is to have a flow system with the media containing the parasite being circulated through the NMR machine from an external reservoir (Thompson et al. 1990).

NMR is relatively insensitive, particularly when whole parasites are being used, its limit of detection typically being 10^{-6} to 10^{-7} mol/ml. Much of the problem is caused by sample inhomogeneity (avoided by using extracts rather than whole parasites) and by the high inorganic ion content of tissue, both of which contribute to high background noise. To detect low levels of metabolites, the signal-to-noise ratio can be improved by long run times, up to several hours; this is possible with tissue extracts, but not with live parasites. Similarly, proton decoupling techniques can cause excessive heating of the sample, unsuitable for live specimens.

The use of NMR in parasite biochemistry has been reviewed by Thompson (1991). In parasites, [^{31}P-NMR] has been used to monitor levels of phosphorus-containing molecules important in energy metabolism (ATP, ADP, glucose-6-phosphate), whilst [^{13}C-NMR] has been used to provide detail about metabolic pathways, in particular carbohydrate metabolism using U-^{13}C-D glucose, ^{13}C$_6$-D-glucose, or NaH^{13}CO$_3$ as substrates (Blackburn et al. 1986; Behm et al. 1987). [^{13}C]-NMR enables the position of the label in the molecule to be determined; this, combined with sequential time scans, allows the disappearance of substrate and the appearance of product to be followed kinetically. It may also be possible to follow sequential changes in pathway intermediates, but the approach is limited by the need to balance repetitive scans (at 5- or 10-min intervals) with the longer scan times required to give high resolution.

There have been relatively few [^{1}H]-NMR studies on parasites, probably due to the complexity of the proton NMR spectra, despite its being more sensitive (Blackburn et al. 1993). Proton NMR does allow detection of a wide range of compounds and has been used for end product and tissue metabolite analysis.

In NMR, peak assignments are based on chemical shifts, and, where ambiguous, checked by the addition of pure compounds to the extracts. Quantitative measurements can be made with NMR by integration of the metabolite peaks relative to an internal intensity standard such as phenyl-phosphoric acid (PPA).

In some instances, a significant fraction of a metabolite may be enzyme-bound and not free in solution. Metabolites bound to macromolecules show broadened resonances and become to a large extent "NMR-invisible". The bound fraction can then be estimated by comparing metabolite levels determined by NMR with enzyme analyses (Matthews et al. 1985).

Hydrogen, carbon and phosphorus are not the only nuclei which can be used for NMR. Others of possible biological interest include ^{15}N, ^{43}Ca, ^{19}F, ^{39}K, ^{25}Mg, ^{23}Na and ^{67}Zn. Fluorine is a particularly useful nucleus, and there are some interesting NMR techniques being developed to measure intracellular calcium and magnesium using fluorine-substituted chelating agents (Kirschenlohr et al. 1988).

HPLC is another technique that can be used to analyze tissue extracts. Four modes of chromatography are available: gel filtration, which separates on the basis of molecular size; adsorption, which has a mobile liquid and a solid stationary phase; partition, where there are two immiscible liquids, one stationary, one mobile; and finally, ion exchange, where compounds are separated on the basis of their charge. The compounds to be analyzed need to be in sufficiently high concentration to be detected, which usually means that some form of pre-purification or concentration is required. Contaminants in the sample can cause the column to loose its resolving power, so guard columns are usually used. With perchloric acid extracts, removing the perchlorate completely can be a major problem; perchlorate gives a very large signal which can obscure minor peaks.

Identification of peaks is usually done from their relative retention times and by "spiking" the sample with known standards. To confirm identification, samples should be run on at least two different column systems and identity confirmed by an independent chemical method such as mass spectroscopy or voltametry.

If a metabolite cannot be measured directly, its concentration may be inferred from other measurements. The most widely used method is to measure the concentration of an abundant metabolite which is in equilibrium with an inabundant target. For example, the concentration of oxaloacetate in cells is extremely low, but its concentration can be estimated from the malate dehydrogenase equilibrium ($K = 2.78 \times 10^{-5}$) if the malate levels and the [NAD]/[NADH] ratio are known. Intracellular pH can be estimated from the chemical shift of Pi in ^{31}P-NMR or from the C_2 resonance of histidine. Free cytosolic [Mg^{2+}] can be estimated from the chemical shifts of the ATP and ADP phosphates, which are a function of the [Mg.ATP]/[ATP] and [Mg.ADP]/[ADP] ratios, respectively (Gupta and Moore 1980; Roberts et al. 1981). The ^{1}H–NMR signal of myoglobin can be used to estimate intracellular pO_2. A number of other intracellular probes have been developed, e.g. the fluorescent calcium indicator quin 2 (Clapper and Lee 1985), but these are more suitable for use with isolated cells or subcellular fractions than whole tissues.

Probes and reporter molecules

1.5
Use of Inhibitors

Inhibitors can be used to study isolated enzymes and metabolic pathways; however, few inhibitors are totally specific for one enzyme, and so results may have to be interpreted cautiously. Inhibitors can be used to:

- Characterize isolated/purified enzymes

- Determine the physiological role of an enzyme

- Demonstrate that a particular metabolic pathway is active

- Confirm the position of an enzyme in a metabolic sequence

With isolated enzymes, inhibitor activity can be quantified in terms of inhibitor constants for reversible inhibitors and rate constants or I_{50} values for irreversible inhibitors. A variety of graphical methods are available to determine the type of inhibition (competitive, non-competitive, uncompetitive or mixed) and also for determination of K_i values (Cornish-Bowden 1974). These are mostly based on plots of 1/v against i at different substrate concentrations or s/v against i; direct linear plots can also be used.

With irreversible inhibitors, an I_{50} value, the concentration of inhibitor giving 50% inhibition in a specified time in the standard assay, is often used. I_{50} values can also be determined graphically by plotting % inhibition against i (Yalcin et al. 1983) or by probit analysis. A better measure of irreversible inhibition is to determine the rate constant for inhibition, which can again be determined graphically by plotting % log remaining activity against time for a series of inhibitor concentrations (Aldridge 1950). Parasite enzymes can differ quite considerably from their mammalian counterparts in their inhibitor profiles.

Inhibiting enzyme activity in vivo can present a number of problems. Ideally, inhibitors used in vitro can be used in vivo, but the organism may be impermeable to the inhibitor, or the inhibitor may be unevenly distributed in the whole organism, or it may be detoxified or become bound to non-enzymic sites in the tissues. Some of these problems can be overcome by using tissue slices or minces, rather than whole organisms. Demonstrating the physiological role of an enzyme and the likely physiological consequences of its inhibition is an important aspect of validating a particular enzyme as a drug target. In order to confirm the in vivo site of action of an inhibitor, the following criteria can be applied.

- Correspondence of inhibitor concentrations, similar levels of inhibitor should cause similar levels of inhibition, in vitro and in vivo. In order to obtain specific effects, optimal concentrations of inhibitor must be used; very high concentrations of inhibitor may lead to non-specific and non-enzymatic effects.

- Use of several different inhibitors; there should be a similar rank order of inhibition in vitro and in vivo.

- Reversal of inhibition by downstream substrates. In the pathway:

$$A \rightarrow B \xrightarrow{\text{X}} C \rightarrow D \rightarrow E$$

If the inhibitor acts at X, then addition of C or D will give rise to an increase in E, the addition of A or B will not. For example, iodoacetamide inhibits glyceraldehyde-3-phosphate dehydrogenase, and iodoacetamide-poisoned tissue oxidizes pyruvate more readily than glucose. Inhibition at X should also lead to an accumulation of B, which may be measurable; however, this is not necessarily the case, as B may be metabolized by alternative pathways.

- Isolation of inhibited enzymes. Following treatment of an organism with an irreversible inhibitor, subsequent homogenization and enzyme assay should show a drop in the specific activity of the target enzyme. Reactivation of the inhibited enzyme is also sometimes possible; for example, inhibition by diethylpyrocarbonate can be specifically reversed by hydroxylamine.

Inhibitors often have characteristic effects on whole organisms, for example on growth or movement, but it is often difficult to correlate these effects with the inhibition of particular enzymes. However, used with specific and rational purpose, inhibitors are a useful adjunct to enzyme assays and metabolite measurements.

References

Aldridge WN (1950) Some properties of specific cholinesterase with particular reference to the mechanism of inhibition by diethyl *p*-nitrophenyl thiophosphate (E605) analogues. Biochem J 46:451–460

Barrett J, Butterworth PE (1982) Carbohydrate and lipid catabolism in the planarian *Polycelis nigra*. J Comp Physiol 146:107–112

Behm CA, Bryant C, Jones AJ (1987) Studies of glucose metabolism in *Hymenolepis diminuta* using ^{13}C nuclear magnetic resonance. Int J Parasitol 17:1333–1341

Blackburn BJ, Hudspeth C, Novak M (1993) [^{1}H]-Nuclear magnetic resonance study of three species of *Hymenolepis* adults. Parasitol Res 79: 334–336

Blackburn J, Hutton HM, Novak M, Evans WS (1986) *Hymenolepis diminuta*: nuclear magnetic resonance analysis of the excretory products resulting from the metabolism of D-[^{13}C$_6$] glucose. Exp Parasitol 62: 231–388

Clapper DL, Lee HC (1985) Inositol triphosphate induces calcium release from non-mitochondrial stores in sea urchin egg homogenates. J Biol Chem 260:13947–13954

Cornish-Bowden A (1974) Principles of enzyme kinetics. Butterworth, London

Crabtree B, Newsholme EA (1972) The activities of phosphorylase, hexokinase, phosphofructokinase, lactate dehydrogenase and the glycerol 3-phosphate dehydrogenases in muscles from vertebrates and invertebrates. Biochem J 130:391–396

Dean RB, Dixon WJ (1951) Simplified statistics for small numbers of observations. Anal Chem 23:636–638

DeLuca MA (1978) Bioluminescence and chemiluminescence. Methods Enzymol 62:3–122

Eisenthal R, Cornish-Bowden A (1974) The direct linear plot. A new graphical procedure for estimating enzyme kinetic parameters. Biochem J 139:715–720

Friedmann HC (1974) Flavin mononucleotide. In: Bergmeyer HU (ed) Methods of enzymatic analysis, 2nd edn. Academic Press, London, pp 2179–2181

Gupta RK, Moore RD (1980) ^{31}P NMR studies of intracellular free Mg^{2+} in intact frog skeletal muscle. J Biol Chem 255:3987–3993

Hort DT, Opperdoes FR (1984) The occurrence of glycosomes (microbodies) in the promastigote stage of four major *Leishmania* species. Mol Biochem Parasitol 13:159–172

Kirschenlohr HL, Metcalfe JC, Morris PG, Rodrigo GC, Smith GA (1988). Ca^{2+}transient, Mg^{2+}and pH measurements in the cardiac cycle by ^{19}F NMR. Proc Natl Acad Sci 85:9017–9021

Knowles WJ, Oaks JA (1979) Isolation and partial chemical characterization of the brush border plasma membrane for the cestode *Hymenolepis diminuta*. J Parasitol 65:715–731

Levi G (1969) Different estimates of tissue extracellular space using [carboxyl-^{14}C]-inulin from different sources. Anal Biochem 32:348–353

Lowry OH, Rosenbrough NJ, Farr AL, Randall RJ (1951) Protein measurement with the folin phenol reagent. J Biol Chem 193:265–275

Matthews PM, Shen L, Foxall D, Mansour TE (1985) ^{31}P-NMR studies of metabolite compartmentation in *Fasciola hepatica*. Biochim Biophys Acta 845:178–188

Raaijmakers JGW (1987) Statistical analysis of the Michaelis-Menton equation. Biometrics 43:793–803

Roberts JKM, Wade-Jardetzky N, Jardetzky O (1981) Intracellular pH measurements by ^{31}P nuclear magnetic resonance. Influence of factors other than pH on ^{31}P chemical shifts. Biochemistry 20:5389–5394

Sedmak JJ, Grossberg SE (1977) A rapid, sensitive and versatile assay for protein using Coomassie brilliant blue G-250. Anal Biochem 79:544–552

Shahkolahi AM, Donahue MJ (1993) Bacterial flora, a possible source of serotonin in the intestine of adult female *Ascaris suum*. J Parasitol 79: 17–22

Thompson SN (1991) Application of nuclear magnetic resonance in parasitology. J Parasitol 77:1-20

Thompson SN, Platzer EG, Lee RWK (1990) In vivo flow FT-NMR spectroscopy of a nematode parasite. Magnet Resonance Med 16:350-356

Wilkinson GN (1961) Statistical estimations in enzyme kinetics. Biochem J 80:324-332

Yalcin S, Jensson H, Mannervik B (1983) A set of inhibitors for discrimination between the basic isoenzymes of glutathione transferase in rat liver. Biochem Biophys Res Commun 114:829-834

Chapter 2

Electrophoresis of Parasite Proteins

Eileen Devaney

2.1
Introduction

Electrophoresis is a widely used method for the characterisation of parasite proteins. The technique relies upon the migration of solubilised proteins in an electric field, and, depending upon the conditions in use, can be used to separate proteins on the basis of size, charge, or both. The rate of migration of a protein within an electric field is determined by the intrinsic characteristics of the protein (size, mass, shape and charge) and by the physical characteristics of the medium (ionic strength, viscosity, pH, temperature). Many different methods of electrophoresis have been described (for detailed descriptions and discussions of the theoretical considerations of the various methods, see Hames 1981), but the most commonly used technique in the field of parasitology is SDS-PAGE, i.e. polyacrylamide gel electrophoresis (PAGE), carried out in the presence of the detergent sodium dodecyl sulphate (SDS). This chapter covers the preparation of parasite extracts for electrophoresis, the principles and practice of SDS-PAGE and some applications of SDS-PAGE to the characterisation of parasite polypeptides.

2.2
Lysis of Parasites and Extraction of Proteins

The first step in the electrophoretic analysis of parasite proteins is the preparation of a soluble extract. A variety of methods can be used to extract parasite proteins, depending upon the organism and the use for which the extract is intended. Large multicellular parasites, such as helminths, are usually broken open by homogenisation; where the parasite is very small (e.g. protozoa or larval stages of helminths such as microfilariae), then it is more efficient to sonicate. For many protozoan species, freeze/thawing is adequate to lyse parasites, but if the nuclei are lysed at the same time, then contaminating DNA can interfere with the subsequent preparation of a soluble supernatant. Once the parasite is lysed, an extract can be made. Depending upon the purpose of the experiment, extracts can be water-soluble, detergent-soluble or total (SDS sample cocktail-soluble). When working with parasites, the amount of material is often limiting, so for many purposes the volumes should be kept to a minimum; e.g. a mini-homogenizer (vol 100 μl) is often more appropriate than a full-size homogenizer, and centrifugation steps are often best carried out in screw-top microtubes in the microfuge rather than in an ultracentrifuge.

2.2.1
Solubilisation in SDS Sample Cocktail

If the extract is prepared solely for the purpose of analysis by SDS-PAGE, then the parasites can be extracted directly into SDS sample cocktail (Sect. 2.3 for recipe). This is also a useful method where the quantity of material is limiting (e.g. the L_3 of filarial nematodes), as the whole procedure can be carried out in a single microtube and there is no loss due to transfer from homogeniser to tube etc. For some parasites, DNA from lysed nuclei can interfere with subsequent sample preparation, and so a detergent-soluble extract should be made (see Sect. 2.2.2)

Resuspend the parasites in a known volume of SDS sample **Procedure**
cocktail in a screw top microtube.
Boil for 3 min and spin out insoluble residue in the microfuge at 13 000 g_{av} for 15 min.
Titrate protein on gel by staining.

2.2.2
Detergent Solubilisation

In many experiments, the preparation of a total SDS soluble extract is inappropriate, e.g. where the extract is also to be used for immunoprecipitation, or where the purpose of the experiment is to extract only a subset of parasite proteins (e.g. water-soluble proteins). The method given below can be used for the sequential extraction of parasite proteins, i.e. preparation of H_2O soluble extract, followed by a detergent extract etc.

Equipment
- Glass/glass homogenizer (e.g. Jencons mini-homogenizer, Cat. no. 361–046)
- Sonicator, best with a microprobe to keep volumes to a minimum (e.g. MSE)
- Microfuge (chilled if possible)
- Boiling water bath, or heating block
- Spectrophotometer or ELISA reader for protein assay
- 0.22-μm spin filters (Millipore or Costar)

Reagents
- 1 M Tris/HCl, pH 8.3, or PBS
- 100-fold concentration of individual protease inhibitors (Table 1); dilute in buffer immediately prior to use
- 5% solution of appropriate detergent (Table 2), add to final concentration of 1%
- Protein assay kit (BioRad or Pearce Chemicals)
- SDS-PAGE sample cocktail (see Sect. 2.3)

Table 1. Commonly used protease inhibitors

Inhibitor	Type	Working conc.	Solvent	m.w.
PMSF	Serine	1 mM	Ethanol	174.2
TPCK	Chymotrypsin, thiol	0.2 mM	Ethanol,DMSO	351.9
TLCK	Trypsin and thiol	0.2 mM	50 mM KPO_4, pH 5.0	369.3
EDTA	Metalloproteases	1.0 mM	H_2O at pH 8–9	380.2
Leupeptin	Serine and thiol	1 μm	H_2O, buffer	475.6
Aprotinin	Serine	0.3 μm	H_2O, buffer	6511

Note: Many protease inhibitors are toxic, take appropriate precautions.

Prepare stock solutions of proteases at ×100 the working concentration. Aliquot and freeze at –70 °C. When a tube is thawed, discard the remaining contents; some protease inhibitors (e.g. PMSF) have a short half-life in aqueous solution. The choice of protease inhibitor is largely empirical; it makes sense to use a cocktail of inhibitors of different specificities.

Table 2. Detergents commonly used for solubilising parasite proteins

Detergent	Type
Na DOC	Anionic
nOG	Non-ionic
TX-100	Non-ionic
TX-114	Non-ionic
CHAPS	Zwitterionic
CTAB	Cationic

The choice of detergent used for solubilisation is often largely empirical. However, detergents do have different properties which can effect the efficiency of solubilisation and the denaturation state of proteins, e.g. detergents such as CHAPS and non-ionic detergents such as nOG and TX-100 are non-denaturing, while CTAB is denaturing. CTAB has been reported to enhance the solubilisation of antigens from filarial nematodes (Freedman et al. 1988) and has been used to "surface strip" cuticular antigens from *Nematospiroides dubius* (Pritchard et al. 1985). In addition, some detergents (e.g. DOC, nOG) are more easily removed by dialysis than others (e.g. Triton X-100), although the inclusion of hydrophobic detergent-binding resins such as Bio Beads SM2 (BioRad), can accelerate the removal of TX-100. DOC has to be prepared in low ionic strength buffer at pH > 8. Triton X-114 has the advantage of phase separation at physiological temperatures, allowing separation into membrane-bound and water-soluble components. For a good review of the properties of different detergents, see Jones et al. 1987.

- 1 M Tris/HCl, pH 8.3.   **Solution**

For 100 ml		
× 100 concentrate		
Tris base		12.1 g
ddH$_2$O		to 70 ml
Adjust pH to 8.3 with HCl	ddH$_2$O	to 100 ml

Store at 4 °C, dilute 1:100 prior to use

1. Wash a known number or concentration of organisms and   **Procedure**
   resuspend in a measured volume of lysis buffer (PBS or

hypotonic lysis buffer such as 10 mM Tris/HCl pH 8.3) containing a cocktail of protease inhibitors (see Table 1). Homogenise with several strokes of the pestle or, alternatively, sonicate on ice (start with six cycles of 30 s duration) in the absence of detergent to prevent foaming; check for breakage/lysis by eye or by microscopy. Leave on ice for 30–60 min with occasional mixing to extract.

2. To efficiently extract membrane proteins, add detergent (see Table 2) to a final concentration of 1% and leave for a further 30–60 min with occasional mixing (these times are empirical, and can be adjusted to suit the organism or life-cycle stage).

3. Spin out the insoluble residue at high speed in a microfuge, 13 000 g_{av} for 15 min, at 4 °C, if a chilled microfuge is available.

4. For some purposes it is useful to pass the soluble supernatant through a 0.22-μm spin filter and then take an aliquot for protein assay. Remember that high concentrations of detergent can interfere with some colorometric protein determinations. Aliquot in appropriate volumes and freeze at –70 °C.

5. The insoluble pellet remaining after centrifugation can be further solubilised in SDS sample cocktail (see Sect. 2.2.1) by boiling for 3 min followed by centrifugation, e.g. to extract cuticular collagens from nematodes. The amount of protein in this extract has to be titrated by SDS-PAGE and staining.

2.3
SDS-PAGE

This is an extremely widely used method which lends itself to many adaptations, a selection of which are given below. Using this technique, proteins are separated essentially on the basis of molecular mass, as the intrinsic charge of the protein is masked by the binding of the anionic detergent SDS, which binds to polypeptides in a constant weight ratio of 1.4 g SDS:1 g polypeptide, with the result that all proteins have an equivalent negative charge. During the preparation of the sample, proteins are broken down to polypeptide subunits and denatured by boiling in the presence of SDS and a reducing agent. Upon the application of an electric field, the proteins in a complex mixture are then separated in the polyacrylamide matrix on the basis of relative molecular mass or M_r. The pore size of the polyacrylamide matrix can be adjusted using different percentages of acrylamide monomer and cross-linker to optimise the separation over a particular M_r range.

SDS-PAGE can be run using continuous or discontinuous buffer systems. In continuous systems, a single separating gel is employed with the same buffer in the gel and the tank. In the discontinuous system (Laemmli 1970), a large-pore stacking gel is prepared using a buffer different from the separating gel, and both buffers have an ionic strength and pH different from the running buffer. The discontinuous system is more widely used because the resolution obtained is better, due to the concentration of proteins into a narrow zone within the stacking gel. The technique given below describes the use of a discontinuous system. A variety of different sorts of apparatus is also available for casting and running gels; most commonly used are vertical slab gels, although horizontal slab gels are useful for isoelectric focusing and tube gels for two-dimensional gel electrophoresis. There are a number of considerations to take into account which are summarised briefly here.

Safety: Acrylamide is a neurotoxin and is toxic by inhalation or skin contact. Wherever possible, it is advisable to buy premade liquid acrylamide; this can be purchased from many different chemical suppliers and obviates the need for weighing acrylamide and bis powder. Polyacrylamide gel is not toxic, but gloves should be worn throughout, as gels can contain unpolymerised acrylamide and contact with fingertips can lead to spurious bands/marks, particularly upon silver staining.

Gel Preparation

Polyacrylamide gels are prepared from a mixture of acrylamide which is polymerised by a cross-linker, the most common of which is N,N'-methylene-bis-acrylamide or bis, for short. The polymerization is catalysed by the addition of a catalyst; ammonium persulphate (APS) is commonly used as the initiator and N,N,N',N'-tetramethylethlene-diamine (or TEMED) as the catalyst. Photochemical, rather than chemical, polymerization is also possible, using riboflavin and TEMED. The molecular sieving effect exerted and therefore the degree of separation achieved depends upon the concentration of acrylamide and cross-linker. The total acrylamide concentration (acrylamide monomer plus bis) in a solution is expressed as %T, while %C denotes the percentage of the total which is bis. Commercially available acrylamide solutions contain both acrylamide and bis and the ratio of one to the other is always stated; if required, both reagents can be purchased separately in liquid form.

For crude mixtures of proteins, gradient gels (e.g. 5–15%) provide optimal separation over a wide molecular size range; in a gradient gel, the pore size is larger at the top of the gel and decreases towards the bottom, and the gradient in pore size also acts to concentrate polypeptides into sharp bands. Pore size can be adjusted either by altering the concentration of acrylamide in the mix, or by altering the concentration of cross-linker. For details of the theoretical considerations underlying the separation of polypeptides by SDS-PAGE, the reader is referred to Hames (1981). Before adding SDS, APS and TEMED, the gel solution is degassed, as dissolved O_2 inhibits polymerization. When the gel is poured, it is immediately overlayed with a solution of water-saturated n-butanol; this ensures that the top of the gel is completely even upon polymerisation and excludes O_2, which could inhibit polymerisation.

Sample Preparation

In preparing the sample for SDS-PAGE, the parasite extract is mixed with an excess of SDS, a reducing agent, usually dithiothreitol (DTT) or 2-mercaptoethanol (2-ME), glycerol and a tracker dye such as bromophenol blue and boiled for 3 min, followed by centrifugation to remove any unsolubilised material. We commonly use DTT, which is more stable than 2-ME. Boiling in the presence of reducing agent and SDS denatures the proteins into subunit polypeptides, and so M_r determinations by SDS-PAGE estimate the size of polypeptide subunits and not native proteins, which may be multimeric. Glycerol (or sucrose) in the sample cocktail increases the viscosity of the mix and ensures that the samples do not float out of the wells when loaded, while the bromphenol blue provides a means of visualising the samples as they migrate through the gel. The mixture of DTT, glycerol, SDS and bromophenol blue is referred to as sample cocktail. For some purposes, it is appropriate to run gels under non-reducing conditions. In this case, the reducing agent is omitted and iodacetamide (1 mg/ml) can be added to the sample cocktail to protect cysteine residues.

Equipment

Most readers will have access to ready-made gel electrophoresis equipment which is available from many different companies, e.g. Pharmacia, Hoefer, BioRad etc. The choice nowadays is between mini-systems and regular gels. Regular gels (usually around 16 cm long) are larger, use more reagents and take longer to run. However, the resolution obtained on a larger gel can be better than on a mini-gel, and large gels are obviously preferable where the object of the exercise is purification of a polypeptide, for example. Mini-gels (7 cm long) are fast, require a fraction of the amount of solutions, and are therefore considerably more economical. Some mini-systems, e.g. Hoefer, have attempted to overcome the problem of resolution by employing a longer separating gel (11 cm long) which can be used on the original mini-apparatus with slight modifications. For both mini- and regular systems, the thickness of the gel can be adjusted using different spacers; the thinner the gel, the less protein can be resolved. Gel apparatus from different companies is generally not interchangeable, so it is best to stick to one manufacturer.

Running Conditions

The running conditions are usually recommended in the data sheets for a particular apparatus; when setting up the power pack, one variable is usually selected, e.g. large gels run during the day at a constant current of 35 mA per gel will take around 5 h to run; as the run progresses, the voltage increases, and it is a good idea to note the starting voltage at a constant current setting, as this will often alert one if something is wrong. For mini-gels (Hoefer), a constant current of 20 mA per gel will take around 1 h. It is important that the gel be cooled during the run, as a considerable amount of heat can be generated which, if not dissipated, can result in aberrant running of samples.

Analysis of Gels

Once the gel has run, it is usually stained or processed for autoradiography. The most commonly used stain is Coomassie blue; during this process the polypeptides in the gel are fixed as well as stained. Destaining removes the stain from the background polyacrylamide matrix, while leaving the polypeptide bands stained. The sensitivity of Coomassie blue staining is around 1 μg of protein per band, and other more sensitive methods of detecting proteins have consequently been developed. Silver staining can detect down to 10 ng of protein, but not all proteins stain well with silver stain. Silver staining kits are avialable from BioRad etc. Once the gel is stained, it can be scanned by densitometry and a permanent record obtained by photography or by simply drying the gel.

M_r Determination

Polypeptides are separated in SDS-PAGE on the basis of their molecular mass, and the M_r of an unknown polypeptide can be calculated by reference to a series of commercially obtainable molecular weight standards analysed in parallel. A standard curve is plotted by measuring the distance moved from the top of the gel (R_f) versus the log M_r of the known markers. A straight line will be obtained only over a particular size range for a single percentage gel, e.g. for low molecular weight standards ranging from 14.4 to 97.4 kDa, a 12.5% gel should separate all standards; lower concentration gels may result in one of the standards running at the dye front. The M_r of the unknown polypeptides can be estimated from the standard curve. The molecular weight determination by SDS-PAGE provides an estimate only; proteins which are heavily glycosylated can behave anomalously, as SDS is bound only to the peptide part of the molecule. Proteins such as histones, which are very basic, also run aberrantly on SDS-PAGE.

Reagents and Storage

All reagents are available in electrophoresis grade from a range of chemical suppliers. Acrylamide should be stored at 4 °C in the dark and stock solutions of gel buffers should be stored at 4 °C. Most protocols advise preparing APS fresh, but we prepare a stock 10% solution, aliquot into 100-μl volumes and store at -20 °C. Once thawed, discard the remaining contents. Aliquots of prepared sample cocktail and of high- and low-range molecular weight markers can also be prepared in advance and stored at -20 °C. Molecular weight markers can be bought prestained or labelled with biotin etc; 10% SDS should be stored at room temperature and running buffer can be made ×10 concentrated, stored at room temperature and diluted prior to use.

– Gel apparatus, including glass plates, spacers, casting stand,   **Equipment**
  combs, upper and lower buffer reservoirs
– Power pack
– Magnetic stirrer and bar

– Acrylamide   **Reagents**
– Separating gel buffer, 1 M Tris/HCl, pH 8.8
– Stacking gel buffer, 0.5 M Tris/HCl, pH 6.8
– 10% APS
– 10% SDS
– TEMED
– Glycerol
– H_2O-saturated n-butanol
– ×10 concentrated running buffer
– Sample cocktail
– Stain
– Destain
– Drying reagent

• Acrylamide solution (30%T, 2.7%C_{bis})*   **Solutions**

For 200 ml	
Acrylamide	59.4 g
Bis	1.6 g
ddH$_2$O	to 200 ml

Filter and store at 4 °C in the dark

*%T refers to the total concentration of acrylamide plus bis; %C_{bis} refers to the percentage bis expressed as percentage of total monomer plus cross-linker

- Separating gel buffer, 1.5 M Tris/HCl, pH 8.8

For 200 ml	
Tris base	36.3 g
ddH$_2$O	to 150 ml
Dissolve on stirrer, adjust pH to 8.8	
with HCl ddH$_2$O	to 200 ml

Store at 4 °C

- Stacking gel buffer, 0.5 M Tris/HCl, pH 6.8

For 50 ml	
Tris base	3.0 g
ddH$_2$O	to 30 ml
Dissolve on stirrer, adjust pH to 6.8	
with HCl ddH$_2$O	to 50 ml

Store at 4 °C

- Running buffer, pH 8.3, (working strength 0.025 M Tris, 0.192 M glycine, 0.1% SDS)

For 5 l	
×10 concentrated	
Tris base	151.15 g
Glycine	721 g
ddH$_2$O	to 4000 ml
Dissolve on stirrer	
SDS	50 g
Check pH, should not require	
adjustment ddH$_2$O	to 5 l

Store at room temperature, dilute 1:10 prior to use

- 10% w/v SDS

For 500 ml	
SDS	50 g
ddH$_2$O	to 500 ml

Store at room temperature

- 10% w/v APS

For 10 ml	
APS	1.0 g
ddH$_2$O	to 10 ml

Aliquot in 250 μl vol and freeze at –20 °C.

- (a) Sample cocktail (0.125 M Tris, 4% SDS, 20% glycerol)

For 100 ml	
0.5 M Tris, pH 6.8	25.0 ml (see Stacking gel buffer above)
10% SDS	40.0 ml
Glycerol	20.0 ml
ddH$_2$O	to 100 ml

Aliquot and store at room temperature.

- (b) 0.1% Bromophenol blue, 1.5 M DTT

For 10 ml	
Bromophenol blue	0.01 g
DTT	2.3 g
ddH$_2$O	to 10.0 ml

Aliquot in 100 μl and store at –20 °C.
For use, mix 650 μl Sample cocktail (a) and 100 μl Bromphenol blue (b).

- Coomassie blue stain

For 5 l	
Coomassie blue	5 g
Methanol	2250 ml
Stir until dissolved	
ddH$_2$O	2250 ml
Acetic acid	500 ml

Filter through Whatman No. 1

- Destain

For 5 l	
Acetic acid	350 ml
Methanol	1000 ml
ddH$_2$O	3650 ml

- Drying reagent

For 1 l	
Acetic acid	100 ml
Glycerol	10 ml
ddH$_2$O	to 1 l

Procedure

1. Wash and clean glass plates, swab with ethanol and polish to remove any grease; assemble the gel cassette following the manufacturer's instructions. Be careful to obtain a good seal between the bottom of the plates and the casting stand, else the gel will leak. A **thin** film of Celloseal or other sealant can be applied to assist the seal. Do not use chipped or cracked glass plates.

2. Prepare gel solution mixing reagents well (see Table 3a for sample recipes) and degas using a vacuum pump. Add SDS, then TEMED folowed by APS, and swirl gently to mix. Pour gel mix into cassette using a 25-ml disposable plastic pipette for large gels, or a 1.0-ml Gilson pipette for mini-gels. Pour the separating gel to leave sufficient space for the stacking gel and comb (a distance of at ~ 2.5 cm on a large system or 1.5 cm, mini-gel). Overlay immediately with H$_2$O saturated n-butanol and leave to set. When the gel is polymerised (30–60 min at room temperature), an interface forms between the gel and the liquid. Multiple gels can be cast at once (some systems allow 12 gels to be poured simultaneously) and stored in the fridge until use.

3. Prior to use, pour off butanol and wash the top of the gel with ddH$_2$O. Tip onto side to dry and insert a piece of blotting paper between the plates to absorb any excess H$_2$O.

4. Prepare stacking gel solution (Table 3b), add SDS, TEMED and APS and pour, using the appropriate comb to form wells. Leave to polymerise, 15–30 min. For mini-gels, do not leave stacking gel too long, as the wells have a tendency to shrink, which can make subsequent loading of samples difficult.

5. Prepare samples for loading. Mix a known volume of sample with at least an equal volume of SDS sample cocktail (more sample cocktail can be used to make up

Table 3a. Recipes for separating gel mix

Separating	5%	7.5%	10%	12.5%	15%
Acrylamide	5.0	7.5	10	12.5	15
Tris, pH 8.8	11.2	11.2	11.2	11.2	11.2
ddH_2O	13.7	11.2	8.7	6.2	0.7
Glycerol	–	–	–	–	3.0
SDS, 10%	300 μl	300 μl	300 μl	300 μl	300 μl
TEMED	20 μl	20 μl	20 μl	20 μl	20 μl
APS, 10%	100 μl	100 μl	100 μl	100 μl	100 μl

Quantity for a single large gel (18×16×1.5 cm); multiply by number of gels required. For mini gels (Hoefer system), use 1/3 of the quantity per gel.

Table 3b. Recipes for stacking gel mix

Stacking	5%	3%
Acrylamide	1.67	1.0
Tris, pH 6.8	1.25	1.25
ddH_2O	7.03	7.7
10% SDS	100 μl	100 μl
TEMED	10 μl	10 μl
10% APS	50 μl	50 μl

volumes), boil for 3 min and cool. Prepare molecular weight markers according to the instructions and boil. Remove the comb from the stacking gel, wash the wells once with ddH_2O and then with running buffer, using a wash bottle, and then fill the wells with running buffer. Samples are most easily loaded with long gel loading tips (available from a variety of suppliers, e.g. Anachem, M-1381) on 10 or 100 μl Gilsons. If possible, load only sample cocktail into the end wells, as samples in these wells can sometimes run aberrantly, due to edge effects.

6. Prepare an appropriate volume of running buffer, depending on apparatus; if system is water-cooled, connect cooling and stand the tank on a magnetic stirrer. Add running buffer to precool. Tank buffer can be reused several times as long as it is not contaminated with radioactivity, for example, but upper reservoir buffer must be used fresh each run.

7. Set up power pack, according to the specifications of the gel apparatus. Sample times are: for Hoefer mini-gel, ~ 60 min at 20 mA constant current, for large system, 5–6 h at 35 mA constant current or overnight at 10 mA. These times are approximate, and it is important to read manufacturer's instruction; times will also vary with percentage of acrylamide, thickness of gel, ambient temperature etc.

8. At the end of the run, when the dye front has just migrated off the end of the gel, switch off power pack, disconnect and remove gel. Prise plates apart with a spacer or other appropriate tool (not a spatula!). Remove gel and place in Coomassie blue or fix for silver staining. Stain for 1–2 h on rocking platform, depending upon how important it is to visualise all the bands (overnight may be preferable). Destain, including a piece of clean sponge to absorb excess dye, on rocker.

9. If the gel is to be dried, dry directly or place in drying reagent for 30–40 min – this is particularly useful for high percentage gels or for gradient gels, where cracking may occur on drying. Remove the gel from the drying reagent or destain and place on a piece of Whatman's 4 M filter paper cut to leave a border of about 50 mm all round. Cover the gel with cling film, but do not attempt to seal the cling film or tuck it underneath the gel, as this will result in trapping moisture as the gel dries. Place on gel drier and apply the vacuum – do not remove the vacuum until the gel is completely dry (around 1 h for a 10% gel dried at 80 °C).

Trouble shooting

- Gel polymerises too quickly
  - Check concentrations of APS and TEMED; cool gel mix at 4 °C prior to adding APS and TEMED, as polymerisation is temperature-dependent. The amount of APS and TEMED should be sufficient to polymerise a gel in 30–45 min.

- Gel does not polymerise
  - Check added APS and TEMED, prepare fresh APS.

- Gel runs with a "smile"
  - Cooling not adequate; check cooling connected during run; limit on voltage rather than current reduces the heat generated, but slows the velocity of the run.

- No sharp bands, streaking of samples
  - Samples overloaded, run less protein.

- Gel breaks/cracks when drying

  - Lower the temperature of drying to 60 °C, soak gel in drying reagent. With the Easy Breeze gel drier (Hofer), the gel is held between two sheets of cellophane in a drying frame and gels can be dried with or without heating.

- Bands/molecular weight standards run in wrong position

  - Check concentration of acrylamide.

- Gels run too fast/too slow

  - Check pH of running buffer and of gel buffers, make fresh if required.

2.4
Modifications of Standard Method

2.4.1
Gradient Gels

To achieve optimal separation of a complex mixture of polypeptide of unknown M_r gradient gels are most useful. A 5–15% gradient will achieve separation of both high and low molecular weights, but clearly the percentage acrylamide can be varied to optimise separation over a particular M_r range.

Equipment

- Gradient maker
- Peristaltic pump with tubing
- Cannula
- Magnetic stirrer with small stirring bar

Procedure

1. Set up casting stand, as above. Connect a piece of tubing to the outlet of the gradient maker and attach to the peristaltic pump. Attach a cannula to the end of the tubing and insert the cannula between the glass plates.

2. Prepare 5 and 15% solutions as in Table 3. Mix and degas each separately. Add SDS, TEMED and APS and swirl gently to mix.

3. Add the heavy solution (15%) to the mixing chamber connected to the outlet; place the stirrer bar in this chamber and mix slowly, so as not to generate air bubbles. Open the stopcock between the two chambers to allow the solution to flow through the channel between the two chambers, but not into the reservoir chamber. Close the stopcock. If any of the heavy solution enters the reservoir chamber, return it to the mixing chamber.

4. Add the light solution to the reservoir chamber, turn on the pump and open the stopcock between the two chambers and pour the gel, lifting the cannula as the level of liquid rises.

5. Overlay the gel as usual and immediately wash out the tubing by pumping through an excess of ddH$_2$O.

6. Prepare stacking gel and run as above.

2.4.2
Tricine Gels for the Analysis of Very Low Molecular Weight Polypeptides (after Hans Hagen, University of Keele)

SDS-PAGE gels cannot provide routine resolution of very small polypeptides (less then 10 kDa), because in the Laemmeli system it is difficult to obtain good separation of small protein-SDS complexes from SDS micelles even when high concentrations of acrylamide are used. This results in streaking at the low molecular weight end of high percentage gels. A number of systems have been described specifically for the separation of peptides in the range of 1.0 to 10 kDa, but some of these rely on the presence of high concentrations of urea in the gels, which can make gels difficult to handle and use reproducibly. The tricine system of Schagger and von Jagow (1987), a modified version of which is described below, utilizes tricine rather glycine as the trailing ion, and relies upon the separation achieved in a series of three gels (the stacking, the spacer and the separating gel), which are poured sequentially.

Equipment – As for standard SDS-PAGE

Reagents – As for standard SDS-PAGE, plus
– Gel buffer
– Anode buffer
– Cathode buffer

- Acrylamide 30% (as above)

- Gel buffer (0.3% SDS, 3.0 M Tris.HCl, pH 8.45)

For 100 ml	
SDS	0.3 g
ddH$_2$O	to 60 ml
Tris	36.34 g
Conc. HCl	~4.0 ml
Adjust pH to 8.45	
ddH$_2$O	to 100 ml

- Anode buffer (lower chamber buffer, working strength 0.2 M Tris/HCl, pH 8.9)

For 5 l	
×10 concentrated	
Tris	1211 g
ddH$_2$O	to 4000 ml
Dissolve, adjust pH to 8.9 with HCl	
ddH$_2$O	to 5000 ml

Dilute 1:10 prior to use

- Cathode buffer (upper chamber, working strength 0.1 M Tris (no HCl), 0.1 M tricine, 0.1% SDS)

For 500 ml	
×10	
concentrated	
Tris	60.5 g
Tricine	89.6 g
SDS	5.0 g
ddH$_2$O	to 400 ml
Check pH is 8.25, should not require adjustment	
ddH$_2$O	to 500 ml

Dilute 1:10 prior to use

1. Assemble the casting stand.

2. Prepare the gel solutions, as described in Table 4, degas and add TEMED and APS.

3. Pour the separating gel to a height of ~11 cm, overlay with ddH$_2$O-saturated n-butanol. It is useful to have marked the

Table 4. Recipes for tricine system

	Separating (11 cm) 16.5%	Spacer (2 cm) 10%	Stacker (3 cm) 4%
Acrylamide, 30%	13.8 ml	3.3 ml	1.3 ml
Gel buffer	8.3 ml	3.3 ml	2.5 ml
ddH$_2$O	–	3.4 ml	6.2 ml
Glycerol	2.9 ml	–	–
TEMED	20 μl	10 μl	20 μl
10% APS	100 μl	50 μl	60 μl

height of each gel on the plates in advance and to pour the gel a little over the mark to allow for shrinkage.

4. When the separating gel has polymerised, wash off the overlay and pour the spacer gel (2 cm); once this has polymerised, pour the stacking gel (3 cm), with the comb in place.

5. Prepare the samples as for SDS-PAGE, using same sample cocktail.

6. Add anode buffer to the tank, place gel in tank and add cathode buffer to upper buffer reservoir.

7. Run at 35 mA constant current through the day or overnight at lower setting.

Schagger and von Jagow recommend adding urea to the separating gel to optimise the separation of very low molecular weight polypeptides. Very low molecular weight standards are available for M$_r$ determination, but Rainbow standards from Amersham are reported to give a good fit (Hagen, pers. comm.).

2.5

Applications of SDS-PAGE to the Characterisation of Individual Polypeptides

While SDS-PAGE is a useful technique for the initial analysis of crude parasite extracts, perhaps its real value is as the end point in a variety of biochemical/immunochemical analyses for characterising individual polypeptides. Immunoblotting is covered in detail in the following chapter, and here I will deal with glycoprotein analysis using specific enzymes, peptide mapping, preparative SDS-PAGE and electroelution and 2D gel electrophoresis, in each of which SDS-PAGE plays an integral part. For all the methods given below, it is advantageous if the protein of interest can be easily localised on a gel, e.g. by the incorporation of radiolabel.

The above techniques are useful for the identification of homologies between specific proteins. Relationships between proteins cannot be defined upon the basis of electrophoretic mobility alone, e.g. polypeptides of different sizes may be related by a precursor-product relationship, or closely related polypeptides may differ in mobility due to the degree of glycosylation. For this reason, techniques such as peptide mapping and 2D gel electrophoresis (covered under Sect. 2.6) are useful for investigating homologies between proteins.

2.5.1
Glycoprotein Analysis Using Specific Enzymes

Although the presence of carbohydrate side chains on specific proteins can be detected using lectin affinity chromatography, for example, digestion with specific endogylcosidases offers a more sensitive and analytical method for dissecting the complexities of glycoprotein structure. Carbohydrate side chains can be either N-linked (linked to the amido N of asparagine) or O-linked (linked to the hydroxyl O of serine, threonine or hydroxylysine). N-linked side chains are more common and have been studied in greater depth than O-linked carbohydrates; some glycoproteins contain both types of linkage. The method given below (modified from Maizels et al. 1991) is for digestion with glycopeptidase F (EC 3.2.2.18) which cleaves intact oligosaccharides (most high mannose and a variety of other structures) from a native or denatured glycoprotein. Information on the availability and specificity of alternative endoglycosidases can be obtained from suppliers such as BCL.

– 10% SDS	**Reagents**
– 2-ME	
– 7.5% NP 40	

- 100 mM 1,10 phenanthroline
- 0.55 M $NaPO_4$ buffer, pH 8.6
- Glycopeptidase F

Solutions • SDS/2-ME (1% SDS, 1.6% 2-ME)

For 1.0 ml	
10% SDS	100 μl
2-ME	16 μl
ddH_2O	884 μl

• 100 mM 1,10 phenanthroline

For 10 ml	
1,10 phenanthroline	198 mg
Absolute ethanol	2.0 ml
ddH_2O	8.0 ml

Aliquot and store at $-70\ ^{\circ}C$

• 0.55 M Na_2PO_4, pH 8.6

	Na_2HPO_4	7.81 g
	ddH_2O	100 ml
	NaH_2PO_4	6.58 g
	ddH_2O	100 ml

Take 50 ml Na_2HPO_4 and add NaH_2PO_4 dropwise until pH 8.6

Procedure 1. Take two aliquots of 5- μl soluble parasite extract, or a known number of TCA precipitable counts if label has been incorporated; make up to 5 μl with ddH_2O, mix with 5 μl SDS/2-ME and boil for 3 min.

2. Cool and add 10.8 μl of 0.55 M $NaPO_4$, 3 μl of 100 mM 1,10 phenanthroline and 5 μl 7.5% NP 40.

3. To one tube add 0.05 units glycopeptidase F, prepared by diluting stock enzyme in double diluted PO_4 buffer

4. To the other tube add an equal volume of double-diluted PO_4 buffer (control)

5. Incubate at $37\ ^{\circ}C$ for 24 h

6. Stop the reaction by boiling for 3 min in an equal volume of SDS sample cocktail and analyse samples on gel.

Denaturation by boiling and treating with SDS increases the deglycosylation rate. It is important to enzyme-treat in the presence of a protease inhibitor (such as EDTA or O-phenanthroline), to guard against trace amounts of metalloproteases, which could be active over the prolonged incubation at 37 °C. Glycopeptidase F is denatured by SDS in the absence of non-ionic detergent, so it is important to add the enzyme after the NP 40. Deglycosylated samples will migrate faster than uncleaved glycoprotein; mobilities will differ, depending on whether a single or multiple side chain/s have been digested.

2.5.2
Peptide Mapping

This is another method of establishing relationships between polypeptides. Peptide mapping, or limited proteolysis, can be carried out by 1- or 2D methods and a variety of enzymes (proteases) or chemicals (e.g.CnBr) can be used to cleave the polypeptide into its constituent peptides. Peptide mapping can be carried out in solution, but in the method given here, adapted from Cleveland (1983), the polypeptides of interest are first located after SDS-PAGE and excised from the gel. The gel slices are placed in the wells of a second 15% SDS polyacrylamide gel together with known amounts of a particular protease. During the second run, the enzyme digests the polypeptide and the resulting peptides are separated on the basis of their molecular mass. The peptide banding pattern is characteristic of the protein substrate and the enzyme used to generate the peptides, so that relationships between different polypeptides are evident by comparing peptide banding patterns. The sensitivity of the method is such that it can distinguish between proteins which are almost identical, e.g. Cleveland (1983) gives the example of hog and chicken tubulins, which differ in only 2 out of 450 amino acids. This minor difference results in an extra band in a peptide fingerprint of the two proteins.

Solution • Gel slice equilibration buffer/protease digestion buffer (0.125 M Tris/HCl, pH 6.8, 0.1% SDS, 10% glycerol, 1 mM EDTA, 0.3% 2-ME)

For 10 ml	
0.5 M Tris/HCl, pH 6.8	2.5 ml
Glycerol	1.0 ml
ddH_2O	6.3 ml
10% SDS	100 μl
0.5 M Na_2EDTA	20 μl
2-ME	30 μl
Bromophenol blue	Trace

Procedure 1. Run the first gel as usual, choosing a % gel which gives optimal separation in the M_r range of interest. If the polypeptide of interest is not radiolabelled, it is important that a single band be excised, as contaminating polypeptides will give rise to numerous peptide fragments, which make interpretation of the subsequent banding pattern very difficult.

2. Locate the polypeptide of interest and excise from the gel using a sharp scalpel. If the polypeptide is radiolabelled, line up the autoradiograph with the dried gel to locate the band.

3. Prepare the second 15% gel, leaving sufficient space for a stacking gel of 5 cm length. Mix twice the volume of stacking gel solution.

4. Remove as much of the backing paper from the dried gel as is possible and place the strip of gel in the well of the 15% gel. Overlay with 20 μl of equilibration buffer and leave for 15–20 min. Add 10 μl of digestion buffer containing varying dilutions of protease. A good protease to start with is *Staphlococcus aureus* protease V8; use dilutions from 0.1 to 5.0 μg per well. Leave some wells without protease.

5. Set the power pack at 60 V constant voltage until the sample has cleared the stacking gel. Once the sample has migrated through the stacking gel, limit on current and run the gel as normal.

6. Visualise the peptides by autoradiography if sample was radiolabelled, or by silver staining if not.

In some protocols, it is recommended that, once the samples have cleared the stacking gel, the power pack is switched off for 30 min to allow the proteolysis to proceed. In our laboratory, running through the longer stacking gel at 60 V provides ample time for proteolysis. V8 protease runs as a doublet around 27 kDa, so it is useful to include a well with enzyme alone to distinguish the enzyme from the peptides generated, particularly if the peptides are to be visualised by staining.

2.5.3
Preparative SDS-PAGE and Electroelution

In this method, a large sample well is formed using an appropriate comb with only two wells, a large well for the sample and an additional regular-sized well for molecular weight standards. Preparative SDS-PAGE is useful for the analysis of large amounts of parasite protein, for example where the aim of the experiment is to purify a protein by SDS-PAGE and electroelution. Proteins can be visualised by staining in Coomassie blue and destaining, as above, but the staining and destaining times should be kept to a minimum. Alternatively, if the protein is radiolabelled, then the gel can be autoradiographed without prior staining and the location of the band determined by the presence of the radiolabel. This is particularly useful when the protein can be surface-labelled with ^{125}I, as the radioactivity provides both a marker for locating the polypeptide in the gel and a means of quantifying the efficiency of the subsequent electroelution.

The application of a current through a slice of gel containing a protein of interest will result in the migration of the protein from the gel. Polypeptides separated by SDS-PAGE are essentially negatively charged, so if the strips of gel are placed in a chamber and an electric current applied across the chamber, the protein will migrate out of the gel towards the positive pole. The protein can then be collected, usually in dialysis membrane with an appropriate pore size, and used directly for immunisation, for example.

Equipment
- SDS-PAGE equipment, as above
- Electroeluter
- Dialysis membrane

Reagents
- As for SDS-PAGE
- 1:10 dilution of working strength running buffer without SDS (2.5 mM Tris/HCl, 19.2 mM glycine), or as recommended for the apparatus

Procedure **Preparative SDS-PAGE**

1. Prepare the separating gel and make a stacking gel using a continuous comb.

2. Prepare the sample; to calculate the amount of material which can be analysed, multiply the usual quantity analysed in each well of a 10-well comb (say 50 μg) by 20 (i.e. the preparative well can contain the equivalent of 20 regular-sized wells). Make the volume up to 1.0–1.5 ml with sample cocktail and prepare the sample as usual. Do not attempt to overload the gel with protein, as this will result in poor separation of individual polypeptides.

3. Visualize the band of interest by staining or by autoradiography, as above.

4. Cut band out using a sharp scalpel.

Equipment for electroelution

An electroelution chamber is available from BioRad, which can be used with the mini-gel tank to elute either protein or DNA from acrylamide or agarose gels. Six samples can be run at once and the eluted sample is collected in a small volume. We have made extensive use of a more basic system available from Genetic Research Instrumentation Ltd., which utilises small chambers to which dialysis membrane is attached. Pieces of gel are placed in one chamber, covered with buffer, current applied and proteins eluted into the other chamber, from where they can be collected. The method given below is for this apparatus. In our laboratory, this apparatus has always been used with SDS-free buffer and has given good results. In the absence of additional SDS, the recovered sample can be used directly as antigen in ELISA or for immunisation.

Electroelution Procedure

1. Set up the electroelution chamber, following the manu-facturer's instructions. Connect the water-cooling system. The whole apparatus can be placed on ice if required. Boil some dialysis tubing and attach to the base of the elution wells.

2. Prepare 500 ml of a 1:10 dilution of running buffer, without SDS, and cool in the fridge.

3. Excise the strips containing the polypeptide of interest from the gel using a sharp scalpel. If the polypeptide was radiolabelled with ^{125}I, count the strips in the γ counter.

4. Cut into small pieces, rehydrate in buffer and remove the paper backing from the dried gel. This is most simply done in a small volume of ddH_2O in a petri dish.

5. Transfer the rehydrated slices to the larger well using a mini-spatula, fill with buffer and apply the voltage. Run at 500 V for 5 h.

6. Switch off the power pack. Pipette off the excess buffer until approximately 500 μl remains in the well containing the eluted sample. Using the remaining buffer, pipette vigorously to remove the sample from the dialysis tubing. Alternatively, the polarity can be reversed for 2–3 min prior to stopping the run to remove sample from the tubing.

7. Count the sample in the γ counter to estimate the efficiency of the elution.

8. The sample is now ready for use.

2.6
2D Gel Electrophoresis

2D gel electrophoresis relies upon the separation of proteins on the basis of their isoelectric point by isoelectric focusing (IEF; O'Farrell 1975), followed by separation on the basis of size by SDS-PAGE. The combination of procedures can sep-arate a complex mixture of proteins into many different components. As with peptide mapping, 2D electrophoresis is a useful method of investigating relationships between different polypeptides. A 2D separation does not guarantee that individual spots represent single polypeptides, but rather minimises the risk of comigration of polypeptides. Each protein has a different pH at which it is electrically neutral – the isoelectric point or pI.

In IEF, proteins are separated according to pI in a gel in which a stable pH gradient has been generated by the addition of ampholytes, extending from a low pH at the anode to a high pH at the cathode. Each protein in the mixture migrates until it reaches its pI, at which point migration ceases and the protein becomes concentrated into a narrow band. IEF is extremely useful for resolving proteins in the pH range 4–7; narrower or broader pH gradients can be utilised, with different sets of ampholines. Basic proteins are not well resolved in IEF, as the pH gradient collapses at the basic end and proteins run off the gel and are lost. In IEF, the gels are prerun to form a pH gradient prior to adding the samples, which are loaded at the basic end of the gel. NaOH is used in the upper chamber and H_3PO_4 in the lower.

Non-equilibrium pH gradient electrophoresis (NEPHGE; O'Farrell et al. 1977) is a modification of IEF which uses a different set of ampholines and separates proteins over a wider pH range. It is perhaps more useful for the initial analysis of a crude mixture of proteins. NEPHGE has the additional advantage that larger sample volumes can be analysed and there is no need for a prerun. In NEPHGE, proteins are loaded at the acidic end of the gel and migrate towards the basic end, but do not reach their pI, so that the spots are not usually so tightly focused on NEPHGE gels as on IEF gels. H_3PO_4 is used in the upper buffer chamber and NaOH in the lower and the electrodes are reversed, i.e. plug the red electrode into the black socket of the power pack and vice versa.

The pH gradient in IEF gels can be measured either by running in parallel commercially available coloured pI markers, or alternatively by running a gel without sample, chopping it into pieces after the run and incubating each piece in ddH_2O. The pH of the solution can then be measured, and is assumed to be similar to that of the gels plus samples. The method for NEPHGE is presented first, followed by the modifications for IEF.

2.6.1
Non-Equilibrium pH Gradient Electrophoresis (NEPHGE)

- Tube gel unit, such as GT 2 (Hoefer), including grommets,   **Equipment**
  stoppers
- Universal tube rack (not essential)
- Glass tubes, 13 cm length, 1.5 mm inner diameter
- Parafilm
- Gel pouring needles (BioRad)
- Gel extrusion needles (BioRad)
- Magnetic stirrer and bar
- Power pack, delivering up to 500 V
- Petri dishes
- SDS-PAGE equipment

- Ultrapure urea   **Reagents**
- 30% acrylamide
- TEMED
- 10% APS
- Ampholines
- 10% NP-40
- DTT
- 0.5 M Tris/HCl, pH 6.8
- 1 M NaOH
- 1 M H_3PO_4
- Overlay buffer
- IEF lysis buffer
- Equilibration buffer

- Gel mix for NEPHGE   **Solutions**
  (Sufficient for 16 tube gels of 9 cm height)

For 5.0 ml	
Ultrapure urea	2.75 g
30%T acrylamide	0.67 ml
10% NP-40	1.0 ml
ddH$_2$O	1.0 ml

pH 3.5–10 ampholines 250 μl
Dissolve the urea on the stirrer, with occasional warming in
the 37 °C water bath;

Note: do not heat, as heating can cause charge artefacts in the
urea.

| 10% APS | 10.0 μl |
| TEMED | 7.0 μl |

Pour gels immediately

- IEF lysis buffer (9.5 M urea, 2% NP-40, 2% ampholines, 50 mM DTT)

For 50 ml	
Urea	28.5 g
DTT	390 mg
NP-40	1.0 ml
Ampholines, pH 3.5–10	2.5 ml
ddH$_2$O	to 50 ml

Aliquot and store at –70 °C; warm gently when thawed to remove urea crystals

- Overlay buffer

| IEF lysis buffer | 100 μl |
| ddH$_2$O | 20 μl |

- Cathode (lower) buffer (working strength 20 mM NaOH)

For 100 ml	
×50 concentrated	
NaOH	4 g
ddH$_2$O	to 100 ml

Dilute 1:50 (to 20 mM) prior to use; for lower chamber, need 1000 ml.
Stock solution keeps 1 week at room temperature.

- Anode (upper) buffer (working strength 10 mM H$_3$PO$_4$)

For 100 ml	
×100 concentrated	
85% H$_3$PO$_4$	6.85 ml
ddH$_2$O	to 100 ml

Dilute 1:100 (10 mM) prior to use; for upper chamber need 500 ml
Stock solution keeps 1 week at room temperature

- Equilibration buffer (0.0625 M Tris/HCl, 2% SDS, 0.05 M DTT, 10% glycerol)

For 250 ml	
0.5 M Tris/HCl, pH 6.8	31.25 ml
10% SDS	50 ml
DTT	1.9 g
Glycerol	25 ml
ddH$_2$O	to 250 ml

Store at room temperature

- Sealing agarose

For 100 ml	
Agarose	1 g
0.5 M Tris, pH 6.8	25 ml
10% SDS	1 ml
ddH$_2$O	to 100 ml

Extracts to be analysed by 2D gel electrophoresis have to prepared in IEF lysis buffer. Take a known number of organisms and add to IEF lysis buffer; gentle homogenisation may help extract parasites such as nematodes. Spin out the insoluble residue at high speed in the microfuge. Sometimes this procedure can lead to the precipitation of protein at the top of the rod gel during the first dimensional run. In such cases, it may be more appropriate to make a detergent soluble extract (see Sect. 2.2.2) and to dilute it 1:1 in double strength lysis buffer.

Sample Preparation for 2D analysis

NEPHGE is most conveniently started first thing in the morning and the first dimension run through the day, with the second dimension SDS-PAGE run overnight.

Procedure, NEPHGE

1. Take clean 13-cm tubes, 1.5 mm inner diameter, and seal one end with Parafilm. Make a mark at 9 cm on the tube. All gels must be poured to the same height. Insert the tubes into the gel pouring stand. Alternatively, attach the tubes around a glass bottle with an elastic band.

2. Prepare the gel mix for NEPHGE (see Solutions), and pour the gels using a gel pouring needle attached to a 5.0-ml syringe. Tap the tubes to release any air bubbles trapped

beneath the surface. Always pour a few extra gels, in case of leaks or trapped air bubbles. Overlay with water.

3. Once the gels have polymerised (30–45 min), remove the water overlay, using a long needle, and load the sample directly; up to 200 μl can be loaded, although resolution is better with smaller volumes. Loading too much protein can result in precipitation at the top of the rod gel. Overlay with overlay buffer (see Solutions) and fill the tubes to the top with 10 mM H_3PO_4 (Anode buffer).

4. Fill the lower reservoir with 20 mM NaOH (Cathode buffer) and connect the cooling. Remove the parafilm from the rod gels, load the gels into the tank and stopper any unused holes. Add the H_3PO_4 to the upper chamber and check for leaks. Connect the electrodes in reverse and run at 400 V constant voltage for 4.5 h. Check the current at the start of the run (around 3 mA), this will drop in time to 0.3 mA.

5. Switch off the power pack and remove the rods one by one. Extrude the gels using a gel extrusion needle attached to a 10-ml syringe containing ddH_2O. Insert the needle between the gel and the tube and gently squirt the ddH_2O out of the syringe, rotating the tube as the needle moves gradually towards the bottom of the tube. As the needle is carefully withdrawn, the gel should slip out of the tube. Extrude the gels directly into small disposable petri dishes, remove any liquid and incubate in 5 ml equilibration buffer for 30 min. Gels are now ready for loading onto the second-dimension SDS-PAGE system or for freezing at -70 °C until analysis.

6. If gels are to be loaded directly, have the separating gel poured. Use a continuous comb with a single well for the stacking gel. Pour the stacking gel while the rod gels are equilibrating.

7. Pick the rod gel up between two glass hooks (made from Pasteur pipettes) and place on a spacer. Gently push the rod gel onto the dried surface of the stacking gel; if the surface is dry, the rod gel will adhere without further manipulation. If a good seal is required, a small amount of molten agar can be added to the stacking gel and the rod gel placed directly on this. Add the molecular weight standards to the well and electrophorese as normal. Try to load the tube gels always in the same orientation. If two

different samples are to be compared directly, then run the second dimension gels back to back. Notched glass plates are available which allow two gels to be poured within a single set of clamps – these "club sandwiches" are useful for analysing four SDS-PAGE gels simultaneously.

8. If the gels are to be frozen, place each gel on a piece of aluminium foil using glass hooks and freeze directly at –70 °C. Write the gel number on the aluminium foil, or alternatively place the tin foil in a labelled tube.

9. To clean the tubes – rinse in ddH_2O and place in chromic acid overnight. Rinse thoroughly in several changes of ddH_2O and then emerse in absolute ethanol. Dry and store in dust-free environment.

Note: Chromic acid is extremely caustic and should be handled with care.

2.6.2
Isoelectric Focusing (IEF)

It is most convenient to run IEF tube gels overnight and to start the SDS-PAGE during the following day. The method is essentially similar to that described for NEPHGE, with the following differences

• The ampholines used are different.

• The concentrations of catalyst and initiator are different.

• The rod gels are prerun to form the pH gradient in advance of loading the sample.

• The volumes which can be focused successfully are much smaller (around 20 μl).

• The polarities are reversed so that the upper chamber contains NaOH and the lower chamber contains H_3PO_4.

See 2.6.1, NEPHGE • Gel mix for IEF (Sufficient for 16 tube gels of 9 cm height)	**Equipment, reageants, solutions**

For 5 ml	
Ultrapure urea	2.75 g
30%T acrylamide	0.67 ml

10% NP-40	1.0 ml
ddH$_2$O	1.0 ml
pH 5–7 ampholines	200 μl
pH 3.5–10 ampholines	50 μl
Dissolve the urea on the stirrer, with occasional warming in the 37 °C water bath;	
Note: do not heat, as heating can cause charge artefacts in the urea.	
10% APS	7.0 μl
TEMED	4.5 μl

Pour gels immediately

- Cathode buffer (upper chamber): 20 mM NaOH, degas prior to use

- Anode buffer (lower chamber): 10 mM H$_3$PO$_4$

Procedure, IEF

1. Pour the gels using the IEF gel mix and overlay with ddH$_2$O.

2. Once the gels have polymerised, remove the ddH$_2$O and replace with 20 μl of IEF lysis buffer, using a gel loading tip on a 20 μl Gilson. Leave for a further 1–2 h.

3. Add H$_3$PO$_4$ (Anode buffer) to tank, attach the cooling, load the tubes into the apparatus and plug any unused holes with a stopper. Add degased NaOH (Cathode buffer) to upper chamber, check for leaks and start the prerun to form the pH gradient. Prerun at 200 V for 15 min, 300 V for 30 min and 400 V for 30 min.

4. Switch off the power, remove the cathode buffer and overlay and load the samples using a gel loading tip on a 20-μl Gilson. Load up to 20 μl. Overlay with 10 μl overlay solution, top up the tubes with NaOH and run overnight (16 h) at a constant voltage of 500 V (8000 V h).

5. At the end of the run, process the rod gels as described above.

References

Cleveland DW (1983) Peptide mapping in one dimension by limited proteolysis of sodium dodecyl sulphate-solubilised proteins. Methods Enzymol 96:222–229

Freedman DO, Nutman TB, Ottesen EA (1988) Enhanced solubilization of immunoreactive proteins from *Brugia malayi* adult parasites using cetyltrimethylammonium bromide. Exp Parasitol 65:244–250

Hames BD (1981) An introduction to polyacrylamide gel electrophoresis. In: Hames BD, Rickwood D (eds) Gel electrophoresis of proteins: a practical approach. IRL Press at Oxford University Press, Oxford, UK

Jones OT, Earnest JP, McNamee MG (1987) Solubilization and reconstitution of membrane proteins. In: Findlay JBC, Evans WH (eds) Biological membranes: a practical approach. IRL Press at Oxford University Press, Oxford, UK

Laemmli UK (1970) Cleavage of structural proteins during the assembly of the head of bacteriophage T4. Nature 227:680–685

Maizels RM, Blaxter ML, Robertson BD, Selkirk ME (1991) Parasite antigens, parasite genes. A laboratory manual for molecular parasitology. Cambridge University Press, Cambridge

O'Farrell PH (1975) High resolution two-dimensional polyacrylamide of proteins. J Biol Chem 250:4007–4021

O'Farrell PZ, Goodman HM, O'Farrell PH (1977) High resolution two-dimensional polyacrylamide of basic as well as acidic proteins. Cell 12:1133–1142

Pritchard DI, Crawford CR, Duce IR, Behnke JM (1985) Antigen stripping from the nematode epicuticle using the cationic detergent cetyltrimethylammonium bromide. Parasite Immunol 7:575–585

Schagger H, von Jagow G (1987) Tricine-sodium dodecyl sulphate-polyacrylamide gel electrophoresis for the separation of proteins in the range from 1 to 100 kDa. Anal Biochem 166:368–379

Chapter 3

Molecular Techniques in Analytical Parasitology

STEVEN HEATH

3.1
Introduction

Molecular biological techniques have revolutionised the field of parasitology over the past 20 years, particularly in the analysis of the parasite genome. However, more importantly, some of these techniques have been adapted for use in identification/diagnostic tests, providing a level of sensitivity and specificity which was otherwise unobtainable. Consequently, diagnostic parasitology has been endowed with new means to detect and identify parasite species, subspecies and even individual isolates.

Molecular parasitologists also ask intriguing questions such as: which genes are expressed at a particular stage in a parasite's life cycle? Are these genes important for the development of the parasite? What is the function of these genes? Molecular biological techniques are being used to answer these questions with a view towards identifying new targets for intervention.

3.2
Isolation of Parasite Nucleic Acid

Nucleic acid has been isolated from a wide range of parasites and the procedures used depend to a large extent on the source of material and whether one is isolating DNA or RNA. In general, extraction procedures use techniques and reagents which are capable of disrupting membranes or teguments and denaturing proteins. Both DNA and RNA are susceptible to nuclease degradation, and so great care must be taken with these procedures. Nuclease contamination can come from the source material itself after disruption or more likely from fingers, hair, dust and sneezes.

3.2.1
Preparation of DNA

- Refrigerated microfuge
- Vacuum dryer
- Water bath
- 1.5-ml sterile microfuge tubes
- DES solution (1 M urea/1 M lithium chloride/0.2% SDS/50 mM Tris. HCl/5 mM EDTA pH 8). Filter solution before addition of SDS
- Proteinase K (20 mg/ml)
- RNAse A (50 μg/ml)
- Propan-2-ol
- Phenol/chloroform (1:1)
- Chloroform
- Ethanol
- 95% ethanol

Equipment and reagents

1. Add 2 ml of DES solution to pellet of parasites.

2. Add 10 μl of proteinase K and shake gently until homogeneous.

3. Incubate at 50 °C for 1 h then cool to room temperature.

4. Add 1.1 ml of propan-2-ol and mix gently by inversion.

5. Centrifuge for 3 min in a microfuge and then remove the supernatant.

6. Dissolve the remaining DNA in 100 μl of TE containing 50 μg/ml of Rnase A.

7. Transfer DNA solution to a sterile 1.5-ml microfuge tube containing 100 μl of DES solution.

8. Add an equal volume of phenol/chloroform and mix thoroughly over a period of 5 min.

9. Centrifuge for 2 min to separate the phases.

10. Remove aqueous phase and repeat the phenol extraction.

Procedure

11. Finally, transfer the aqueous phase to a clean microfuge tube and precipitate the DNA by adding an equal volume of 95% ethanol.

12. Pellet the DNA by centrifugation for 10 min.

13. Wash the pellet in 95% ethanol twice and finally dry it under vacuum.

14. Dissolve pellet in TE buffer over several hours.

3.2.2
Preparation of RNA

Equipment and reagents
- Refrigerated microfuge
- Vacuum dryer
- 1.5 ml sterile microfuge tubes
- Solution D (4 M guanidinium thiocyanate/25 mM Sodium citrate pH 7/0.5% sarcosyl/0.1 M 2 mercaptoethanol)
- β-Mercaptoethanol
- 2 M sodium acetate pH 4
- Water-saturated phenol
- Chloroform/isoamyl alcohol (24:1)
- Isopropanol
- Diethylpyrocarbonate (DEPC)
- DEPC-treated water
- 75% ethanol (made up with DEPC-treated water)

Procedure

1. Add 20 μl of β-mercaptoethanol to 2.8 ml of solution D.

2. Add 0.5 ml of this solution to a pellet of parasite in a 1.5-ml microfuge tube.

3. Mix thoroughly and leave to stand at room temperature for 2 min.

4. Add 50 μl of 2 M Na acetate and mix thoroughly.

5. Add 0.5 ml of water-saturated phenol and mix by inversion.

6. Add 100 μl of chloroform/isoamyl alcohol and shake vigorously for 10 s and chill on ice for 15 min.

7. Centrifuge for 20 min and then transfer the upper aqueous phase to a sterile microfuge tube.

8. Add an equal volume of isopropanol and leave at –20 °C for 1 h.

9. Centrifuge at 10 000 g for 20 min at 4 °C and then discard the supernatant.

10. Dissolve the pellet in 150 μl of solution D and add 300 μl of isopropanol.

11. Chill at –20 °C for 1 h and then centrifuge again at 10 000 g for 20 min at 4 °C.

12. Discard the supernatant and wash the pellet in 75% ethanol.

13. Dry the pellet under vacuum and then dissolve it in an appropriate volume of DEPC-treated water.

3.2.3
Quantification of Nucleic Acid

DNA and RNA absorb in the ultraviolet part of the spectrum between 250 and 270 nm. Using the known molar extinction coefficients for the four bases at 260 nm, the concentration of solutions of nucleic acids can be calculated. Thus, using a 1 cm cuvette with a 1-cm path length, an absorbance of 1 is indicative of a concentration of double-stranded DNA of 50 $\mu g/ml$ and for RNA or single-stranded DNA of 40 $\mu g/ml$. The ratio of absorbance at 260 to 280 nm is a good indication of purity. For DNA solutions, values of 1.8 to 1.9 and RNA solutions, 1.9 to 2 are perfectly acceptable as a measure of purity. The presence of protein and or phenol will decrease this ratio significantly.

3.3
Gel Electrophoresis and Blotting of Nucleic Acid

3.3.1
Gel Electrophoresis

The ability to resolve different-sized fragments of nucleic acid is a very powerful technique for the molecular biologist. The exact size of fragments can be determined accurately by direct observation of gels which have had the fluorescent compound ethidium bromide added. Fractionation of DNA by gel electrophoresis is a very versatile technique and can be applied to DNA, ranging from a few bases to chromosome-sized fragments. DNA can be fractionated in either agarose or polyacrylamide, but for simple analysis and blotting procedures, agarose is the gel material of choice.

Near neutral pH, DNA is negatively charged and migrates from cathode to anode, its mobility dependent upon fragment size. Small linear DNA fragments migrate faster than larger ones, and DNA duplexes ranging from 70 bp (3% agarose) to 800 000 bp (0.1% agarose) can be easily separated in non-denaturing agarose gels. Single-stranded DNA can also be separated by agarose gel electrophoresis; however, a denaturant (e.g. formaldehyde) has to be added to the gel prior to electrophoresis. Techniques for fractionation of DNA can equally be applied to RNA. As most RNA encountered is single-stranded, a denaturing agarose gel technique is required for successful separation. It is important to note that gel electrophoresis of single-stranded RNA, in particular MRNA, should be performed using equipment and reagents that have been treated in order to inactivate or remove nucleases. Some laboratories even keep certain pieces of equipment for RNA use only.

Resolution of DNA in agarose is a function of the concentration of dissolved agarose. The migration rate is also dependent on molecular size, conformation and voltage gradient.

Equipment
- Horizontal agarose electrophoresis tank
- Transformer/power pack
- Agarose powder
- TAE buffer (40 mM Tris-acetate/1 mM EDTA pH 8)
- RNA running buffer (1×): 0.04 M morpholinopropanesulphonic acid (MOPS), pH 7/10 mM sodium acetate/1 mM EDTA
- Gel loading buffer: 30% Ficoll/1 mM EDTA/0.25% bromophenol blue/0.25% xylene cyanole FF
- Formaldehyde (37% in water, 12.3 M)

Preparation of a non-denaturing gel

1. Add powdered agarose to TAE buffer to give the correct percentage. The agarose is dissolved completely by heating, ensuring that the solution is completely homogeneous.

2. Allow solution to cool to approximately 50–60 °C before pouring and add ethidium bromide to a final concentration of 0.5 μg/ml.

3. Pour the gel into the mould and insert the well-forming comb.

4. After setting (~30 min) the apparatus is topped up with buffer until the gel surface is just submerged.

5. Run agarose gel at 5 V/cm.

Samples are loaded onto the gel by addition of a concentrated Ficoll/dye, EDTA solution. The limits of detection of a DNA band in agarose using ethidium bromide is 1–5 ng of double-stranded DNA. For analysis of genomic DNA, approximately 5–10 μg of DNA should be loaded per lane.

Formaldehyde-agarose gels are routinely used to fractionate RNA prior to transfer to hybridisation membranes such as nitrocellulose and nylon. Formaldehyde solutions and gels should always be handled under a fume hood.

Preparation of a denaturing gel

1. Formaldehyde (12.3 M) and 5× RNA running buffer are added to the desired percentage of molten agarose in water at 50–60 °C, to give final concentrations of 1× RNA buffer and 2.2 M formaldehyde.

2. Up to 20 μg of RNA in 5 μl of sterile water is heated to 55 °C for 15 min and then 2 μl of 5× RNA buffer is added along with 10 μl of freshly deionised formamide.

3. Finally, 3.5 μl of 12.3 M formaldehyde is added along with 2 μl of sterile gel loading buffer.

4. In terms of resolution, best results are achieved by running the agarose gels at <5 V/cm.

Nucleic acids that have been denatured using formaldehyde are not visualised efficiently with ethidium bromide. A useful tip for sizing RNA transcripts is to cut off and isolate the marker lanes and remove formaldehyde by washing in water (4 × 30 min) and 0.1 M ammonium acetate (2 × 30 min) and staining for 1 h in 0.1 M ammonium acetate, 0.1 M β-mercaptoethanol containing 0.5 μg/ml ethidium bromide.

Technical hint

3.3.2
Blotting of Nucleic Acid

One of the most powerful tools of the molecular biologist is the ability to fractionate nucleic acids and to determine which of the fragments has complementary sequences to a DNA or RNA probe. The method of transferring size-fractionated DNA from a gel to an inert support matrix followed by hybridisation to a labelled probe is referred to as a Southern blot. The identical procedure using RNA is referred to as a Northern blot.

The Southern blotting technique for transferring fractionated DNA to nitrocellulose produces a faithful replica of a high-resolution gel by a passive diffusion process. The mechanism of binding of nucleic acids to nitrocellulose is unknown, but it is believed to be a non-covalent interaction. Nylon is also used for blotting purposes; however, nitrocellulose remains the material of choice even though it does have some disadvantages, for example it binds fragments of 200–300 bases poorly.

Preparation of blots for Southern transfer

1. Fractionate DNA in agarose gel as previously described using 10–40 μg of restriction enzyme digested DNA.

2. Float nitrocellulose (cut to the approximate size of the gel) on water to wet it thoroughly and then soak in blotting buffer (20× SSC).

3. If fragments of >8 kb are to be transferred, the gel should be soaked in 0.25 M HCl for 10 min.

4. Rinse the gel with water to remove excess acid and then denature by soaking it in 1.5 M NaCl/0.5 M NaOH for 15 min. This step should be repeated.

5. Neutralise the gel by soaking in 3 M NaCl/0.5 M Tris-HCl (pH 7.4) for 15 min. This step should be repeated.

6. Arrange three sheets of Whatmann 3 MM filter paper in a dish (cut to 3 mm larger than gel). Saturate paper with blotting buffer and place the gel on top. Ensure all exposed paper surfaces with plastic wrap.

7. Place the nitrocellulose on top of the gel and remove air bubbles. Add one sheet of filter paper saturated with

blotting buffer on top of the nitrocellulose followed by a stack (1–2 inches) of paper towels and a light weight to keep the blot compressed.

8. Allow the transfer to proceed overnight. The efficiency of transfer can be checked by restaining the gel. As a rule of thumb, when the thickness of the gel has reduced to about 1 mm, the transfer can be stopped, as the concentration of gel at this stage is too great to allow any further transfer.

9. Place blots between two sheets of thick filter paper and bake in a vacuum oven at 80 °C for at least 2 h when using nitrocellulose.

1. After fractionation of RNA in a denaturing gel system, the marker lanes can be removed and stained as previously described.

2. Transfer the gel to nitrocellulose overnight as described for Southern blotting using either 10 or 20× SSC.

3. Dry the blot as described for Southern transfer.

4. If attempting to identify rare mRNAs (< 0.01%), it is important to use 1–10 μg of Poly (A)$^+$ RNA or an equivalent amount of total cellular RNA [assuming Poly (A)$^+$ RNA is 1–5% of total].

Preparation of blots for Northern transfer

3.4
Labelling and Hybridisation of DNA Probes

3.4.1
Radiolabelling of DNA Probes

There are a large number of techniques for introducing ^{32}P into DNA and RNA, such as end labelling RNA with ATP, filling in staggered DNA ends, primer extensions on DNA and RNA templates, to name but a few. However, by far the commonest methods for labelling DNA using DNA modifying enzymes are

- Nick translation (Rigby et al. 1977) and

- Random hexamer priming

which are discussed in this chapter. Both of these techniques can be modified to incorporate, for example, biotin-labelled nucleotides, although the consensus of opinion is that non-radioactive probes are less sensitive than their isotopically labelled counterparts and would be unable to identify rare genes or transcripts.

Labelling DNA by Nick Translation

Equipment and reagents
- 600 μl sterile microfuge tubes
- 1 ml syringes
- dNTPs (20 mM)
- [α^{32}P] dATP (3000 Ci/mmol, 10 μci/μl, Amersham)
- DNase (60 pg/μl in 50 mM Tris. HCl pH 7.4, 10 mM MgCl$_2$, 1 mg/ml DNase-free BSA)
- 10× Nick translation buffer (NT buffer: 50 mM MgCl$_2$, 250 μM dNTPs, 50 mM Tris. HCl pH 7.4)
- E. coli DNA polymerase I (0.8 units/μl)
- 250 mM EDTA
- Sephadex G-50
- NET (10 mM Tris-HCl pH 8, 50 mM NaCl, 0.1 mM EDTA pH 8)

Nick translation

1. Add 2 μl of 10× NT buffer to 8 μl of DNA (100–400 ng).

2. Then add 1 μl of DNase and distilled water up to 19 μl.

3. Add 1 μl of DNA polymerase (Pol I, 0.8 units/μl) and incubate at 15 °C for 1–2 h.

4. Add 0.1 vol of 250 mM EDTA to stop the reaction.

5. Remove unincorporated nucleotides by passing the reaction mixture over a Sephadex G-50 column in 10 mM Tris-HCl pH 8, 50 mM Nacl, 0.1 mM EDTA pH 8 in a 1 ml syringe.

Technical hints Failure of this labelling reaction is invariably due to poor quality DNA. This can sometimes be eleviated by performing a phenol: chloroform extraction. If this does not solve the problem, then the amounts of DNase and Pol I can be increased.

Labelling DNA by Random Hexamer Priming

This procedure produces probes of very high specific activity and is optimised for probes in the size range of 600–800 nucleotides (Feinberg and Vogelstein 1983).

- Boiling water bath
- Klenow fragment (3 units $/\mu l$)
- Random primer buffer (0.17 M Tris. HCl/17 mM $MgCl_2$/ 1.33 mg/ml BSA/18 OD_{260} units/ml random primers/pH 6.8)
- $[\alpha^{32}P]$ dATP (3000 Ci/mmol, 10 $\mu ci/\mu l$, Amersham)
- dNTPs (20 mM)
- 250 mM EDTA
- 1 ml syringes
- Sephadex G-50
- NET (10 mM Tris-HCl pH 8, 50 mM NaCl, 0.1 mM EDTA pH 8)

Equipment and reagents

1. Denature 25 ng of DNA dissolved in 5–20 μl of distilled water in a 600 μl microfuge tube by heating in a boiling water bath for 5 min. Immediately cool on ice.

2. Mix together 2 μl of dATP, dGTP, dTTP, 15 μl of random primers buffer mixture, 5 μl (50 μci) of $[\alpha^{32}P]$ dCTP (3000 Ci/mmol) and then add distilled water to a total volume of 49 μl.

3. Add 1 μl of Klenow fragment and mix thoroughly but gently.

4. Incubate at 25 °C for 1 h.

5. Stop the reaction by adding 0.1 vol EDTA.

6. Remove unincorporated nucleoside triphosphate by passing down a 1 ml column of Sephadex G-50 in 10 mM Tris-HCl pH 8, 50 mM NaCl, 0.1 mM EDTA pH 8.

Random hexamer priming

Incubation times longer than 1 h may give higher specific activities; however, the level of incorporation tails off after approximately 2 h at 25 °C.

Technical hints

3.4.2
Non-Isotopic Labelling of DNA Probes

The use of non-radioactive DNA detection systems is becoming increasingly popular in molecular parasitology research. At present, the most popular methods involve either fluorescence, colorimetric staining or chemiluminescence. This section describes the labelling of DNA with a biotinylated nucleotide and subsequent detection using chemiluminescence (for review see Beck and Koster 1990). DNA probes can be labelled with biotin via a number of different enzymatic procedures by simply replacing one nucleotide with a biotinylated derivative which are now commercially available. The detection regime would normally involve a "streptavidin bridge" comprising a streptavidin-labelled alkaline phosphatase. Synthetic oligonucleotide probes can be 3′ end-labelled by using a terminal transferase adding one or more biotinylated nucleotides (Moyzis et al. 1988). Cloned DNA probes can be labelled with biotin by either random hexamer priming using Klenow fragment or nick translation using DNA polymerase I (see above). Using these procedures dTTP is normally replaced with biotin-11-dUTP in standard reactions. A significant advantage of using biotinylated probes over ^{32}P labelled probes is that the former are stable for 6–12 months.

Equipment and reagents

- Water bath
- 600 μl sterile microfuge tubes
- 1 ml syringes
- Biotin-11-dUTP (0.4 mM in 100 mM Tris. HCl (pH 7.5))
- Reagents for random hexamer priming (see above)
- Reagents for Nick translation (see above)

Procedure See procedure labelling by Nick translation or by random hexamer priming.

3.4.3
Hybridisation and Signal Detection

Southern or dot blots should be prepared by standard protocols using nylon membranes. For optimum signal-to-noise ratios, the target DNA should first be either cross-linked to the membrane by ultraviolet light or fixed by baking membranes at 80 °C.

<table>
<tr><td>

- Heat-sealable plastic bags
- Heat sealer
- Water bath
- Orbital shaker
- Nylon membrane
- X-ray film (Fuji medical), film cassette and film development facility
- 10× SSC (1× SSC= 0.15 M NaCl, 0.015 M sodium citrate)
- 10% (w/v) SDS
- 0.25 M disodium phosphate pH 7.2
- 0.25 M EDTA
- 5% BSA
- Tween 20
- Substrate buffer (0.1 M diethanolamine, 1 mM $MgCl_2$ pH 10)
- AMPPD (1,2-dioxetane-3, 2tricyclo [3-3-1-1] decan]-4-yl) phenyl phosphate

</td><td>

Equipment and reagents

</td></tr>
</table>

1. Wet the nylon membrane in 5× SSC.

2. Incubate in hybridisation buffer: 0.5% (w/v) polyvinyl-pyrrolidone (PVP), 1 mM EDTA, 1 M NaCl, 5% (w/v) dextran sulphate, 0.2% heparin, 4% (w/v) SDS, 50 mM Tris. HCl pH 7.2 for 1 h at 55 °C.

3. Heat-denature the biotinylated probe in hybridisation buffer (5–20 fmol/ml) by boiling in a water bath for 3–5 min.

4. Add the probe-hybridisation mixture to the membrane in a heat-sealable bag and incubate overnight at 65 °C.

5. Wash the membrane as follows: twice for 5 min each in 2× SSC, 1% (w/v) SDS, twice for 15 min in 0.1× SSC, 1% SDS at 65 °C, twice for 5 min each at room temperature in 1× SSC.

6. Chemiluminescent detection of bound probe is performed by initially washing the membrane twice for 5 min in blocking buffer: PBS containing 2% (w/v) BSA, 0.1% (w/v) Tween 20.

7. Incubate the membrane in the same blocking buffer for 30 min.

Hybridisation with biotinylated DNA probe

8. Then incubate the membrane with streptavidin-alkaline phosphatase (1: 5–10 000) in blocking buffer minus Tween 20 for 30 min at room temperature.

9. Wash the membrane for 5 min in blocking buffer and four times (5 min each) in 0.3% (w/v) Tween 20 in PBS.

10. Then wash the membrane twice (5 min each) in substrate buffer.

11. The membrane is finally incubated for 5–10 min in 0.25 mM AMPPD in substrate buffer.

12. After removing excess substrate from the membrane, it is sealed in a heat-sealable bag and exposed to film for 10–20 min.

13. The procedure for hybridisation and detection of ^{32}P-labelled probes is as above (steps 1–5) and the membrane is finally placed wet into a heat-sealable bag and exposed to Fuji X-ray film.

3.5
Polymerase Chain Reaction (PCR)

The PCR is an in vitro procedure for synthesising enzymatically, specific sequences of DNA (Mullis and Faloona 1987). This technique utilises two oligonucleotide primers that flank the region of interest and hybridise to opposite strands of the target DNA. DNA amplification is accomplished by a series of repetitive cycles involving template denaturation, primer annealing and the extension of the annealed primers using the thermostable *Taq* DNA polymerase. The result of this amplification is the exponential accumulation of DNA target sequence whose ends are defined by the 5′ termini of the primers. As amplification product from one cycle can be utilised as template in further cycles the number of copies of the target DNA doubles every cycle. This can result in a million-fold increase in the number of copies of target DNA.

The PCR technique is a relatively complicated biochemical reaction of molecules which has constantly changing kinetic interactions, and it is probably impossible to describe a single set of conditions which will provide the desired product in all situations. There are, however, certain guidelines which should be adhered to when attempting to perform the technique. Firstly, primer selection is extremely important. Where possible, primers should be selected from regions which possess a GC content which is similar to that of the fragment being amplified. They should not possess complementary regions, particularly at their 3' ends. This will prevent the formation of "primer dimers". Finally, primers should not be selected from regions that possess significant secondary structure. Computer programmes are available that can identify such areas (e.g. Sqiggles – University of Wisconsin).

Alterations in the PCR reaction buffer can dramatically alter the outcome of a reaction. In particular, the concentration of $MgCl_2$ can have a significant effect on the specificity and yield of a product. Generally, concentrations of 1.5 mM are optimal but if circumstances dictate, then Mg^{2+} may need to be added. Higher concentrations may lead to formation of non-specific amplification products by stabilising the primer-template complex. The concentrations of deoxynucleotide triphosphates are normally present in the reaction at 200 μM each. Higher concentrations may lead to misincorporations by the polymerase. *Taq* polymerase is typically used in PCR reactions at about 2.5 units per 100 μl; however, if amplification reactions involve DNA with a high sequence complexity it may be necessary to increase the enzyme concentration up to 4 units per 100 μl.

In a typical cycling reaction, double-stranded DNA is first denatured briefly at 90–95 °C; the primers are allowed to anneal to their complementary target sequence by rapidly cooling to 40–60 °C. The annealed primers are extended with *Taq* polymerase by heating to 70–75 °C. The melting temperature of the primers is defined by their GC content and, as a rough guide, can be calculated as follows: T_m= (2 °C for AT+4 °C for GC). The times of incubation periods will vary according to the length of the target sequence being amplified.

3.5.1
Reverse Transcriptase – PCR (RT-PCR)

At any instant, only a proportion of all the genes in an organism's genome will be active, and consequently RNA will be transcribed only from this group of genes. A useful way to compare developmental gene expression is to isolate messenger RNA from a specific developmental form and make a complementary copy (cDNA) using the enzyme reverse transcriptase, and then perform a standard PCR amplification using the cDNA as a template. In this way, identification of an amplification product is indicative of the presence of an active gene within a particular developmental form.

Equipment and reagents
- Automated thermal cycler
- 600 μl sterile microfuge tubes
- PCR buffer: 50 mM KCl/10 mM Tris HCl (pH 8.3)/2.5 mM MgCl$_2$/0.1 mg/ml gelatin
- dNTPs (20 mM)
- Random hexamer primer
- Oligo dT primer
- Upstream and downstream primers
- MuLV reverse transcriptase (BRL)
- RNasin (1 unit/μl)
- *Taq* DNA polymerase
- Sterile dH$_2$O
- Ice bath

Basic technique

1. Heat 1–5 μg of total RNA in a microfuge tube to 65 °C for 2 min in PCR buffer together with 100 pmol of random hexamer primer.

2. Then add each dNTP to a final concentration of 1 mM, 1 μl of RNasin and 200 units of MuLV reverse transcriptase. Adjust the volume to 20 μl with sterile dH$_2$O.

3. Incubate at room temperature for 10 min and then at 42 °C for a further 60 min.

4. Stop the reaction by heating to 95 °C for 5 min and then quickly chilling on ice.

5. Then add 80 μl of PCR buffer containing 1 μM of each upstream and downstream primer and 2.5 units of *Taq* polymerase. Overlay the reaction mixture with 75 μl of liquid paraffin to reduce evaporation.

6. Denature at 95 °C for 5 min, anneal at T$_m$–5 °C for 5 min and extend at 72 °C for 5 min. Peform this "low stringency" routine for two cycles.

7. Perform "high stringency" cycles by decreasing the incubation times to 1 min and increasing the primer annealing temperature to either T_m or T_m+5 °C. This routine comprised 30 cycles.

8. After completing the amplification, the aqueous phase containing DNA can be removed by briefly vortexing with 100 μl of chloroform to remove the liquid paraffin. An aliquot (10 μl) of the sample can then be run on a NuSieve agarose (1–3%) gel for analysis.

In order to obtain optimum amplification, the above procedure should be performed using a range of template and primer concentrations. The magnesium ion concentration may effect: primer annealing, formation of primer dimer artefacts, enzyme activity and fidelity (Innis et al. 1988). *Taq* DNA polymerase requires free magnesium in addition to that bound by template DNA, primers and DNTPS. A range of Mg^{2+} concentrations (in the form of $MgCl_2$) up to 6 mM should be assessed for the ability to produce specific amplification of target DNA. The temperature and length of time for primer annealing should be optimised for each primer-template combination. An applicable starting annealing temperature is T_m-5 °C.

Technical hints

3.5.2
Touchdown PCR

Mispriming of either one or both oligonucleotide primers usually gives rise to spurious smaller bands in the amplification product. Unfortunately, these products can become the major species in a reaction due to a stochastic advantage that shorter misprimed products have over the longer correct amplification product during cycling. This problem appears to be worse when there is only a very small amount of template, and can sometimes be alleviated by adjusting the Mg^{2+} concentration. Don et al. (1991) have devised a modified cycling procedure which greatly reduces the chances of these misprimed products appearing in the final amplification reaction. The reaction begins at or above the expected annealing temperature. The annealing temperature is then reduced by 1 °C every second cycle from 65 °C to a "touchdown" temperature of 55 °C, at which point ten cycles are carried out. It has been estimated that a 5 °C difference between the correct and incorrect T_m can lead to a 4^5 (1024)-fold advantage of correct over spurious product.

References

Beck S, Koster H (1990) Anal Chem 62, 2258
Don RH, Cox PT, Wainwright BJ, Baker K, Mattick JS (1991) Nucl Acids Res 19(14), 4008
Feinberg AP, Vogelstein B (1983) Anal Biochem 132, 6
Innis MA, Mayambo KB, Gelfand DH, Brown MAD (1988) Proc Natl Acad Sci USA 85, 9436
Moyzis RK, Buckingham JM, Cram LS, Dani M, Deaven LL, Jones MD, Meyne RL, Ratliff RL, Wu JR (1988) Proc Natl Acad Sci USA 85, 6622
Mullis KB, Faloona F (1987) Meth Enzymol 155:335
Rigby PWJ, Dieckmann M, Rhodes C, Berg P (1977) J Mol Biol 113, 237

Chapter 4

The Production and Analysis of Helminth Excretory-Secretory (ES) Products

ALAN BROWN, GARY GRIFFITHS, PETER MICHAEL BROPHY,
BARBARA ANNE FURMIDGE, and DAVID IDRIS PRITCHARD

4.1
Introduction

Parasitic helminths secrete and excrete a wide range of molecules during culture in vitro. Given the limitation of these culture conditions, one must presume that the molecules secreted have a biological function, and a role to play in maintaining the homeostasis of the host/parasite relationship. In the present chapter we have attempted to review the conditions already used to maintain a range of parasites considered to be of medical, economic and biological importance. In doing so, we have also attempted to draw the attention of the reader to some of the more chemically defined molecules found in culture supernates or excretory-secretory (ES) products. Particular importance has been placed on molecules considered to be essential to the maintenance of the life cycle of these parasites, as these molecules probably represent an "Achilles heel" of many of the parasite species described.

It should also be noted that some of the parasites considered in this chapter produce molecules of potential therapeutic importance, and two examples are worthy of further illumination. The integrin blocker, NIF, produced by *A. caninum*, is being currently considered as an anti-inflammatory agent for use in many disease states of man. Similarly, it is highly likely that the human hookworm *Necator americanus* produces an integrin blocker which prevents human platelet aggregation.

In this sense, improvements in parasite culture and further chemical characterisation of the contents of culture supernatants will, in time, lead to the exploitation of the parasite by man, a classic case of biological role reversal.

4.2
Collection of Excretory-Secretory (ES) Products

When collecting ES products from helminths, a number of important factors must be taken into consideration to ensure that the supernates collected contain only material excreted or secreted by the parasite, and is free from host and bacterial contamination.

Removal of the Parasite from the Host

In removing the parasite from the host, care has to be taken to avoid contamination with host material. This is particularly important when the parasite is attached to areas such as the ileum, where removal by forceful methods would result in parasite damage and contamination of subsequent cultures with material from the host's intestine. Various strategies have been developed to overcome this, most of which involve allowing the parasite to detach voluntarily from the gut into a culture medium, thus avoiding the removal of a bolus of host tissue (Hotez and Cerami 1983; Knox and Jones 1990; Blackburn and Selkirk 1992; Brown and Pritchard 1993). In the case of *S. mansoni*, adult worms may be removed from the liver of infected animals by perfusion with Dulbecco's balanced salt solution (Rotmans et al. 1981) or 0.85% sodium chloride/0.75% sodium citrate solution (Dresden et al. 1981). Parasites should then be washed thoroughly and cultured for a "sterilising period" to allow any host material adsorbed onto the cuticle, or ingested by the parasite to detach or be secreted prior to the main culture period (see below).

Choice of Media

The media chosen must be capable of supporting the stage of the helminth life cycle concerned in a viable condition under (as far as possible) stress-free conditions. Usually, this is a tissue culture medium containing various supplements according to the individual parasite's requirements. This also means that a suitable method of assessing viability is required. In most cases, this is motility (Rotmans et al. 1981;

Hill et al. 1993; Griffiths and Pritchard 1994) although, in some cases, parasite secretions (AChE) have been monitored as a measure of parasite "well being" (Burt and Ogilvie 1975). Another important consideration in the choice of culture medium is the purpose for which the ES products are required. For example, culture media containing serum will contain some protease inhibitors, e.g. α_2 macroglobulin and intrinsic acetylcholinesterase activity (e.g. foetal calf serum is an extremely rich source of AChE), and will, therefore, be unsuitable for either the purification or monitoring of parasite AChE and proteolytic activities.

Monitoring Possible Bacterial Contamination

Once removed from the host, every effort must be made to avoid bacterial and fungal contamination of subsequent ES cultures. The key to this is extensive washing of the parasite (at least 1 h) with either a suitable physiological salt solution, e.g. Hanks' balanced salt solutuion (HBSS) or culture medium, supplemented with antibiotic (penicilin, streptomycin) and antifungal agents (amphotericin B, nystatin). The concentrations used will vary according to the parasite species involved but should be high enough to kill contaminating bacteria and fungi without causing damage to the parasite itself. Only after this period of "surface sterilisation" should parasites be cultured for ES products. After the culture period, the ES products collected should be tested for bacterial contamination by streaking a small aliquot onto either nutrient agar or LB agar (Brown and Pritchard 1993; Hill et al. 1993) containing no antimicrobial or antifungal agents. The plates should be incubated at 37 °C overnight. Any cultures showing signs of contamination should be discarded immediately.

After collection, ES products may be filtered through a 0.22-μm filter and, if necessary, concentrated by freeze drying or by using a centrifugal concentrator, e.g. Centricon (Amicon) with a 10-kDa cutoff, although the reader should be aware that some molecules of interest may be smaller than 10 kDa. Centrifugal concentrators with smaller molecular weight cutoffs are available. Finally, ES products should be stored at −20 to −40 °C until required.

Table 1. Conditions and media published for some common Helminth species

Helminth	Media	Time and temperature	Special conditions	Reference
Ancylostoma duodenale	RPMI 1640 supplemented with 100 i.u./ml penicillin, 100 μg/ml streptomycin and 2 mM L-glutamine	37 °C for 24 h	None	Pritchard (unpubl.)
Ancylostoma ceylanicum	RPMI 1640 supplemented with 100 i.u./ml penicillin, 100 μg/ml streptomycin and 2 mM L-glutamine	37 °C for 24 h	None	Garside et al. (1989)
Ancylostoma caninum	Eagles minimal essential medium supplemented with 200 i.u./ml penicillin and 100 μg/ml streptomycin	37 °C for 24 h	5% carbon dioxide	Carrol et al. (1984)
Brugia malayi (adults)	RPMI 1640 supplemented with 100 i.u./ml penicillin, 100 μg/ml streptomycin 25 mM Hepes and 1% glucose	37 °C for up to 7 days	5% carbon dioxide	Kwan-Lim et al. (1989)
Brugia malayi (larvae)	Hanks balanced salt solution (HBBS)	28 °C for 24 h	None	Kharat et al. (1989)
Fasciola hepatica (adults)	NCTC-135 supplemented with 25 i.u./ml penicillin and 25 mg/ml streptomycin	37 °C for 24 h	5% carbon dioxide	Fairweather et al. (1987)

Fasciola hepatica (juveniles)	RPMI 1640 supplemented with 0.2% glucose, 2% foetal calf serum, 25 i.u./ml penicillin and 25 mg/ml streptomycin	37 °C for 24 h	5% carbon dioxide	Smith and Clegg (1981)
Haemonchus contortus (adults)	Phosphate-buffered saline (PBS) supplemented with 1% glucose, 500 i.u./ml penicillin and 5 mg/ml streptomycin	37 °C for 8 h	None	Knox and Jones (1990)
	Dulbecco's modified eagle medium (DMEM) supplemented with 100 i.u./ml penicillin and 0.1 mg/ml streptomycin	37 °C for 22 h	None	Karanu et al. (1993)
Haemonchus contortus (larvae)	Phosphate buffered saline (PBS) supplemented with 1% glucose, 500 i.u./ml penicillin and 5 mg/ml streptomycin	37 °C for 8 h	None	Knox and Jones (1990)
Heligmosomoides polygyrus	RPMI 1640 supplemented with 100 i.u./ml penicillin, 100 µg/ml streptomycin and 2 mM L-glutamine	37 °C for 24 h	None	Lawrence and Pritchard (1993)

Table 1 (*Contd.*)

Helminth	Media	Time and temperature	Special conditions	Reference
Necator americanus (adults)	RPMI 1640 supplemented with 100 i.u./ml penicillin, 100 μg/ml streptomycin and 2mM L-glutamine	37 °C for 24 h	None	Brown and Pritchard (1993)
	Hanks balanced salt solution supplemented with 100 i.u./ml penicillin, 100 μg/ml streptomycin and 0.5% hydrolysed lactalbumin (HLac)	37 °C for 24 h	None	Burt and Ogilvie (1975)
Necator americanus (larvae)	RPMI 1640 supplemented with 100 i.u./ml penicillin, 100 μg/ml streptomycin and 2mM L-glutamine	37 °C for 72 h	Exsheathment of larvae required prior to culturing	Kumar and Prichard (1992)
Nippostrongylus brasiliensis	RPMI 1640 supplemented with 100 i.u./ml penicillin, 100 μg/ml sreptomycin, 2% glucose and 2mM L-glutamine	37 °C for 72 h	5% carbon dioxide	Blackburn and Selkirk (1992)
Schistosoma mansoni (adults)	H-199 medium supplemented with 100 i.u./ml penicillin, 100 μg/ml streptomycin, 25 μg/ml chloramphenicol, miconazol, 20 mM Hepes, 5 mM sodium bicarbonate and 30 mM glucose	37 °C	8% carbon dioxide. Cultures incubated in the dark	Rotmans et al. (1981)

Species	Medium	Conditions	Comments	Reference
Schistosoma mansoni (schistosomula)	Hanks balanced salt solution	37 °C for 16 h	Mechanical or skin preparation of shistosomula from cercariae required	Verwaerde et al. (1988) Ramalho-Pinto (1974) Clegg and Smithers (1972)
Schistosoma mansoni (cercariae)	*Biomphalaria glabrata* embryo culture medium supplemented with 20% foetal calf serum, 100 i.u./ml penicillin, 100 μg/ml streptomycin and 110 μg/ml gentamycin	26 °C	None	Lodes and Yoshino (1989)
Trichinella spiralis	RPMI 1640 supplemented with 100 i.u./ml penicillin, 100 μg/ml streptomycin and 2mM L-glutamine	37 °C for 24 h	None	Jarvis and Pritchard (1992)
Trichostrongylus colubriformis (larvae)	Tris saline buffer pH 7.6 (3.03 g/l Tris; 0.04 gl EDTA; 8.5 g NaCl/l)	41 °C for 4–6 h	None	O'Donnell et al. (1989)
Trichostrongylus colubriformis (adults)	*Trichostrongylus* basal medium (yeast extract 50 g/l, glucose 25 g/l, K_2HPO_4 4 g/l, KH_2PO_4 4 g/l, pH 7.4) supplemented with penicillin 1.2 g/l, streptomycin 2 g/l S-fluorcytosine 0.01 g/l and amphtericin B 0.01 g/l	37 °C for 96 h	None	Griffiths and Pritchard (1994)

Table 1 (*Contd.*)

Helminth	Media	Time and temperature	Special conditions	Reference
Trichuris suis	RPMI 1640 supplemented with 500 i.u./ml penicillin, 500 μg/ml streptomycin, 1.25 μg/ml amphotericin B, 350 μg/ml chloramphenicol and 1% glucose	37 °C for 9–10 days	5% Carbon dioxide	Hill et al. (1993)
Wuchereria bancrofti (microfilariae)	Equal proportions of Leibovitz 15, NCTC 135 and IMDM	37 °C for 24h	5% Carbon dioxide	Subrahmanyam (1990)
Wuchereria bancrofti (larvae)	Hanks balanced salt solution	28 °C for 48 h	None	Kharat et al. (1989)

Culture Conditions Recommended for Some Common Helminths

Table 1 lists the conditions and media published for the culture of some common helminth species.

4.3
Acetylcholinesterase

Acetylcholinesterase (AChE) (EC 3.1.1.7) is actively secreted (Rathaur et al. 1987) by many nematodes (Ogilvie et al. 1973), including: *Trichostrongylus* spp. (Rothwell et al. 1973; Jones and Knox 1990), *N. americanus* (Pritchard et al. 1991), *H. polygyrus* (Mallett 1989; Lawrence and Pritchard 1993), *N. brasiliensis* (Blackburn and Selkirk 1992) and *Oesophagostomum* spp. (Bremner et al. 1973). A definitive function for secretory AChE in the host-parasite relationship has yet to be discovered, although there are many suggested functions based on the known properties and functions of vertebrate AChE. Possible roles include: modulation of the host's immune response (Kaliner and Austen 1974), inhibition of mucus secretion by goblet cells (Philipp 1984) or a possible detoxification role (Sutherland and Lee 1993). These, and other possible roles, are reviewed elsewhere (Pritchard 1993).

AChE is a highly polymorphic enzyme. Simply stated, AChE forms consist of multiples of a basic catalytic subunit (approx. 75 kDa), associated with or without a collagen-like "tail". Six major forms of AChE have been identified in vertebrates and these can be classified into either asymmetric or globular forms.

Globular forms consist of monomers, dimers or tetramers of catalytic subunits, termed G_1, G_2 and G_4 respectively. These forms range from totally soluble (hydrophilic) species to species with hydrophobic glycophospholipids attached to the catalytic subunit (amphiphilic). Asymmetric forms consist of 1, 2 or 3 groups of catalytic tetramers, disulphide-linked to a collagen-like tail. These are known as A_4, A_8 and A_{12} forms, respectively. The structural aspects of AChE are considerably more complex and subtle than described here, and more comprehensive reviews may be found by consulting the following: Taylor et al. (1987); Toutant and Massoulie (1987); Maelicke (1991); Taylor (1991). In vertebrates, this heterogeneous collection of molecules is generated from a single gene, by alternative splicing of transcripts, and by post-translational modification of the precursors. In contrast, only globular forms have been described so far in helminths; for example the major AChE form in the trematode S. mansoni is a G_2 dimer attached to the tegument by a phosphatidylinositol anchor (Camacho et al. 1994; Espinoza et al. 1988) The nematodes *N. americanus* and *T. colubriformis* secrete a hydrophillic G_2 dimer (Griffiths and Pritchard 1994; Pritchard et al. 1994). It has been shown that AChE production by *Caenorhabditis elegans* is coded by three different genes (ace-1, ace-2 and ace-3; Johnson and Russell 1983). This genetic heterogeneity apparently results in three classes of catalytic subunit (A, B and C), differing in their inhibitor and substrate specificities.

Kinetic analysis, which consists of: substrate affinity, inhibitor sensitivity, detergent sensitivity and thermolability (thermal inactivation), reveals that AChE forms can be categorised into one of three distinct kinetic classes termed A, B or C.

Class A AChEs have a lower K_m, i.e. higher substrate specificity, are most sensitive to inhibitors and are markedly sensitive to inhibition with 0.05% Triton X-100 Class B AChEs are the most common and have a higher K_m, i.e. lower substrate specificity, and are less sensitive to inhibitors and Triton X-100. Class C AChEs have a very low K_m, i.e. show a very high affinity for acetylcholine, are highly insensitive to carbamate and organophosphate inhibitors, and are markedly sensitive to thermal inactivation. The latter form has been identified as a minor component (5%) in only a few species, e.g. *C. elegans* and *H. glycines*. (Johnson and Russell 1983; Chang and Opperman 1991, 1992).

The following methods describe how to identify, purify and characterise AChE from parasite ES products.

4.3.1
Quantitation of Acetylcholinesterase Activity

AChE activity may be detected by monitoring the release of radiolabelled choline from acetylcholine iodide (Johnson and Russell 1975). However, the most commonly used method involves monitoring colormetrically the hydrolysis of acetylthiocholine iodide to thiolcholine and acetic acid (Ellman et al. 1961). Released thiolcholine reacts with 5,5′-dithiobis-(2-nitro-benzoic acid) (DTNB) to give a yellow coloration that can be followed spectrophotometrically.

– Spectrophotometer or ELISA plate reader capable of **Equipment**
  reading at 412 nm
– I ml plastic or silica cuvette

a) 50 mM sodium phosphate buffer pH 7   **Solutions**

b) 75 mM acetylthiocholine iodide (21.67 mg/ml) in 50 mM sodium phosphate buffer pH 7

c) 10 mM DTNB (39.6 mg/ml) in 50 mM sodium phosphate buffer pH 7

d) 10% Triton TX-100 in 50 mM sodium phosphate buffer pH 7

– The working solution consists of
  20 ml of solution a
  700 μl of solution c
  200 μl of solution d

– Triton TX-100 is added to prevent aggregation of amphi-phillic forms of AChE. However, Triton TX-100 will also inhibit some forms of AChE, e.g. class A AChE (see above), and so care must be taken when deciding whether to add detergent or not.

1. Pipette 793 μl of the working solution into the cuvette.   **Procedure**

2. Add 7 μl of solution b.

3. Start the reaction by adding 20 μl of the sample under test.

4. Monitor the change in absorbance at 412 nm over a given time period.

It is important to carry out both positive and negative control reactions. The negative control may consist of either 20 μl of buffer or 20 μl of the sample completely inhibited by eserine (1 mM). A number of AChEs of known activity are available commercially and may be used as a positive control. Under these conditions:

$$c = A/ \in l,$$

where A = change in absorbance at 412 nm, l = light path of 1 cm, $\in$ = molar extinction coefficient = 1.36×10^4/(cm. M) and c = concentration of product, taking into account the natural breakdown of acetylthiocholine iodide at pH 7 (0.0016 ab/min)

$$c = A - 0.0016/1.36 \times 10^4 = moles/l/min.$$

If the protein concentration of the sample is known, then the specific activity of the sample per mg or μg of protein can also be calculated. AChE activity is often expressed in units. Typically, 1 unit of AChE is defined as being the activity required to hydrolyse 1 μM of substrate per minute at 25 °C, although other definitions are also in use. Specific AChE activities are expressed as units per mg of protein.

Notes • It is important to define any activity detected as true AChE activity. This may be achieved by applying a number of criteria (Toutant 1989).

- Cholinesterase activity is distinguished from non-specific esterase activity by its sensitivity to 10 μM eserine.
- AChE is typically inhibited by 10 μM bis(4-allyldimethyl-ammoniumphenyl) pentan-3-one dibromide (BW 284C51) and is insensitive to the butrylcholinesterase inhibitor, tetramonoisopropylpyrophosphotetramide (isoOPMA) at concentrations of 10 μM.
- AChE hydrolyses acetylcholine and acetylthiocholine iodide faster than propionylcholine or propionylthiocholine iodide and has little effect on butyrylcholine or butylrylthiocholine iodide.
- AChE is inhibited by high substrate (acetylthiocholine) concentrations (> 1 mM). Butyrycholinesterases are not inhibited by excess substrate.

• While the Ellman assay is useful for determining the biochemical properties of AChE, the soluble nature of the reaction product renders it unsuitable for the visualisation of AChE activity on polyacrylamide gels or the localisation of AChE activity on histological sections. A suitable method for the staining of AChE was developed in 1964 by Karnovsky and Roots. In this method, thiocholine released by hydrolysis of acetylthiocholine iodide reduces a ferricyanide ion to ferrocyanide. This subsequently combines with Cu^{2+} to form insoluble brown deposits of copper ferrocyanide at the site of AChE activity.

4.3.2
Detection of AChE Activity
Following Polyacrylamide Gel Electrophoresis

- Vertical or flat-bed electrophoresis equipment, e.g. BioRad **Equipment** Protean II or LKB Multiphor II

- 130 ml 0.1 M sodium acetate pH 6.4   **Solutions**
- 10 ml sodium citrate
- 20 ml distilled water
- 200 mg acetylthiocholine iodide
- 20 ml potassium ferrricyanide

The reagents must be added in the above order for the stain **Staining** to function correctly.

Gel electrophoresis is carried out under denaturing conditions (Laemmli 1970) or non-denaturing conditions in the absence of sodium dodecyl sulphate (SDS). Following electrophoresis, the gel is washed in distilled water for 30 min to leach out the SDS if present, allowing renaturation of the enzyme. Gels are then incubated in Karnovsky's stain (Karnovsky and Roots 1964) for up to 24 h or until brown bands of AChE activity are fully developed. For further characterisation of AChE activity detected following electrophoresis, inhibitors (10× stock solution) may be incorporated into the stain in place of 20 ml of distilled water. This stain may also be used on histological sections that have been unfixed and sectioned on a cryostat.

4.3.3
Purification of AChE

While it is possible to carry out many of biochemical characterisations of AChE directly on ES products, some techniques, i.e. kinetic studies, N-terminal sequencing, assessment of immune responses to the enzyme and the raising of specific antisera, require the purified enzyme. AChE has been purified by a number of methods, the majority of which involve affinity chromatography on different matrices, e.g. acridinium-Sepharose 4B (Rosenberry and Scoggin 1984) and 9-[N$^\beta$-(ε-aminocaproyl)-β-aminopropylamino]-acridine-Sepharose (Blackburn and Selkirk 1992), the latter being difficult to synthesise. One of the simplest methods of AChE purification is affinity chromatography using epoxy-activated Sepharose 6B to which the specific AChE inhibitor edrophonium chloride has been coupled (Hodgson and Chubb 1983). This method is described in detail below.

Preparation of edrophonium Sepharose Wash epoxy-activated Sepharose 6B (pharmacia) as directed by the manufacturer with distilled water and 0.2 M sodium borate buffer pH 10. Edrophonium chloride is coupled to the epoxy-activated Sepharose by incubating the Sepharose with 0.2 M edrophonium chloride pH 10 at 50 °C overnight.

Affinity chromatography

1. Wash the edrophonium Sepharose in sequence with 10 volumes each of
   0.1 M sodium acetate pH 4.5
   0.012 M sodium borate pH 10
   distilled water

2. Pack the gel slurry into a column and equilibriate with starting buffer (50 mM phosphate buffer pH 8) and adjust the flow rate to 12 ml/h (using a protein purification system, e.g. the Bio-Rad Econo system).

3. Apply the ES products to the column and remove unbound material with seqential washes of
   50 mM phosphate buffer pH 8
   50 mM phosphate buffer, pH 8, containing 0.5 M NaCl
   Finally, bound AChE is eluted with 50 mM phosphate buffer, pH 8, containing 0.5 M NaCl and 12 mM edrophonium chloride.

4. 1–2-ml fractions are collected and assayed for AChE activity and protein content.

The purification of AChE from the ES products and somatic extracts of parasites using this method has been achieved with *N. americanus* (Pritchard et al. 1991), *N. brasiliensis* and *H. polygyrus* (Griffifths 1994). However, yields vary a great deal between parasite species, which may indicate differences

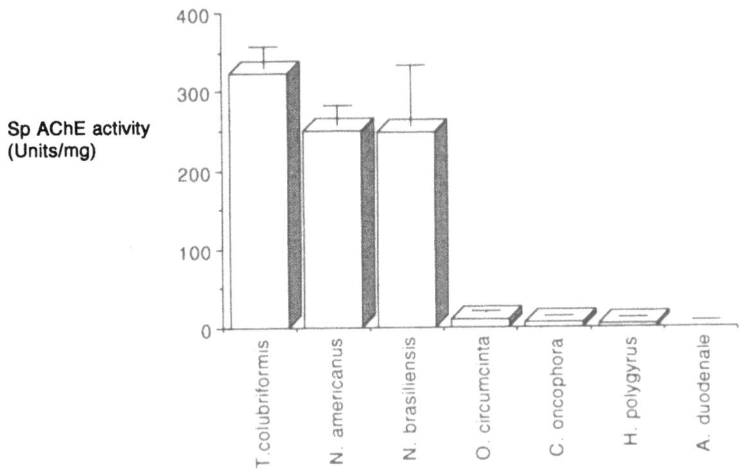

Fig. 1. The specific AChE activities of a number of helminths following 24 h of culture. A number of helminths were cultured for ES products over a period of 24 h. The resulting culture supernates were assayed for both AChE activity and protein content and the specific AChE activity (units/ mg) calculated. However, it must be borne in mind that the specific activities may not simply reflect the amount of AChE secreted but the ratio of active AChE to other proteins present in ES products For example, *Trichostrongylus colubriformis* may secrete AChE very little else, while *Necator americanus* may secrete more AChE than *T. colubriformis*, but may also secrete greater amounts of other proteins, thus lowering the observed specific activity. The diagram also shows that, while AChE secretion may be common among nematodes, they may be split into two groups having high and low specific activities

in binding or stability between the respective AChEs. Similarly, AChE is produced in differing amounts by different parasites (Fig. 1) and this may also affect the amount of starting material required for successful purification.

This can be initially estimated by assessing the specific activity (ratio of AChE activity to protein content) of the samples and by analysing purified fractions on 10% SDS-PAGE, 10% native (non-denaturing) and preformed IEF gels, pH 2-10 (Pharmacia). Proteins can be stained using Coomassie blue or the BIO-RAD silver stain. Confirmation of the presence of AChE is achieved using the Karnovsky activity stain (Karnovsky and Roots 1964) on native and IEF gels, although it must be borne, in mind that AChE, that resolves

Assessment of purity

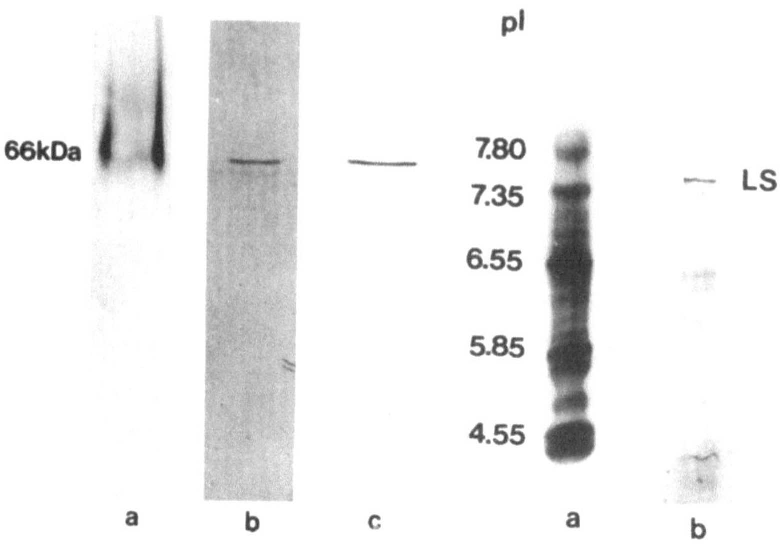

Fig. 2. *Left*: Polyacrylamide gel analysis of *Trichostrongylus colubriformis* AChE following purification on edrophonium chloride Sepharose. a 10% native (**b, c**) 10% SDS-PAGE gels of purified AChE fractions **a** Karnovsky activity stain; **b** Coomassie stain; **c** silver stain. Each gel shows one major band at 66 kDa, indicating the protein possesses AChE activity that there are no other proteins present in the purified fractions. **b** IEF gels of purified *Trichostrongylus colubriformis* AChE. *Right*: **a** Karnovsky stain. A wide range of bands can be seen ranging from pI 4. 55–7.80; **b** silver stain. The strongest bands are at pI 6.55 with other faint bands at pI 7.35 and pI 4.55. Each silver–stained band has a corresponding activity band. In both cases, samples were loaded at position LS (loading site)

as a single band on SDS-PAGE, may still give rise to multiple bands on IEF gels (Fig. 2).

4.3.4
Identification and Separation of AChE Molecular Forms

Molecular characterisation can be achieved on the basis of size, using various biochemical techniques including: gel filtration (Griffiths and Pritchard 1994), centrifugation using Centricon C-100, sucrose density gradients (Pritchard et al. 1994) and native gels (with activity staining). Sucrose density gradients can also be used to identify amphiphilic and hydrophilic forms by running in the presence and absence of 1% Triton X-100 and noting any shift in sedimentation (a sedimentation shift indicates the presence of amphiphilic forms). The instability of AChE to purification, dilution and physical processes is well documented, and a substantial loss of AChE activity is to be expected during many of these procedures.

4.3.5
Kinetic Properties of AChE

By varying the conditions of the Ellmann assay, the biochemical properties of AChE may be examined. Inhibitors may be incorporated into the working solution to determine IC 50 values (inhibitor concentration required to decrease AChE activity by 50% compared with uninhibited enzyme).

Similarly, by varying the substrate concentration, kinetic constants (V_{max} and K_m) which are also important in defining AChE classes may also be determined. However, it is important that for each change in the test conditions a suitable blank assay is carried out, as the natural rate of decay of acetylthiocholine iodide varies with changing substrate concentration and pH.

The assay described above uses a final volume of 1 ml. However, the assay may be scaled down to take advantage of the growing number of ELISA multiplate readers capable of monitoring reaction rates in individual wells of 96-well plates.

4.4
Proteolytic Enzymes

The secretion of proteolytic enzymes by helminths is well documented (McKerrow and Doenhoff 1988; McKerrow 1989) and they undoubtedly play vital roles in the maintenance of the parasite life cycle. These roles include: feeding, digestion of host tissue (Dresden and Deelder 1979; Hotez et al. 1985), prevention of blood coagulation (Loeb and Smith 1904; Hotez and Cerami 1983; Carrol et al. 1984), the facilitation of entry into the host (Dresden and Asch 1972; Matthews 1975, 1982; Hotez et al. 1990) and evasion of the host immune response (Auriault et al. 1981; Kumar and Prichard 1992; Pritchard et al. 1990a).

Helminth proteolytic enzymes are worthy of study for these and other reasons. They are also being considered as candidate vaccines, a concept first proposed by Chandler in 1932 and later applied to hookworms by Thorson in 1956. More recently, Wijffels et al. (1994) have shown that vaccination with a purified cysteine proteinase from *Fasciola hepatica* decreases worm fecundity. Similarly, it has been shown that antibody responses against *N. americanus* ES products (which also contain a mixture of proteinases) reduce worm weight and fecundity (Pritchard et al. 1990b). Proteases have also been considered as serodiagnostic antigens (Zerda et al. 1987) and the possibility of using non-toxic protease inhibitors as chemotherapeutic agents is also under active investigation (McKerrow 1989).

However, it is not proposed to describe the characterisation of helminth proteolytic enzymes here as enzyme classes and techniques suitable for their assay in vivo will be discussed in detail in Chapter 5.

4.5
Anticoagulants, Blood Coagulation Assays

One of the more serious aspects of intestinal parasitism is the blood loss and subsequent anaemia caused by the feeding behaviour of some parasites. This is particularly true in the case of the human hookworms *Necator americanus* and *Ancylostoma duodenale* (Hotez 1989). Adult parasites attach themselves to the gasterointestinal mucosa and, during this process, intestinal capillaries of the *lamina propria* are damaged. Under normal circumstanses, the exposure of collagen following tissue damage would trigger haemostasis. It was postulated as early as 1904 that hookworms may secrete anticoagulants to facilitate feeding and promote blood loss (Loeb and Smith 1904). Subsequently, a number of investigators have reported the presence of antihaemostatic activities (Hotez and Cerami 1983; Carrol et al. 1984). The effects of these agents include the prolongation of clotting times and anti-platelet and fibrinogenolytic activities, although to date many of the molecules responsible for these activities have not been fully characterised. Recently, however, a 16-kDa polypeptide was isolated from the dog hookworm *Ancylostoma caninum*, which prolonged both the prothrombin time (PT) and the activated partial thromboplastin time (APTT). Using chromogenic and clotting time assays, the anticoagulant activity was attributed to a specific inhibitor of Factors VIIa and Xa (Capello et al. 1993). This molecule shows homology with the human tissue factor pathway inhibitor (TFPI) and a similar activity would also appear to be secreted by the human hookworm *Necator americanus*.

Methodology

Coagulation of plasma involves the interactions of three enzymatic pathways known as the intrinsic, extrinsic and common pathways (Fig. 3), resulting in the conversion of fibrinogen to fibrin and the aggregation of platelets to form a plug at the site of injury. Both the intrinsic and extrinsic pathways result in the activation of Factor X (a serine protease) which, in the presence of lipids, Factor V and Ca^{2+} ions, converts prothrombin into active thrombin, which in turn catalyses the conversion of fibrinogen to fibrin. Events following the activation of Factor X are known as the common pathway.

The intrinsic pathway is initiated by the contact of Factor XII with an abnormal surface, e.g. exposed collagen. Activation of Factor XII results in an enzymatic cascade, the activated form of one factor catalysing the activation of the next factor. The end result is the activation of Factor X and the formation of fibrin.

In contrast, the extrinsic pathway is initiated by trauma to a blood vessel which causes the release of a lipoprotein called tissue factor. A complex of tissue factor and Factor VII then catalyses the activation of Factor X.

Helminth anti-coagulatory products may affect the intrinsic, extrinsic or common pathways of coagulation and a number of assays have been developed to examine the effects of ES products on these pathways.

4.5.1
The Collection of Plasma

To carry out these assays, a supply of human blood is required. This must be taken (according to the relevant establishment's local rules) by venepuncture from healthy volunteers who have been questioned to exclude the ingestion of any drug which may interfere with platelet function.

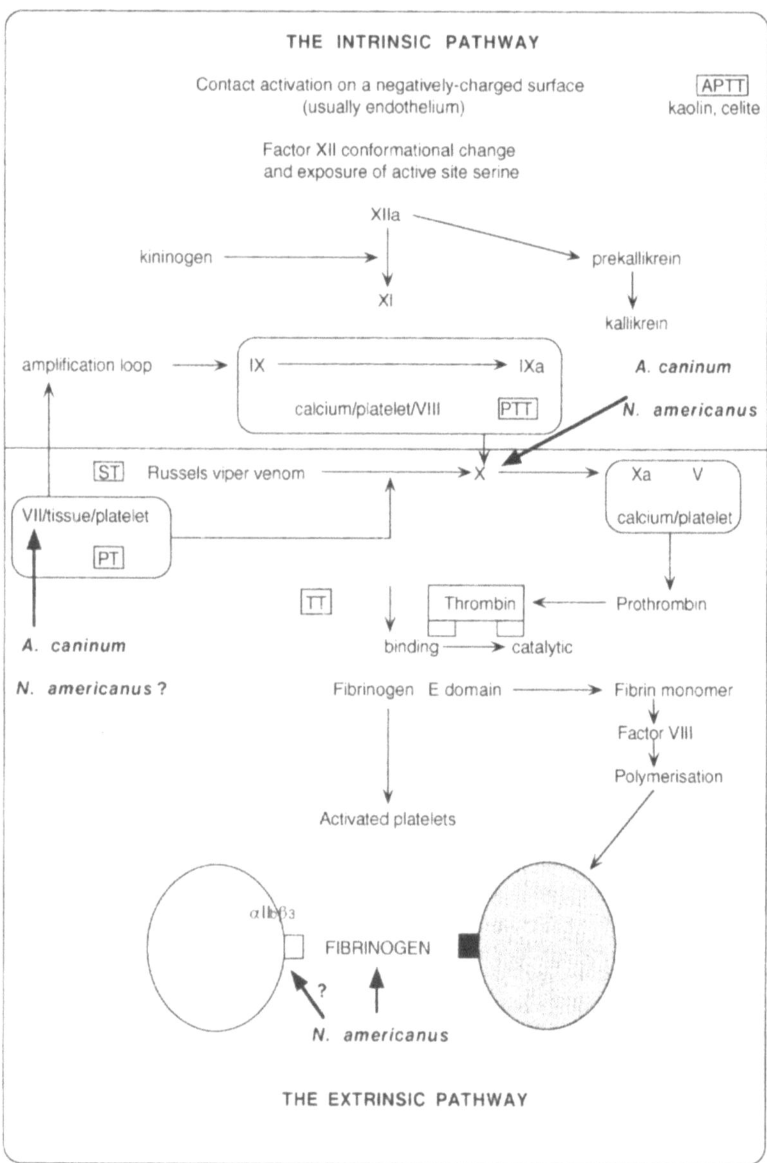

Fig. 3. The intrinsic and extrinsic clotting pathways involved in the clotting of plasma and the events leading up to fibrinogen formation and the activation of platelets The sites of inhibition by *Ancylostoma caninum* and *Necator americanus* are indicated by *bold arrows*. Platelet activation occurs when exposed collagen activates phospholiphase C, which in turn exposes $\alpha II_b\beta_3$

Blood should be taken under conditions of minimum stress and trauma to avoid adrenaline secretion.

The blood obtained is immediately placed into plastic tubes containing 0.08 M trisodium citrate (9:1 blood: citrate). To each 100 ml of citrated blood, 5 g of Hepes buffer [N-(2-hydroxyethyl) piperazine-N'-(2 ethane sulphonic acid)] is added to maintain pH and improve stability of the samples on storage (if the blood obtained is to be used for platelet aggregation assays, then this step is omitted).

Citrated blood samples are then chilled to 4 °C and centrifuged in closed containers at 200 g for 10 min to obtain platelet-rich plasma (PRP) or 2000 g for 15 min to obtain platelet-poor plasma (PPP).

Plasma is then pipetted into plastic storage containers using wide-bore pasteur pipettes (coated with silicone to prevent activation of the plasma via the intrinsic pathway) and stored at –20 °C until required.

Note: Trisodium citrate prevents plasma coagulation by removing Ca^{2+} ions. During tests to determine clotting time, Ca^{2+} ions are returned as $CaCl_2$. In all the following assays, clean unused glass tubes must be used.

4.5.2
Recalcification Time Test (Intrinsic Pathway)

Platelet-rich plasma (PRP) contains all the factors necessary to activate prothrombin by the intrinsic pathway (with the exception of Ca^{2+} ions). The rate of clotting is a measure of the overall coagulant activity and this rate will be decreased if there is any inhibition of any factor or factor complex.

Equipment

- Water bath
- Stop clock

Materials

- Platelet rich plasma, kept on ice
- 0.85% sodium chloride, held at 37 °C in a water bath
- 0.025 M calcium chloride, held at 37 °C in a water bath
- ES products of known protein concentration

Procedure
1. Add a known volume of ES products to a glass tube and make up to 100 μl with 0.085 M sodium chloride.

2. Add 100 μl of PRP (control incubations contain 100 μl of PRP and 100 μl of sodium chloride).

3. Incubate for 4 min at 37 °C.

4. After 4 min add 100 μl of 0.025 M calcium chloride and at the same time start the clock. Every 10 s remove the tube from the water bath and tilt it so the contents run halfway down the side of the tube.

5. Note the time of appearance of fibrin threads.

This test, however, is relatively crude, as both the number of platelets and the extent of exposure to glass may vary. A more quantitative assay of intrinsic pathway inhibition is the activated partial thromboplastin test (APTT).

4.5.3
Activated Partial Thromboplastin Test (Intrinsic Pathway)

In this assay platelet substitute is added in the form of a phospholipid emulsion to platelet-poor plasma (PPP). Optimal glass activiation is achieved by incubating the plasma with a suspension of Celite (glass particles).

Equipment
– Water bath
– Stop clock

Materials
– Platelet-poor plasma
– 4% Celite in 0.85% sodium chloride (mix well)
– Rabbit brain encephalin (RBC)-Sigma stock suspension
– 0.025 M calcium chloride, held at 37 °C in a water bath
– ES products diluted in 0.85% saline if required

Procedure
1. 50 μl of PPP and ES products are incubated together in glass test tubes for 6 min at 37 °C in a water bath (control incubations contain an equal volume of saline in place of ES products).

2. Add 25 μl of RBC and 25 μl of celite and incubate the tubes for a further 6 min with gentle agitation.

3. Add 50 μl of warmed calcium chloride. Tilt the tube every few s and measure the time taken for the formation of a fibrin clot.

4.5.4
Prothrombin Time Test (Extrinsic Pathway)

In this assay, used to monitor inhibition of the extrinsic pathway tissue, thromboplastin, a phospholipoprotein, is added to platelet-poor plasma to activate Factor VII and to provide the tissue phospholipid needed for Factor X activation.

Equipment

- Water bath
- Stopclock

Materials

- Platelet-poor plasma
- 0.85% sodium chloride
- Rabbit thromboplastin (Sigma) + 0.025 M calcium chloride (mixed in equal volumes and held at 37 °C in a water bath)
- ES products (diluted in saline if required)

Procedure

1. Mix 50 μl of PPP and ES products in a glass test tube and incubate in water bath at 37 °C for 6 min (control incubations contain an equal volume of saline in place of ES products).

2. Add 100 μl of warmed thromboplastin mixture and commence timing the reaction.

3. After the initial 10 s, tilt the tube every few seconds and note the time of clot formation.

4.5.5
Stypven Clotting Time Test (Common Pathway)

In the presence of brain phospholipid, Russell viper venom acts directly on Factor X. Stypven clotting time is the accelerated clotting time of recalcified plasma when mixed with Russell viper venom. If parasite ES products prolong the

Stypven clotting time, then this shows that coagulation inhibition is occurring below Factor X in the clotting cascade. If plasma Factor V and prothrombin levels are normal, then anticoagulant activity may be due to Factor Xa inhibition or anti-thrombin activity.

Equipment
- Water bath
- Stop clock

Materials
- Platelet-poor plasma
- Russell viper venom (RVV) in rabbit brain encephalin (Sigma diagnostics); 1 vial is dissolved in 3 ml of 0.85% sodium chloride
- 0.025 M calcium chloride
- ES products (diluted in 0.85% sodium chloride if required)

Procedure
1. Warm 50 μl of PPP in a glass test tube.

2. Add 25 μl of RVV solution and incubate for 3 min.

3. Add ES products to the mixture (control incubations contain an equal volume of saline in place of ES products) and incubate for a further 10 min.

4. Add 50 μl of calcium chloride and note the time taken for clot formation.

Prolongation of clotting time indicates inhibition of the common pathway by ES products.

Results Results of these assays typical for *N. americanus* ES products are shown in Fig. 4.

4.5.6
Platelet Aggregation

Equally important for parasite survival is the ability to interfere with platelet activation. Primary haemostasis is achieved by platelet adherence to the subendothelium, followed by the formation of a platelet plug at the site of vascular injury. During the formation of this plug, platelet aggregation, thromboxane synthesis and the secretion of platelet granule contents take place (Steen and Holmsen

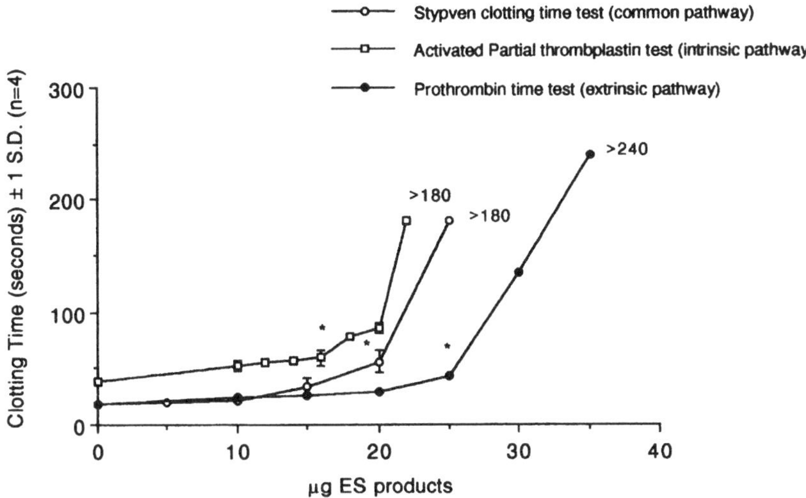

Fig. 4. The effects of *N. americanus* ES products on PT, APTT, and Stypven clotting time tests (using conditions described in the text) are shown * Denotes very small clot formation

1987). Furthermore, by providing a phospholipid surface, platelets contribute to and enhance the activity of the prothrombinase, and "tenase" complexes. (Miletich et al. 1977). Recently, major advances have been made in the understanding of platelet interaction in the subendothelium and platelet/platelet interaction. The platelet glycoprotein $\alpha II_b \beta_3$ is the key receptor involved in platelet/platelet interaction during aggregation (McEver et al. 1983). When platelets are activated, IIb-IIIa undergoes a conformational change allowing it to bind fibrinogen resulting in platelet cross-linking (Shatill 1993).

Inhibition of platelet aggregation has been shown by ES products from *A. caninum* (Carrol et al. 1984) and, more recently, we have shown that ES products from *N. americanus* potently inhibit platelet aggregation induced by ADP, collagen and platelet-aggregating factor (PAF).

The effects of hookworm ES products on platelet aggregation (in the presence of agonists to induce aggregation) using platelet-rich plasma (PRP) have been studied using a platelet aggregometer. This is a modified spectrophotometer that measures light transmission changes in PRP as an index of platelet aggregation in vitro (Born and Cross 1963).

Preparation of PRP

1. Venous blood is taken from a healthy donor and centrifuged, as previously described to give PRP.

2. The number of platelets is then estimated using a Coulter counter. A background count is made in triplicate, using Isoton buffer. Then, using a positive placement pipette, add 6.7 μl of PRP to 20 ml of Isoton buffer and count the number of platelets present. The number of platelets/μl is obtained by subtracting the mean background count from the mean sample count and multiplying by 6.

3. The ideal PRP count for use in the aggregometer is 3×10^4. PRP can be adjusted to this platelet concentration by the addition of platelet-poor plasma (PPP).

Measurement of platelet aggregation

1. Prewarm the aggregometer to 37 °C.

2. Calibrate the aggregometer to 100% transmission using PPP.

3. Add 0.2 ml of PRP to the cuvette (stirred magnetically at 1200 rpm).

4. Add 20 μl of the agonist under study. Platelet aggregations are formed and the transmission of the PRP increases relative to PPP. Platelet sensitivity to different aggregating agents may vary. Consequently, various concentrations of each agonist should be monitored to determine the concentration that induces optimal aggregation. Platelet aggregation assays should be carried out as quickly as possible, as platelet viability declines after 2–3 h.

5. Once the optimal agonist concentration has been determined a concentration of agonist giving sub-optimal aggregation should be chosen for inhibition assays using ES products.

6. Incubate a range of ES product concentrations with PRP at 37 °C for 30–40 min (during which time the agonist dose response assays should be carried out). If necessary, assay volumes may be made up with 0.85% saline.

7. After incubation, the aggregating agonist is added at a suitable concentration and any inhibition of platelet aggregation noted. *N. americanus* ES products have been shown to inhibit platelet aggregation in the presence of the

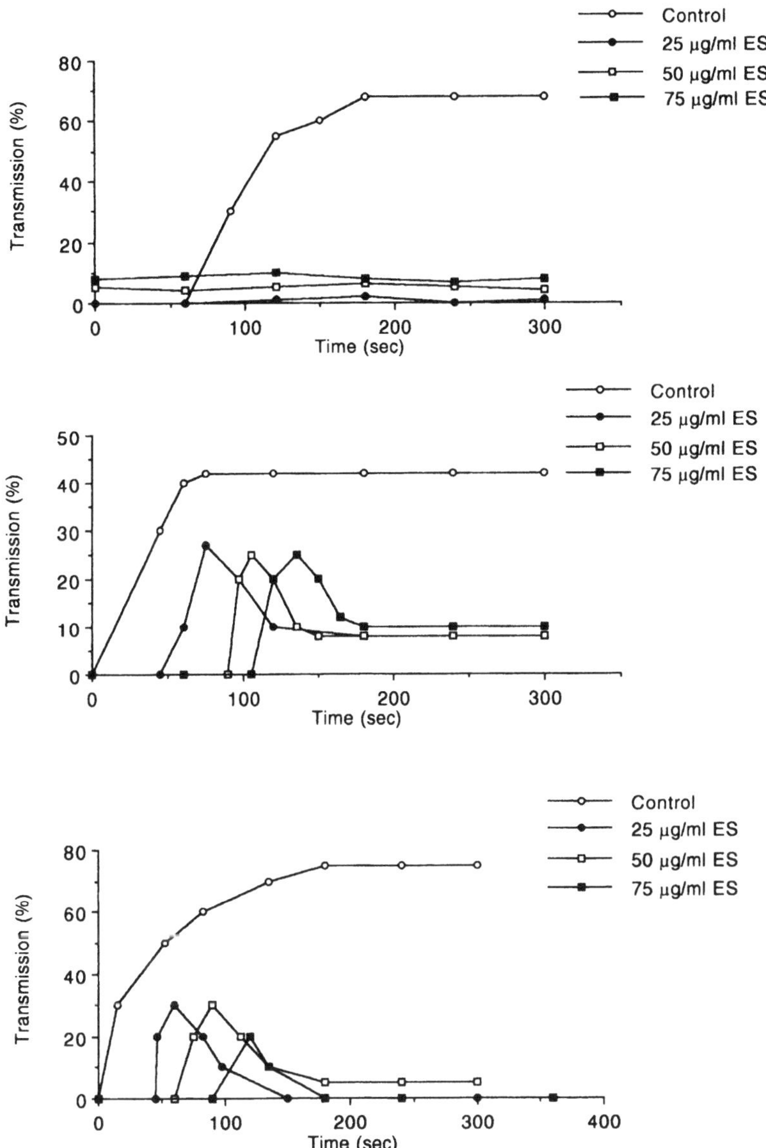

Fig. 5a–c. The effects of *N. americanus* ES products on platelet aggregation in human plasma mediated by collagen, adenosine diphosphate (ADP) and platelet–activating factor (PAF). Platelet-rich plasma was preincubated at 37 °C with 0, 25, 50, 75 µg/ml ES products for 40 min; 20 µl of agonist (**a** 2 mg/ml collagen, **b** 2.5 µM ADP, **c** 0.06 µg/ml PAF) was added to 200 µl of each PRP sample. In all cases, *N. americanus* ES products inhibited platelet aggregation

agonists collagen, adenosine diphosphate (ADP) and platelet-activating factor (PAF).

Typical results are shown in Fig. 5a–c.

4.6
Potential Immunomodulators

4.6.1
Glutathione s-Transferases

Glutathione s-transferases (EC 2.5.1.18) are multi-functional phase II broad-range detoxification proteins distributed throughout the animal kingdom (Ketterer et al. 1989), and GST activity is detectable in a range of helminth extracts (Brophy and Barrett 1990b). The GSTs can function as enzymes via their cofactor glutathione or as binding/transport proteins (Ketterer et al. 1989). In mammals, GSTs are expressed as isoenzymes that can be grouped into four species-independent soluble superfamilies (Alpha, Mu, Pi and Theta) with a subunit size of 24–26 kDa, a 13-kDa family and two membrane-bound families (Mannervik et al. 1985; Blocki et al. 1992)

Immunotherapy- and chemotherapy-directed research programmes have independently identified parasitic helminth GSTs as potential targets (Smith et al. 1986; Balloul et al. 1987; Brophy and Barrett 1990b). Parasitic helminth GSTs may function as chemotherapeutic defence proteins by neutralising anthelmintics or as immune defence proteins by detoxifying the secondary products of lipid peroxidation that can arise from immune initiated free radical attack on lipid membranes (Brophy and Pritchard 1992; Brophy et al. 1994a).

Helminth GSTs also exist as isoenzymes, but the enzymes cannot be placed into one of the mammalian GST families based on the combined criteria of biochemical activity, antibody cross-reactivity and primary amino acid sequence (Brophy et al. 1995a). In fact, specific interspecies helminth GST superfamilies may be expressed: the schistosome 28-kDa family, the digenean/cestode 25–26-kDa family and at least two parasitic nematode families (Brophy et al. 1995a, b).

ES and surface GSTs are expressed in the digeneans *Schistosoma* and *Fasciola* and in the hookworm *Necator americanus* (Davern et al. 1991; Panaccio et al. 1992; Brophy et al. 1995b). The parasite ES GSTs could function as "molecular dumping" proteins for toxic ligands or act as anti-inflammatories by neutralising lipid peroxidation products at the mucosal surface.

As was the case with acetylcholinesterase, glutathione s-transferase activity may be detected both in liquid assays and following polyacrylamide gel electrophoresis.

4.6.1.1
Spectrophotometric Assay of Glutathione s-Transferase

A number of subsrates may be used to assay for glutathione s-transferase activity, e.g. the natural substrate *trans*-2-nonenal or the model substrate 1-chloro-2, 4-dinitrobenzene (CDNB; Habig et al. 1974).

- Spectrophotometer capable of measuring absorbance at   **Equipment**
  225 nm
- 1-ml quartz cuvette

- 2 mM *trans*-2-nonenal   **Materials**
- 0.1 M potassium phosphate buffer pH 6.5 (pH may vary
  with differing substrates)
- 50 mM glutathione

1. Pipette 500 μl of buffer into a quartz cuvette and add 20 μl   **Spectrometric**
   of glutathione (final concentration 1 mM).   **Assay**

2. Add ES products or distilled water (negative control) to a
   final volume of 975 μl.

3. Allow the contents to preincubate for 5 min before starting
   the reaction by adding 25 μl of the *trans*-2-nonenal stock
   solution (final concentration 50 μM).

4. The activity is followed by monitoring the decrease in
   absorbance at 225 nm over a known time period.

5. The amount of product formed can be calculated using an
   extinction coefficient of 19.2 mM^{-1} cm^{-1}.

6. A unit of GSH transferase activity is defined as the amount of enzyme catalysing the formation of 1 μM of product per minute under the conditions of the assay. Specific activity is defined as the units of enzymic activity per mg of protein.

If other substrates are to be used, then the pH chosen should be one at which the non-enzymatic product formation is minimal.

4.6.1.2
Detection of Glutathione s-Transferase Activity on Polyacrylamide Gels

Several methods for the detection of glutathione s-transferase have been described, some of which are based on the development of the blue starch-iodine colour due to the conjugation product formed between glutathione and aromatic compounds (Board 1980; Clark 1982). Others are based on measuring the spectrophotometric absorption of the reaction product following gel electrophoresis (Kenney and Boyer 1981). These methods have a number of limitations including low sensitivity, spreading of the coloured bands and the need for spectrophotometric gel scanning apparatus. A more convenient and simple method is based on the the fast reduction of nitroblue tetrazolium (NBT) to blue insoluble formazan when incubated with reduced glutathione (GSH) and phenazine methosulphate (PMS). Where glutathione s-transferase activity is present, the rate of NBT reduction is depressed owing to the enzymatic conjugation of the sulphydryl group. Thus, glutathione transferase activity is visualised as clear bands on a blue background (Ricci et al. 1984). Glutathione transferase activity may be detected following native PAGE or isoelectric focusing.

Detection on polyacrylamide gels

1. Following electrophoresis, the gel is washed at 37 °C with gentle agitation in 0.1 M potassium phosphate buffer pH 6.5 containing 4.5 mM GSH, 1 mM 1-chloro-2, 4-dinitrobenzene (CDNB) and 1 mM NBT (10 min).

2. Wash the gel in distilled water.

3. Finally, incubate the gel in 0.1 M Tris-HCl buffer pH 9.6 containing 3 mM PMS. Blue insoluble formazan appears in about 3–5 min except in areas where GsT activity is present.

4. For storage purposes, the gels may be washed with distilled water and placed in 1 M NaCl. Under these conditions the bands remain tightly defined for up to 1 month.

4.6.1.3
Purification of Glutathione s-Transferase

GSTs can be purified by a combination of glutathione-affinity chromatography and chromatofocusing.

Glutathione-affinity resins are now available from commercial sources (Pharmacia, Sigma), or the resins may be synthesised by the method of Simons and Van der Jagt (1977).

Gluta thioneaffinity chromato- graphy

Glutathione s-transferase may be purified in a manner similar that for acetylcholinesterase. A relatively crude extract (10 000 g supernatant) can be loaded onto a glutathione resin with GSTs retained on the column and subsequently eluted with 5 mM glutathione (Brophy et al. 1994a). However, it must be borne in mind that some helminth GSTs will not bind to a glutathione resin unless they are partially purified by another matrix such as hydroxylapetite (BioRad) or chromatofocusing (reviewed by Brophy and Barrett 1990a).

For most analytical purposes, a relatively rapid glutathione-affinity batch method can be used to isolate a small quantity of GSTs.

1. Wash approximately 250 µl of glutathione-agarose gel with 10 vol phosphate buffered saline (PBS) in an Eppendorf tube by centrifugation at low speed for 10 s in a bench microcentrifuge.

2. Somatic extract or ES products are mixed with the gel for 30 min at 4 °C.

3. Centrifuge at high speed for 10 s in a bench microcentrifuge.

4. Remove the supernatant and wash the gel matrix with 20 gel vol PBS.

5. Centrifuge at high speed for 10 s in a bench microcentrifuge.

6. Elute bound GSTs by washing the gel with 50 mM Tris-HCl pH 9.6 buffer containing 5 mM reduced glutathione.

Chromato-focusing Glutathione affinity resins will separate GSTs as a pool of possible isoenzymes. The different GST forms can be separated by charge-based methods such as ion-exchange techiques or chromatofocusing. The latter has been used to separate GSTs from cestodes and digeneans (Brophy et al. 1989; O'Leary and Tracy 1988). In chromatofocusing, a pH gradient is established in the column following application of the sample. This may be achieved manually or by programming a low-pressure chromatography system (e.g. BioRad Econosystem). Different forms of GST will elute from the column at their respective pI (i.e. neutral charge). In this way, GSTs differing in pI of only 0.1 unit may be separated, provided the column is not overloaded and a shallow and narrow pH range is used to elute the enzymes from the matrix. Chromatofocusing can also be employed to analyse GST isoenzyme profiles in unpurified extracts (Brophy and Barrett 1990a).

4.6.1.4
Cloning GST Sequences

GSTs have been cloned from digeneans by the classical methods of probing cDNA libraries with either DNA or antibodies (Taylor et al. 1988). However, polymerase chain reaction (PCR) strategies can also be used to isolate helminth GST sequences. The following is a brief outline of the strategy used to isolate a GST sequence from the nematode *Heligmosomoides polygyrus*. Many of the methods used are standard and can be found in any good molecular biology textbook, e.g. (Maniatis et al. 1989).

Cloning Procedure Native GSTs were purified by glutathione – affinity chromatography and two polypeptides of 23- and 24-kDa were resolved by SDS-PAGE, blotted onto PDVF membrane and their N-termini sequenced. Twelve amino acids (VHYKL-TYFNGRG, one-letter code) and six amino acids (VHYKLT) were sequenced from the N-terminus of the 23- and 24-kDa subunits respectively.

A degenerate PCR primer, CA(TC) TA(TC) AA(AG) TC/TT (inosine) AC(inosine) TA(TC) TT(TC) AA, was synthesised to the N-terminal sequence of the 23-kDa subunit using inosine incorporation to reduce the degeneracy. Messenger (m)RNA was isolated from *H. polygyrus* using a Quickprep mRNA purification kit (Pharmacia). First-strand cDNA synthesis was catalysed from mRNA using Moloney Murine Leukemia Virus (MMLV) reverse transcriptase. The first-strand cDNA synthesis was primed using an oligo (dT)-based primer (5'-d[AACTGGAAGAATTCGCGGCCGCAGGAAT$_{18}$]-3'. This primer was also used as the downstream PCR primer.

A PCR reaction was carried out using

- 0.1 μg heat-denatured cDNA,
- 20 pmol of each upstream and downstream primer,
- 2 mM MgCl$_2$,
- 0.2 mM dNTPs,
- 2.5 units Taq polymerase (Hoffman-La Roche) and reaction buffer.

Thirty cycles were completed in a thermal cycler (Techne), each cycle consisting of 1 min at 95 °C, 1 min at 50 °C and 1 min at 72 °C.

The reaction product was analysed by agarose gel electrophoresis and a 800-base pair product was shown to be amplified under these conditions.

The 800-base pair product was excised from the gel purified by Sephaglas (Pharmacia) blunt-ended by Klenow fragment and T$_4$ polynucleotide kinase treatment and cloned into the plasmid pUC18 (Pharmacia). The product was sequenced by the method of Sanger et al. (1977). The predicted protein sequence had an overall homology of 30% to the mamalian alpha GST family.

This techique should be useful for isolating other GST sequences from limited starting tissue provided there are sufficient protein sequence data available for primer construction and relatively stable or abundant mRNA with poly-adenylation at the 3' end.

4.6.2
Superoxide Dismutases (SODs)

SOD expression is probably ubiquitous in parasitic helminths (Callahan et al. 1988). SOD has been purified from several helminths including *Trichinella spiralis*, *Dirofilaria immitis*, *Onchocerca cervicalis* and *Taenia taenaeformis* (Rhoads 1983;

Leid and Suquet 1986; Callahan et al. 1988). SOD forms are also secreted by a number of helminths (Leid and Suquet 1986; Knox and Jones 1992;) and a secretory SOD has been cloned from *Schistosoma mansoni* (Simurda et al. 1988).

In mammals, SOD, in concert with catalase activity or possibly glutathione peroxidase, neutralises the superoxide anion produced as a by-product of cellular metabolism [reactions (1–3)]. Furthermore, activated leukocytes employ reactive oxygen species such as the superoxide anion as part of their armory for anti-parasitic activity (Callahan et al. 1988).

$$O_2^- + O_2^- + 2H^+ \text{ (superoxide dismutase)}$$

$$\longrightarrow H_2O_2 + O_2 \tag{1}$$

$$2H_2O_2 \quad \text{(catalase)} \longrightarrow 2H_2O + O_2 \tag{2}$$

$$2GSH + H_2O_2 \text{ (GSH peroxidase)} \longrightarrow GS\text{-}SG + 2H_2O \tag{3}$$

Therefore, it has been suggested that SOD is expressed by parasites as a part of an enzymatic defence against this immune effector mechanism. However, in the majority of helminths, glutathione peroxidase and especially catalase activity is relatively low (Callahan et al. 1988).

This finding highlights a possible biochemical enigma, as the hydrogen peroxide produced by SOD activity is itself a good diffusable oxidant that can be converted to the highly toxic hydroxyl radical (Halliwell and Gutteridge 1992). Therefore, a novel pseudocatalase could be expressed in helminths to neutralise the hydrogen peroxide produced by SOD acivity. In contrast, it has been postulated that in hookworms, the secretion of SOD in the absence of a hydrogen peroxide neutralising enzyme can be described as a pro-active parasite mechanism (Brophy, Patterson and Pritchard, unpubl. observations). Secretory hookworm SOD would utilise the reactive oxygen species produced by immune accessory cells in order to produce hydrogen peroxide for counter-attacking host tissues. This would provide a suitable surface for hookworm attack via for example proteolytic enzymes.

**Spectrophotometric Assay for the Detection
of Superoxide Dismutase Activity**

The protocol here is based on the method of (Beyer and
Fridovitch 1987).

- Spectrophotometer capable of reading at 560 nm   **Equipment**
- Fluorescent GroLux tube (Sylvania)

a) 50 mM potassium phosphate buffer pH 7.8   **Materials**

b) 30 mg/ml L-methionine

c) 1.41 mg/ml nitro blue tetrazolium (NBT)

d) 1% Triton X-100

e) 4.4 mg/100 ml riboflavin

A working solution is made up of 27 ml of solution a, 1.5 ml
of solution b, 1 ml of solution c and 0.75 ml of solution d.

1. Add 0.9 ml of working solution to 90 μl of sample (or   **Spectro-**
   buffer for negative control).   **photometric**
                                                              assay
2. Add 10 μl of riboflavin.

3. Measure the absorbance at 560 nm.

4. Illuminate under a fluorescent tube for 7 min.

5. Measure the absorbance again at 560 nm.

One unit of superoxide dismutase will cause a 50% inhibition
in the rate of NBT reduction.

4.6.3
Detection of Immunomodulation of Antibody Production

Many species of parasite are capable of supressing host immunity (Playfair 1982). This may be achieved by interfering
with lymphocyte proliferation (Raybourne et al. 1983), inhibiting cytotoxic T cells (Mazingue et al. 1983), depressing
IgE responses (Langlet et al. 1984), reducing B cell numbers
(Cross and Klesius 1989) or stimulating supressor cell activity (Liew et al. 1987).

The parasite products responsible for immunomodulatory activity have been partially characterised for the nematodes *Ostertagia ostertagi* (Cross and Klesius 1989) and *Heligmosomoides polygyrus,* (Monroy et al. 1989; Pritchard et al. 1994). In the case of *Heligmosomoides polygyrus,* immunosupression was measured by monitoring the inhibition in vitro of the production of antibodies to keyhole limpet haemocyanin (KLH).

Equipment
– Sterile laminar flow cabinet
– Centrifuge
– ELISA plate washer (optional) and plate reader

Cell preparation

1. Balb/c mice are primed with 0.1 ml keyhole limpet haemocyanin (10 mg/ml in saline) given intraperitoneally.

2. Four weeks later, the animals are killed and their spleens removed under sterile conditions and placed in Hanks' balanced salt solution (HBSS). Under sterile conditions the spleens are teased apart using a scalpel and forceps until a single cell suspension is formed. This suspension is then filtered through a wire mesh into a 50-ml conical tube (maximum six spleens/tube).

3. Centrifuge (1500 rpm for 10 min) and resuspend the cell pellet in Tris-ammonium chloride pH 7.2 (maximun 2 ml/spleen) to lyse any remaining red blood cells.

4. Centrifuge (1500 rpm for 10 min), resuspend the cells in HBSS and filter the cells again.

5. Centrifuge (1500 rpm for 10 min) and resuspend the cells in CTCM (RPMI 1640, 10% NuSerum, 20 mM L-glutamine, 100 i.u./ml penicillin G and 100 μg/ml streptomycin) maximum 5 ml/3 spleens.

6. An aliquot of the cell suspension is stained with Nigrosin (1 vol cells: 3 vol saline: 1 vol Nigrosin) and counted. The cell suspension is adjusted to 2.5×10^7/ml by the addition of RPMI 1640.

Sample and KLH preparation

Parasite ES products to be tested are filter sterilised (0.22 μm) and a doubling dilution series prepared in CTCM.

A solution of KLH was prepared in CTCM (0.0166 mg/ml) and filter sterilised.

Sterile 96-well ELISA plates are used for the assay   **Plate layout**

1. 50 μl of the required dilution of the sample under test are added to the sample wells. Control wells contain 50 μl of CTCM in place of the sample.

2. 100 μl of KLH is added to every well except for the negative control wells, to which a further 100 μl of CTCM are added.

3. 100 μl of the cell suspension is added to each well.

4. Incubate the plates for 6 days at 37 °C in 5% CO_2.

5. Centrifuge the plates (100 rpm for 5 min) and remove 150 μl of the culture supernatant and replace with 150 μl of fresh medium. Repeat.

6. Add 200 μl of fresh medium and incubate for a further 24 h.

7. Harvest the supernates and assay for the presence of anti-KLH antibodies by ELISA (at this point the culture supernates may be stored at –40 °C until required).

A typical plate layout is shown in Fig. 6.

	1	2	3	4	5	6	7	8	9	10	11	12
A			Sample A						Dilution 2			
B			Dilution 1						Dilution 3			
C			Dilution 2						Dilution 4			
D			Dilution 3						Dilution 5			
E			Dilution 4						Positive control			
F			Dilution 5						Positive control			
G			Sample B						Negative control			
H			Dilution 1						Negative control			

Fig. 6. A typical ELISA plate layout to detect the presence of anti KLH antibodies. ELISA plates are coated with KLH at 10 μg/ml and bound antibodies visualised with mouse alkaline conjugated anti-mouse Ig G. Results are expressed as the Abs 405 nm of the ELISA plate, or as the percentage inhibition of antibody production compared with control cultures containing medium alone

ELISA
to detect
of anti KLH
antibodies

1. Coat a 96-well polystyrene plate with 150 μl KLH (10 μg/ml) in 0.05 M sodium carbonate/bicarbonate buffer pH 9.6. Incubate for 3 h at room temperature.

2. Wash plates four times with phosphate-buffered saline/0.05% Tween 20 (PBS/Tween).

3. Add 150 μl of culture supernatant under test (diluted 1:1) per well and incubate for 2 h.

4. Wash plates four times with PBS/Tween 20.

5. Add 150 μl rabbit anti-mouse IgG conjugated to alkaline phosphatase (dil 1: 1000) per well. Incubate overnight at 4 °C.

6. Wash plates four times with PBS/Tween 20.

7. Antibody binding is revealed by the addition of 150 μl of alkaline phosphatase substrate (5 mg p-nitrophenyl phosphate/5 ml diethanolamine buffer pH 9.8).

8. After approximately 30 min, measure the absorbance at 405 nm using an automated ELISA plate reader (prior to reading the plate 25 μl of 1 M sodium hydroxide may be added to halt the colormetric reaction).

The results of this assay may be expressed as the absorbance at 405 nm of the ELISA plate or as the percentage inhibition of antibody production compared with control cultures containing medium alone. This assay is capable of detecting concentrations of immunomodulator as low as 50 ng/ml, and has proved to be more reliable than Mishell-Dutton culture systems previously used to demonstrate immuno-modulatory activity from *H. polygyrus* (Pritchard et al. 1984). Data obtained using this assay to monitor immunosuppression by *H. polygyrus* is shown in Fig. 7a and b.

4.6.4
Neutrophil Inhibitory Factor (NIF)

The survival of many chronic endoparasites depends on their ability to evade the immune response of the host. It is proposed that parasites may accomplish this by releasing factors that attenuate leukocyte function (Maizels et al. 1993). Recently, a 41-kDa glycoprotein has been purified from the

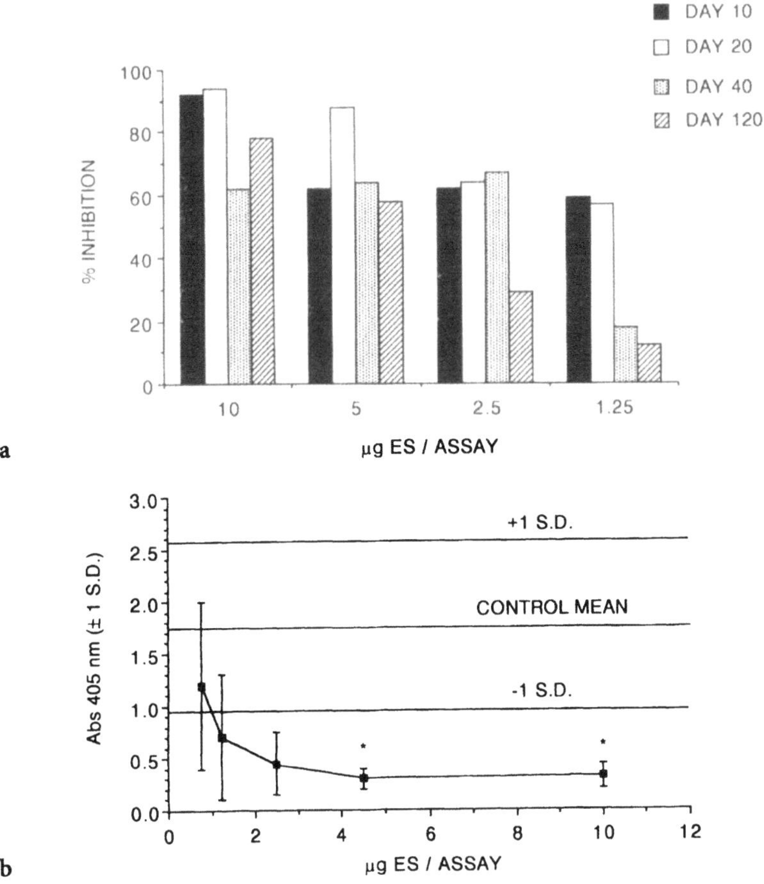

Fig. 7a,b. Standarisation of the procedures for the collection assay of ES products for *II. polygyrus*. **a** The inhibition of antibody to keyhole limpet cyanin (KLH) in vitro using ES products taken from worms taken 10–120 days post–infection. **b** The effect of ES products taken 24 h following the culture of 10–20-day-old worms in on vitro antibody production to (KLH). Antibody levels are expressed as the absorbance reading at 405 nm obtained in ELISA in the presence of increasing concentrations of ES products. Control (-ES) levels are shown ± 1 SD[*] indicates significant inhibition of antibody production: P <0.01 in Students t–test. The data shown in **b** are the results of few separate experiments

canine hookworm *A. caninum* that inhibits the adhesion of activated human neutrophils to vascular endothelial cells, and the release of hydrogen peroxide from activated adherent neutrophils (Moyle et al. 1994). This molecule has been designated neutrophil inhibitory factor (NIF), and appears to bind with high affinity to the integrin CD11/CD18 (Mac-1, Mol, $\alpha_M\beta_2$, CR3) that is present on the neutrophil surface.

The CD11/CD18 family of integrins are essential for normal leukocyte trafficking and inflammatory functions. However, excessive or inappropriate activation of leukocytes leads to local pathological inflammatory leisions and tissue injury. In some model systems of acute inflammation, monoclonal antibodies to CD11/CD18 have been shown to mitigate leukocyte-mediated injury (Harlan et al. 1992). Therefore, molecules that target CD11/CD18 integrins (as secreted by *A. caninum*) may prove useful in the treatment of acute inflammatory disease.

The presence of NIF may be detected by monitoring the inhibition of binding of human neutrophils to human umbilical vein endothelial cell (HUVEC) monolayers.

Isolation of neutrophils Neutrophils may be isolated from heparinised venous blood (taken as before under stress-free conditions) by using a one-step Ficoll-Hypaque gradient (e.g. Mono-poly, ICN Biomedicals).

1. Layer 5 ml of blood onto 3 ml Mono-poly in a 16×100 mm glass tube.

2. Centrifuge at 300 *g* for 60 min at 20 °C.

3. Collect the cell layer containing neutrophils using a Pasteur pippette and resuspend in 10 vol cold DME media (Gibco).

4. Centrifuge at 200 g for 10 min at 4 °C.

5. Resuspend in 5 ml of cold ACK buffer (155 mM NH_4Cl, 10 mM $KHCO_3$, pH 7.4) and incubate for 5 min at room temperature to lyse contaminating red blod cells.

6. Centrifuge at 200 *g* for 10 min at 4 °C.

7. Resuspend in Hanks' balanced salt solution (approx 10^7 cells/ml).

8. Determine cell viability by tryptan blue exclusion.

Neutrophil-HUVEC adhesion assays The adherence of human neutrophils to HUVEC monolayers is monitored by using cells that are preloaded with the fluorescent dye calcein AM (molecular probes)

1. Pelleted neutrophils (above) are resuspended in Hanks' balanced salt solution (10^7 cells/ml) containing a final concentration of 10 μg/ml calcein AM (stock solution 10 mg/ml in DMSO, stored at –20 °C until required).

2. Allow labelling to continue for 10 min with occasional mixing.

3. Centrifuge at 200 g for 10 min at 4 °C.

4. Resuspend in cold HSA buffer (RPMI without sodium phosphate, 1% human serum albumin, 1.2 mM $CaCl_2$, 1 mM $MgCl_2$, 10 mM Hepes, pH 7.3 at a concentration of 1.32×10^7 cells/ml. Store at 4 °C until use.

5. Preincubate 175 μl of labelled neutrophils with 175 μl of test ES products in the presence of 100 ng/ml PMA (phorbol 12-myristate 13-acetate, stock solution 1 mg/mlinDMSO, stored at –70 °C until required) for 10 min at 20 °C.

6. Add 100 μl of the cell suspension to a confluent monolayer of primary HUVEC (Clonetics) grown in a 96-well micro-titer plate. Incubate at 37 °C for 30 min.

7. After 30 min remove non-adherent cells by centrifugation of the inverted, sealed plates for 3 min at 75 g.

8. Add 100 μl 0.1% Triton TX-100 (in 50 mM TrisHCl, pH 7.4) to lyse adherent neutrophils.

9. Measure the fluorescence emission of each well (excitation 485 nm, emission detection 530 nm) on a fluorometric plate reader.

Data points should be performed in triplicate. Under these conditions, 40% of the total input cells bind to the HUVEC monolayer in the absence of inhibitor. Cell number is calculated from an internal standard curve using calcein-labelled neutrophils.

A cDNA-encoding NIF has been isolated from an *A. caninum* cDNA library and the recombinant protein expressed in *Pichia pastoris*. The mature protein is a polypeptide of 257 amino acids and has no significant sequence homologies to any previously reported protein. Thus, NIF may represent a novel class of leukocyte function inhibitors.

4.7
Concluding Remarks

Having given the reader an insight into the management of parasites and the handling of their ES products, we must now consider future developments. Firstly, the designation of proteins as "functional", in a sense that they defend the parasite or are required for parasite metabolism, should be tested in relevant vacination systems. Secondly, the potential therapeutic value of molecules purified from ES products can be tested in appropriate assays. Finally a plethora of material in parasite ES products remains unresearched or un- discovered. As an example we include a two-dimensional

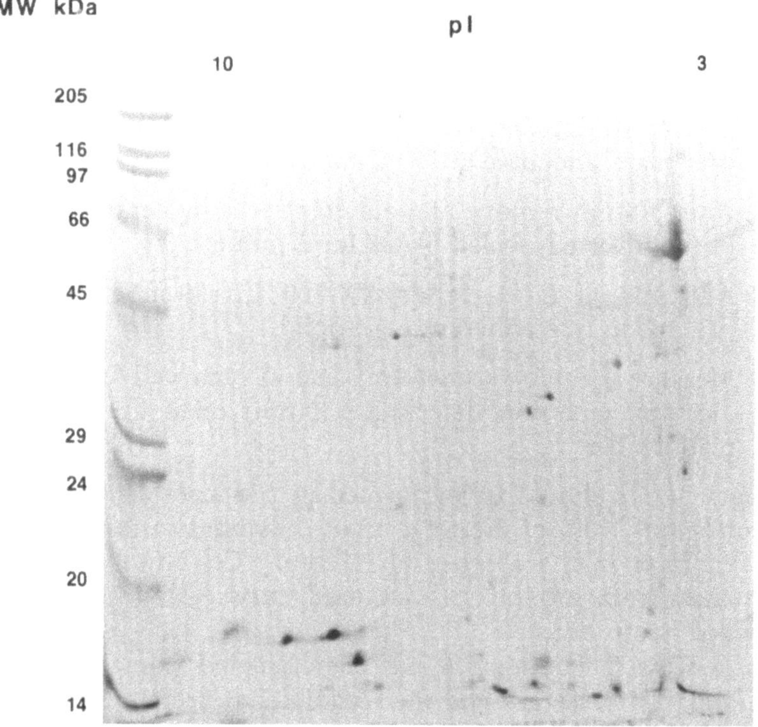

Fig. 8. Two-dimensional SDS-PAGE of *Necator americanus* ES products. 100 μg of *Necator americanus* ES products were separated by tube gel isoelectric focusing (4% gel, ampholyte range pH 3–10) then analysed by SDS-PAGE (12% gel, reducing conditions). Molecular weights and pI points are as indicated

SDS-PAGE gel (Fig. 8) depicting the ES products of the human hookworm *Necator americanus*. We know the indentity of perhaps four of these proteins. The purpose of the remainder awaits discovery.

References

Auriault C, Ouaissi MI, Torpier G, Eisen H, Capron A (1981) Proteolytic cleavage of IgG bound to the Fc receptor of *Schistosoma mansoni* schistosomula. Parasite Immunol 3:33–44

Balloul JM, Sondermeyer P, Dreyer D, Capron M, Grzych J-M, Pierce RJ, Carvallo D, Lecocq J-P, Capron A (1987) Molecular cloning of protective antigen of schistosomes. Nature (Lond) 326:149

Beyer WF, Fridovitch I (1987) Assaying for superoxide dismutase activity: Some large consequences of minor changes in conditions. Anal Biochem 161:559–566

Blackburn CC, Selkirk ME (1992) Characterisation of the secretory acetylcholinesterases from adult *Nippostrongylus brasiliensis*. Mol Biochem Parasitol 53:79–88

Blocki FA, Schlievert PM, Wackell LP (1992) Rat liver protein linking chemical and immunological detoxification. Nature (Lond) 360:269–270

Board PG (1980) A method for the localisation of glutathione-s-transferase isoenzymes after starch gel electrophoresis. Anal Biochem 105:147–149

Born GVR, Cross MJ (1963) The aggregation of blood platelets. J Physiol 168:178–195

Bremner KC, Ogilvie BM, Keith RK, Berrie DA (1973) Acetylcholinesterase secretion by parasitic nematodes – III. *Oesophagostomum* spp. Int J Parasitol 3:609–618

Brophy PM, Barrett J (1990a) Blocking factors and the isolation of glutathione transferases from *Hymenolepis diminuta* (Cestoda: Cyclophyllidea). Parasitol 100:137–141

Brophy PM, Barrett J (1990b) Glutathione transferase in helminths. Parasitol 100:345–349

Brophy PM, Pritchard DI (1992) Metabolism of lipid peroxidation products by the gastro-intestinal nematodes *Necator americanus, Ancylostoma ceylanicum* and *Heligmosomoides polygyrus*. Int J Parasitol 22(7):1009–1012

Brophy PM, Southan C, Barrett J (1989) Glutathione transferases in the tapeworm *Moniezia expansa*. Biochem J 262:939–946

Brophy PM, Ben-Smith A, Brown A, Behnke JM, Pritchard DI (1995a) Differential expression of glutathione S-Transferase by adult *Heligmosomoides polygyrus* during primary infection in fast and slow responding hosts. Int J Parasitol 25(5):641–645

Brophy PM, Patterson LH, Brown A, Pritchard DI (1995b) *Necator americanus*: GST expression: possible role of excretory secretory forms of GST. Acta Tropica 59(3):259–263

Brown A, Pritchard DI (1993) The immunogenicity of hookworm (*Necator americanus*) acetylcholinesterase (AChE) in man. Parasite Immunol 15:195–203

Burt JS, Ogilvie BM (1975) In vitro maintenance of nematode parasites assayed by acetylcholinesterase and allergen secretion. Exp Parasitol 38:75–82

Callahan HL, Crouch RK, James ER (1988) Helminth anti-oxidant enzymes. A protective mechanism against host oxidants. Parasitol Today 4(8):218–225

Camacho M, Tarrab-Hazdai R, Espinosa B, Arnon R, Agnew A (1994) The amount of acetylcholinesterase on the parasite surface reflects the differential sensitivity of schistosome species to metrifonate. Parasitol 108:153–160

Capello M, Clyne LP, McPhedran P, Hotez PJ (1993) *Ancylostoma* Factor Xa inhibitor: Partial purification and its identification as a major hookworm derived anticoagulant in vitro. J Infect Dis 167:1474–1477

Carrol SM, Howse DJ, Grove DI (1984) The anticoagulant effects of the hookworm, *Ancylostoma caninum*: observations on human and dog blood in vitro and infected dogs in vivo. Thromb Haemostasis 51(2):222–227

Chandler AC (1932) Susceptibility and resistance to helmithic infections. J Parasitol 18:135–152

Chang S, Opperman CH (1991) Characterisation of acetylcholinesterase molecular forms of the root-knot nematode, *Meloidogyne*. Mol Biochem Parasitol 49:205–214

Chang S, Opperman CH (1992) Separation and characterisation of *Heterodera glycines* acetylcholinesterase molecular forms. J Nematol 24(1):148–155

Clark AG (1982) A direct method for the visualisation of glutathione-s-transferase activity in polyacrylamide gels. Anal Biochem 123:147–150

Clegg JA, Smithers SR (1972) The effects of immune Rhesus monkey serum on schistosomula of *Schistosoma mansoni* during cultivation in vitro. Int J Parasitol 2:79–98

Cross DA, Klesius PH (1989) Soluble extracts from larval *Ostertagia ostertagi* modulating immune function. Int J Parasitol 19:57–61

Davern KM, Wright MD, Herrman VR, Mitchel GF (1991) Further characterisation of *Schistosoma japonicum* protein Sj 23, a target antigen of an immunodiagnostic monoclonal antibody. Mol Biochem Parasitol 48(1):67–75

Dresden MH, Asch HL (1972) Proteolytic enzymes in extracts of *Schistosoma mansoni* cercariae. Biochem Biophys Acta 289:378–384

Dresden MH, Deelder AM (1979) *Schistosoma mansoni*: thiol proteinase properties of adult worm hemoglobinase. Exp Parasitol 48:190–197

Dresden MH, Rutledge ML, Chappell CL (1981) Properties of the acid thiol proteinase from *Schistosoma mansoni* adults. Mol Biochem Parasitol 4:61–66

Ellman GL, Courtney KD, Andrew V, Featherstone RM (1961) A new and rapid colorimetric determination of acetylcholinesterase activity. Biochem Pharmacol 7:88–95

Espinoza B, Tarrab-Hazdai R, Silman I, Arnon R (1988) Acetylcholinesterase in *Schistosoma mansoni* is anchored to the membrane via covalently attached phospatidylinositol. Mol Biochem Parasitol 29:171–179

Fairweather I, Anderson HR, Baldwin TMA (1987) *Fasciola hepatica* – tegumental surface alterations following treatment in vitro with the deacetylated (amine) metabolite of diaphenethide. Z Parasitenkd 73:99–106

Garside P, Behnke JM, Rose RA (1989) The immune response of male DSN hamsters to a primary infection with *Ancylostoma ceylanicum*. J Helminthol 63(3):251–260

Griffifths G (1994) Secretory acetylcholinesterase from the nematode *Trichostrongylus colubriformis*. PhD, University of Nottingham, Nottingham

Griffiths G, Pritchard DI (1994) The biochemical characterisation of secretory acetylcholinesterase from *Trichosrongylus colubriformis*. Parasitology (in press)

Griffifths G, Pritchard DI (1994) Purification and biochemical charactertisation of acetylcholinesterase (AChE) from the ES products of *Trichostrongylus colubriformis*. Parasitol 108(5):579–586

Habig W, Pabst MJ, Jakoby WB (1974) Glutathione S-transferases. The first step in mercapturic acid formation. J Biol Chem 249(22):7130–7139

Halliwell B, Gutteridge JMC (1992) Biologically relevant metal ion-dependent hydroxylradical generation – an update. FEBS Lett 307(1):108–112

Harlan JM, Winn RK, Vedder NB, Doerschuk CM, Rice CL (1992) In vivo models of leucocyte adherence to endothelium. In: Harlan JM, Liu DY (eds) Adhesion: its role in inflammatory disease. Freedman, New York, pp 117-150

Hill DE, Gamble HR, Rhoads ML, Fetterer RH, Urban JF (1993) *Trichuris suis*: a zinc metalloprotease from culture fluids of adult parasites. Exp Parasitol 77:170–178

Hodgson AJ, Chubb IW (1983) Isolation of the secretory form of acetylcholinesterase by using affinity chromatography on edrophonium sepharose. J Neurochem 41:654–662

Hotez P, Haggerty J, Hawdon J, Milstone L, Gamble HR, Schad G, Richards F (1990) Metalloproteinases of infective *Ancylostoma* hookworm larvae and their possible functions in tissue invasion and ecdysis. Infect Immun 58(12):3883–3892

Hotez PJ (1989) Hookworm disease in children. J Pediatr Infect Disease 8:516–520

Hotez PJ, Cerami A (1983) Secretion of a proteolytic anticoagulant by *Ancylostoma* hookworms. J Exp Med 157:1594–1603

Hotez PJ, Le Trang N, McKerrow JH, Cerami A (1985) Isolation and characterisation of a proteolytic enzyme from adult hookworm *Ancylostoma caninum*. J Biol Chem 260(12):7343–7348

Jarvis LM, Pritchard DI (1992) An evaluation of the role of carbohydrate epitopes in immunity to *Trichinella spiralis*. Parasite Immunol 14:489–501

Johnson C, Russell RL (1975) A rapid radiometric assay for cholinesterase, suitable for multiple determinations. Anal Biochem 64:229–238

Johnson CD, Russell RL (1983) Multiple molecular forms of acetylcholinesterase in the nematode *Caenorhabditis Elegans*. J Neurochem 41:30–46

Jones DG, Knox DP (1990) Evidence for the presence of nematode-derived acetylcholinesterase in sheep infected with *Trichostrongylus colubriformis*. Res Vet Sci 48:136–137

Kaliner M, Austen KF (1974) Cyclic nucleotides and modulation of effector systems of inflamation. Biochem Pharmacol 23:763–771

Karanu FN, Rurangirwa FR, McGuire TC, Jasmer DP (1993) *Haemonchus contortus*: identification of proteases with diverse characteristics in adult worm excretory-secretory products. Exp Parasitol 77:362–371

Karnovsky MJ, Roots L (1964) A "direct-coloring" thiocholine method for cholinesterases. J Histochem Cytochem 12:219–221

Kenney WC, Boyer TD (1981) Detection of glutathione-s-transferase isoenzymes after electrofocusing in polyacrylamide gels. Anal Biochem 116:344–348

Ketterer B, Meyer DJ, Clark AG (1989) Soluble glutathione transferase isoenymes. In: (Sies H, Ketterer B eds) Glutathione conjugation: its mechanism and biological significance. Academic Press, London, pp 73–135

Kharat I, Cheirmaraj K, Prasad GBKS, Harinath BC (1989) Antigenic analysis of excretory-secretory products of *Wuchereria bancrofti* and *Brugia malayi* infective larval forms by SDS-PAGE. Indian J Exp Biol 27:681–684

Knox DP, Jones DG (1990) Studies on the presence and release of proteolytic enzymes (proteinases) in gastro-intestinal nematodes of ruminants. Int J Parasitol 20(2):243–249

Knox DP, Jones DG (1992) A comparison of superoxide dismutase (SOD, EC-1. 15. 1. 1.) distribution in gasterointestinal nematodes. Int J Parasitol 22(2):209–214

Kumar S, Prichard DI (1992) The partial characterisation of proteases present in the excretory/secretory products and exsheathing fluid of the infective (L3) larva of *Necator americanus*. Int J Parasitol 22(5):563–572

Kwan-Lim GE, Gregory WF, Selkirk ME, Partono F, Maizels RM (1989) Secreted antigens of filarial nematodes: A survey and characterisation of in vitro excreted/secreted products of adult *Brugia malayi*. Parasite Immunol 11:629–654

Laemmli UK (1970) Cleavage of structural proteins during the assembly of the head of bacteriophage T4. Nature (Lond) 227:680–685

Langlet C, Mazigue C, Dessaint JP Capron A (1984) Inhibition of primary and secondary IgE response by a schistosome-derived inhibitory factor. Int Arch Allergy Appl Immunol 73:225–230

Lawrence CE, Pritchard DI (1993) Differential secretion of acetylcholinesterase and proteases during the development of *Heligmosomoides polygyrus*. Int J Parasitol 23(3):309–314

Leid RW, Suquet CM (1986) A superoxide dismutase of metacestodes of *Taenia taeniaeformis*. Mol Biochem Parasitol 18(3):301–311

Liew FY, Scott MT, Lim DS, Croft SL (1987) Suppressive substance produced by T cells from mice chronically infected with *Trypanosoma cruzi*. J Immunol. 139:2452–2457

Lodes MJ, Yoshino TP (1989) Characterisation of excretory-secretory proteins synthesised in vitro by *Schistosoma mansoni* primary sporocysts. J Parasitol 75(6):853–862

Loeb L, Smith AJ (1904) The presence of a substance inhibiting the coagulation of the blood in Anchylostoma. Proc Pathol Soc Philadelphia 7(6):6–11

Maelicke A (1991) Acetylcholinesterase: the structure. Trends Biochem 16:355–356

Maizels RM, Bundy DAP, Selkirk ME, Smith DF, Anderson RM (1993) Immunological modulation and evasion by helminth parasites in human populations. Nature (Lond) 365:797–804

Mallett S (1989) In vitro acetylcholinesterase secretion by male or female *Heligmosomoides polygyrus*. Ann Rech Vét 20:107–110

Maniatis T, Fritsch EF, Sambrook J (1989) Molecular cloning, a laboratory manual. Cold Spring Harbor Laboratories. Cold Spring Harbor, New York

Mannervik B, Alin P, Guthenberg C, Jennson H, Kalim Tahir M, Warholm M, Jornvall H (1985) Identification of three classes of cytosolic glutathione transferase common to several mammalian species: correlation between structural data and enzymatic properties. Proc Natl Acad Sci USA 82:7207–7206

Matthews BE (1975) Mechanisms of skin penetration by *Ancylostoma tubaeforme* larvae. Parasitology 70:25–38

Matthews BE (1982) Skin penetration by *Necator americanus* larvae. Z Parisitenk 68:81–86

Mazingue C, Dessaint JP, Schmitt-Verhulst AM, Cerottini JC, Capron A (1983) Inhibiting cytotoxic T lymphocytes by a schistosome-derived inhibitory factor is dependent of an inhibition of the production of interleukin. Int Arch Allergy Appl Immunol 72:22–29

McEver RP, Bennett EM, Martin MN (1983) Identification of two structurally and functionally distinct sites on human platelet membrane glycoprotein IIb-IIIa using monoclonal antibodies. J Biol Chem 258(8):5269–5275

McKerrow JH (1989) Parasite proteases. Exp Parasitol 68:111–115

McKerrow JH, Doenhoff MJ (1988) Schistosome proteases. Parasitol Today 4(12):334–340

Miletich JP, Jackson CM, Majerus PW (1977) Interaction of coagulation Factor Xa with human platelets. Proc Natl Acad Sci USA 74:4033–4036

Monroy FG, Dobson C, Adams JH (1989) Low molecular weight immunosuppressors secreted by adult *Nematospiroides dubius*. Int J Parasitol 19:125–127

Moyle M, Foster DL, McGrath DE, Brown SM, Laroche Y, De Meutter J, Stanssens P, Bogowitz CA, Fried VA, Ely JA, Soule HR, Vlasuk GP (1994) A hookworm glycoprotein that inhibits neutrophil function is a ligand of the integrin CD11b/CD18. J Biol Chem 269(13):1008–1015

O'Donnell IJ, Dineen JK, Wagland BM, Letho S, Dopheide TAA, Grant WN, Ward CW (1989) Characterisation of the major immunogens in the excretory-secretory products of exsheathed third stage larvae of *Trichostrongylus colubriformis*. Int J Parasitol 19(7):793–802

Ogilvie BM, Rothwell TLW, Bremner KC, Schnitzerling HJ, Nolan L, Keith RK (1973) Acetylcholinesterase secretion by parasitic nematodes – I. Evidence for the secretion of the enzyme by a number of species. Int J Parasitol 3:589–597

O'Leary KA, Tracy JW (1988) Purification of three cytosolic glutathione S-transferases from adult *Schistosoma mansoni*. Arch Biochem Biophys 264:1–12

Panaccio M, Wilson LR, Crameri SL,Wijffels GL, Spithill TW (1992) Molecular characterisation of cDNA sequences encoding glutathione s-transferases of *Fasiola hepatica*. Exp Parasitol 74:232–237

Philipp M (1984) Acetylcholinesterase secreted by intestinal nematodes: a reinterpretation of its putative role of "biochemical holdfast". Trans R Soc Trop Med Hyg 78:138–139

Playfair JHL (1982) Workshop report: supressor cells in infectious disease. Parasite Immunol 4:299–304

Pritchard DI (1993) Why do some parasitic nematodes secrete acetylcholinesterase (AChE)? Int J Parasitol 23(5):549

Pritchard DI, Ali NMH, Behnke JM (1984) Analysis of the mechanism of immunodepression following heterologous immunity by soluble adult antigens of *Nematospiroides dubius*. J Helminthol 59:251–256

Pritchard DI, McKean PG, Tighe PJ (1990a). Recent and predicted advances in hookworm biology. In: Kennedy MW (ed) Parasite genes, membranes and antigens. Taylor and Francis, London

Pritchard DI, Quinnell RJ, Slater AFG, McKean PG, Dale DDS, Raiko A, Keymer AE (1990b) Epidemiology and immunology of *Necator americanus* infection in a community in Papua New Guinea: humoral reponses to excretory-secretory and cuticular collagen antigens. Parasitol 100:317–326

Pritchard DI, Leggett KV, Rogan MT, McKean PG, Brown A (1991) *Necator americanus* secretory acetylcholinesterase and its purification from excretory-secretory products by affinity chromatography. Parasite Immunol 13:187–199

Pritchard DI, Lawrence CE, Appleby P, Gibb IA, Glover K (1993) Immunosuppressive proteins secreted by the gasterointesinal nematode parasite *Heligmosomoides polygyrus* Int J Parasitol (in Press)

Pritchard DI, Brown A, Toutant J-P (1994) The molecular forms of acetylcholinesterase from *Necator americanus* (Nematoda), a hookworm parasite of the human intestine. Eur J Biochem 219:317–323

Ramalho-Pinto G, Gazinelli G, Howells RE, Mota-Santos TA, Figueiredo EA Pelligrino J (1974) *Schistosoma mansoni*: defined system for stepwise transformation of cercariae to schistosomula in vitro. Exp Parasitol 36:360

Rathaur S, Robertson BD, Selkirk, ME, Maizels RM (1987) Secretory acetylcholinesterase from *Brugia malayi* adult and microfilarial parasites. Mol Biochem Parasitol 26:257–265

Raybourne R, Desowitz RS, Kliks MM, Deardorff TL (1983) *Anisakis simplex and Terranova sp*: inhibition by larval excretory-secretory products of mitigen-induced rodent lymphblast proliferation. Exp Parasitol 55:238–298

Rhoads ML (1983) *Trichinella spiralis*: identification and purification of superoxide dismutase. Exp Parasitol 56(1):41–54

Ricci G, Bello ML, Caccuri AM, Galiazzo F, Federici G (1984) Detection of glutathione tranferase activity on polyacrylamide gels. Anal Biochem 143:226–230

Rosenberry TL, Scoggin DM (1984) Structure of human erythrocyte acetylcholinesterase. J Biol Chem 259(9):5643–5652

Rothwell TLW, Ogilvie BM, Love RJ (1973) Acetylcholinesterase secretion by parasitic nematodes–II *Trichostrongylus spp* Int J Parasitol 3:599–608

Rotmans JP, Van der Voort MJ, Looze M, Mooij GW, Deelder AM (1981) *Schistosoma mansoni*: characterisation of antigens in excretions and secretions Exp Parasitol 52:171–182

Sanger F, Nicklen S, Coulson AR (1977) DNA sequencing with chain terminating inhibitors. Proc Natl Acad Sci 74:5463–5467

Shatill SJ (1993) Regulation of platelet anchorage and signalling by integrin $\alpha II_b \beta_3$. Thromb Haemostasis 70(1):224–228

Simons PC, Vander Jagt DL (1977) Purification of glutathione S-transferase from human liver by glutathione affinity chromatography. Anal Biochem 82:334–341

Simurda MC, Vankeulen H, Rekosh DM (1988) *Schistosoma mansoni*: identification and analysis of a messenger–RNA and a gene encoding S-peroxide dismutase (Cu/Zn). Exp Parasitol 67(1):73–84

Smith MA, Clegg JA (1981) Improved culture of *Fasciola hepatica* in vitro Z. Parasitenk 66:9–15

Smith DB, Davern KM, Board PG, Tiu WU, Garcia EG, Mitchell GF (1986) M_r 26000 antigen of *Schistosoma japonicum* recognised by resistant WEHI 129/J mice is a parasite glutathione s–transferase. Proc Natl Acad Sci USA 83:8703–8707

Steen VM, Holmsen H (1987) Current aspects on human platelet activation and responses. Eur J Haematol 38:383–399

Subrahmanyam D, Parab PB, Rajasekariah GR, Rudin W, Betschart B, Weiss N (1990) Interactions of monoclonal antibodies with cuticular antigens of filarial parasites *Brugia malayi* and *Wucheriera bancrofti*. Acta Trop 47:381–390

Sutherland IA, Lee DL (1993) Acetylcholinesterase in infective–stage larvae of *Haemonchus contortus, Ostertagia circumcincta and Trichostrongylus colubriformus* resistant and susceptible to benzimadazole anthelmintics. Parasitology 107:553–557

Taylor JB, Vidal A, Torpier G, Meyer D J, Roitsch C, Balloul JP, Southan C, Sondermeyer P, Pemble S, Lecocq JP, Capron A, Ketterer B (1988) The glutathione transferase activity and tissue distribution of a cloned M_r28K protective antigen of *Schistosoma mansoni*. EMBO J 7(2):465–472

Taylor P (1991) The cholinesterases. J Biol Chem 266(7):4025–4028

Taylor P, Schumacher M, MacPhee-Quigley K, Friedman T, Taylor S (1987) The structure of acetylcholinesterase: relationship to its function and cellular disposition. TINS 10: 93–95

Thorson RE (1956) The stimulation of acquired immunity in dogs by injections of extracts of the esophagus of adult hookworms. J Parasitol 42:501–504

Toutant J-P (1989) Insect acetylcholinesterase: catalytic properties, tissue distribution and molecular forms. Prog Neurobiol 32:423–446

Toutant J-P, Massoulie J (1987) Acetylcholinesterase In: Turner KA (ed) Mammalian ectoenzymes. Elsevier, Amsterdam, pp 289–328

Verwaerde C, Aurialt C, Neyrinck JL, Capron A (1988) Properties of serine proteases of *Schistosoma mansoni* schistosomula involved in the regulation of IgE synthesis. Scand J Immunol 27:17–24

Wijffels GL, Salvatore L, Dosen M, Waddington J, Wilson L, Thompson C, Campbell N, Sexton J, Wicker J, Bowen F, Friedel T, Spithill TW (1994) Vaccination of sheep with purified cysteine proteinases of *Fasciola hepatica* decreases worm fecundity. Exp Parasitol 78:132–148

Zerda KS, Dresden MH, Damian RT, Chappell CL (1987) *Schistosoma mansoni*: anti SMw32 proteinase response in vaccinated and challenged baboons. Am J Trop Med Hyg 37(2):320–326

Parasite Proteinases

Michael J. North

5.1
Introduction

The importance of parasite proteolytic enzymes in host-parasite relationships and pathogenesis and their role in processes such as nutrition, the invasion of host cells and tissues and the countering of host defence processes is now well established (McKerrow 1989; North et al. 1990a; North 1991; McKerrow et al. 1993; North and Lockwood 1995). One driving force behind research on these enzymes has been the possibility that they might prove useful as targets for anti-parasite chemotherapy. Much work has been aimed at establishing the characteristics of the enzymes, in particular their substrate specificity and sensitivity to inhibitors, with a view to using this information in the design of novel anti-parasitic agents. This chapter deals with procedures for detecting, assaying and purifying proteolytic enzymes, and for determining specificity and inhibitor sensitivity, and so reflects these priorities. Less emphasis is given to methods concerned directly with elucidating the function of the enzymes. However, gaining an understanding of their role will be dependent on the methods described, as these should point to likely physiological substrates and highlight promising inhibitors whose effect on parasite processes can be tested.

Proteinase Nomenclature and Terminology

By definition, proteolytic enzymes (peptidases) catalyse the cleavage of peptide bonds in proteins and peptides. Proteolytic enzymes possess endopeptidase activity (cleavage of internal bonds) or exopeptidase activity (cleavage of bonds at or adjacent to the end of a peptide chain), and some have both. The term proteinase is used here for any enzyme which has endopeptidase activity. According to the nomenclature of Schechter and Berger (1967), amino acid residues neighbouring the scissile bond are called P_1, P_2, P_3, P_4 etc. in the direction of the amino terminus, and P_1', P_2', P_3', P_4' etc. in the direction of the carboxy terminus (there are complementary subsites on the enzyme called S_1, S_2 etc.). Each enzyme will have a requirement for particular residues or types of residue in one or more of these positions, and this determines its specificity.

Proteinase Classes

The majority of proteolytic enzymes, whether endopeptidases or exopeptidases, fall into one of four classes according to their catalytic mechanism. Aspartic (carboxyl), cysteine (thiol) and serine peptidases are so called because of the involvement of the respective amino acids in the mechanism, while metallopeptidases have a metal atom, usually zinc, at the active site. Each class is further divided into clans and families based on sequence similarities which reflect common evolutionary origins (Rawlings and Barrett 1993). Proteinase class is usually apparent from the effect on enzyme activity of a group of diagnostic inhibitors (see Sect. 5.4). The class and family to which an enzyme belongs are relevant to the choice of assay conditions, substrates and inhibitors. Knowledge of an enzyme's class may allow reasonable predictions to be made about some properties, e.g. pH dependence. Others, such as specificity, may be less predictable, but knowing the class does provide some clues to the types of substrate worth testing.

Occurrence of Proteinases and Proteinase Inhibitors in Parasites

Proteinases of all four classes have been described in both protozoa and helminths (North and Lockwood 1995). This has been established from their sensitivity to inhibitors and from sequence determinations, usually of the corresponding genomic DNA or cDNA. Their occurrence is summarised in Table 1, which also includes some information on putative proteinases encoded by genes in the free-living nematode *Caenorhabditis elegans*. Most parasite enzymes for which

Table 1. Occurrence of proteinases in protozoa and helminths

Proteinase class	Occurrence	Peptidase families (see Rawlings and Barrett 1993)	Comments
Protozoa			
Aspartic	*Plasmodium, Eimeria*	A1 (pepsin)	Well-characterized enzymes reported only in sporozoa
Cysteine	Widely reported	C1 (papain)	Found in most species during at least one stage of their life cycle Most are closer to cathepsin L in structure than cathepsin B
Metallo-	*Leishmania* and some other trypanomastids	M8 (unique family)	Metalloproteinases have been detected in other species but have not yet been characterised
Serine	*Eimeria* *Plasmodium p68* Trypanosomes	–	Some trypsin-like features in specificity

Table 1 (*Contd.*)

Proteinase class	Occurrence	Peptidase families (see Rawlings and Barrett 1993)	Comments
Helminths			
Aspartic	*Dirofilaria immitis*	–	Possible activity in other species but not yet characterised
Cysteine	Widely reported	C1 (papain)	The majority are cathepsin B-like in structure but some are cathepsin L-like, especially in trematodes
	Schistosoma mansoni Sm32	C13 (asparaginyl proteinase)	Other members of this family have only recently been reported in plants, e.g. legumain
		C2 (calpain, same clan as C1)	Gene sequence
	Caenorhabditis elegans ced-3	C14 (interleukin-1b converting enzyme)	Involved in apoptosis
Metallo-	Widely reported	–	Found in all groups of helminth
Serine	Widely reported		Found in all groups of helminth
	S. mansoni	S1 (chymotrypsin)	Cercarial elastase
	C. elegans	S8 (subtilisin)	Gene for prohormone convertase
	C. elegans	S10 (serine-type carboxypeptidase)	Gene sequence
		–, no sequence information available	

sequence information is available are members of the more common eukaryotic peptidase families. Noteworthy exceptions are the *Leishmania* surface proteinase (leishmanolysin, gp63) and the *Schistosoma mansoni* haemoglobinase, Sm32. Most of the reported protozoan proteinases are intracellular, although there are examples of enzymes located on the cell surface, and secretion of cysteine proteinases occurs in a few species. Helminth proteinases have been found in whole parasites and in excreted/secreted material at various life-cycle stages.

This chapter also provides some information on methods for analysing endogenous proteinase inhibitors. These have been reported quite frequently in helminths, in which some are major antigens, but they have only recently been detected in parasitic protozoa (North and Lockwood 1995). A number of genes encoding helminth inhibitors have now been cloned, including some for cystatins (cysteine proteinase inhibitors), serpins (serine proteinase inhibitors) and pepsin inhibitors (aspartate proteinase inhibitors). This has confirmed that at least some of them are related to similar proteins produced by mammalian cells.

Selection of Methods

Studies on parasite proteolytic enzymes have tended to focus largely on endopeptidases, and the majority of the methods given here were designed for detecting and analysing proteinases. The methods presented relate to the enzymatic properties of proteinases. All have been used with one or more enzymes from parasites. Many of the parasite proteinases appear to be typical of the family and class to which they belong (see above), and this has allowed the direct application of techniques developed for analysing similar enzymes from other organisms, including those of the hosts. It has been possible to include only some of the procedures in the form of detailed protocols, and the information presented is intended as much as a guide to the available methods which would be suitable for a range of parasite enzymes. The literature cited should be consulted where necessary for additional details. Other techniques which may be suitable are described in books devoted to proteolytic enzymes (Beynon and Bond 1989; Barrett 1994, 1995), and some further details on protozoan cysteine proteinases are available in other recent articles (North 1994; Scholze and Tannich 1994). General methods for enzyme or protein analysis will, of course, also be appropriate. For example, immunological approaches will often be important, as a number of parasite proteolytic enzymes are major antigens whose enzymatic properties were not always known or recognised initially.

5.2
Sample Preparation

Most proteolytic enzymes are robust, and special procedures for sample preparation are not normally required. Parasite material can be extracted by standard procedures such as lysis with non-ionic detergent (eg 0.1–0.25% Triton X-100 and 0.5% Nonidet P-40), or disruption with glass beads, by grinding, by freeze-thawing, by homogenisation in a glass-Teflon homogenizer or in a French pressure cell, as appropriate. The composition of the extraction buffer may reflect the optimal assay conditions or the requirements for the first purification step, but few components are essential and for some samples extraction into non-buffered solutions, e.g. 0.25 M sucrose, is adequate. For cysteine proteinase analysis, samples may be prepared in buffer containing a reducing agent, such as DTT (1–4 mM), and EDTA (1–4 mM), but this is not normally essential. Some serine proteinases are stabilised by divalent cations and calcium chloride (1–5 mM) should be included. For substrate-SDS-PAGE analysis (see Sect. 5.5), extracts may be prepared directly in electrophoresis sample buffer. Some purification schemes involve the initial isolation by density gradient centrifugation of a proteinase-containing compartment such as secretory vacuoles or lysosomes (see Sect. 5.6).

Autoproteolysis can be a problem in preparations containing proteinases. Proteinase inhibitors can be used to eliminate proteolytic problems (North 1989), with the obvious proviso that the enzyme(s) of interest can still be detected. Irreversible inhibitors can be added to eliminate the activity of unwanted proteinases, but reversible ones will be needed to suppress the activity of those which are of interest. For example, 1 mM $HgCl_2$ may be added to buffers to inhibit cysteine proteinases: its effect is reversed by the presence of excess reducing agent in the assay/detection system.

A few examples of membrane-bound proteinases are known. These can be separated from cytosolic enzymes by partitioning in Triton X-114 (based on Bordier 1981). Washed cells are extracted in 1% (v/v) Triton X-114 in TBS (Tris-buffered saline: 10 mM Tris pH 7.5, 150 mM NaCl) for 15 min on ice. The suspension is centrifuged at 10 000 g for 5 min at 4 °C. The pellet is discarded and the supernatant removed to a fresh tube and incubated at 37 °C for 10 min followed by centrifugation at 1000 g for 5 min. The aqueous (upper) and detergent (lower) phases are recovered and re-extracted with detergent and buffer, respectively.

Many helminth proteinases are major components of excretory/secretory material, and parasites must be incubated under suitable conditions to obtain this. A few protozoa, e.g. trichomonads and *Entamoeba histolytica*, secrete proteinases, and spent growth medium can be used as a source of enzymes. However, serum proteins, if present, can interfere with proteinase assays and cause problems with purification. Extracellular samples will often need to be dialysed and/or concentrated before being analysed. Ultrafiltration filters with cutoffs of 10 kDa or lower should be used, as a number of proteinases are small enzymes with molecular weights of less than 30 kDa.

5.3
Substrates and Proteinase Assays

5.3.1
Optimal Conditions for Proteinase Activity

With most proteolytic enzymes, the major factors in selecting assay conditions will be the pH and substrate. The majority of proteinases are active over a broad range of pH, and at a particular pH the choice of buffer can be flexible. Traditional buffers such as glycine, acetate, citrate, phosphate and borate are still widely used along with Tris and "good" buffers like MES, PIPES and HEPES. As a rough guide, aspartic proteinases tend to be most active at low pH (use pH 2–5), cysteine proteinases just below neutral pH (pH 5–7) and metalloproteinases and serine proteinases just above neutral pH (pH 7–9). Cysteine proteinases require a reducing agent for full activity and DTT (1–10 mM) or cysteine (5–20 mM) should be included. EDTA (1–4 mM) may also be added, but this will inhibit some metalloproteinases. Divalent cations stabilise some serine proteinases and $CaCl_2$ (1–2 mM) may be needed. Zinc ions can be included when analysing metalloproteinases, but are not normally essential unless the enzyme has been pretreated with a chelating agent.

Some of the assays described below are not selective enough to measure the activity of single enzymes in samples containing multiple proteinases. By adding appropriate inhibitors (see Sect. 5.4), it may be possible to determine the contribution of a particular enzyme or group of enzymes to the total activity. For example, the relative contributions of cysteine and aspartic proteinases to haemoglobin digestion by *Plasmodium falciparum* proteinases have been determined by including 10 μM E-64 to eliminate the activity of the former (Gluzman et al. 1994).

5.3.2
Protein and Peptide Substrates

Not surprisingly, the hydrolysis of proteins has been the basis for many detection and assay methods, and a variety of proteins have been used with parasite enzymes (Table 2). The advantages of using proteins as substrates include the following. First, with the exception of enzymes with very narrow specificity, most proteinases will be able to cleave a number of bonds in a given protein allowing the same substrate to be used for a range of enzymes of different class and specificity. Second, the use of proteins as substrates more closely matches the physiological situation. Third, many protein substrates are cheap and available in quantity. There are, however, some disadvantages. Varying the assay conditions (e.g. pH) may affect the substrate as much as the enzyme, detailed kinetic analysis is not usually possible and the activities of individual enzymes may be difficult to distinguish in a standard assay.

Choice of substrate depends on the degree of specificity and sensitivity required and on whether or not the products need to be identified and analysed. Proteins can be selected because they are known or potential physiological targets for

Table 2. Proteins used as substrates for assaying parasite proteinases

Native proteins	Proteins with chromophores
Haemoglobin	Azocasein (366 nm)
Gelatin	Azocoll (540 nm)
Fibrinogen	Elastin-orcein (550 nm)
Immunoglobulins	Hide powder azure (595 nm)
Complement C3	Keratin azure (595 nm)
Matrix proteins including:	
Collagens	The wavelength for spectrophotometric
Laminin	measurement is indicated
Elastin	
Spectrin	
Radiolabelled proteins	**Other labelled proteins**
Haemoglobin	Biotinylated-gelatin
Bovine serum albumin	FITC-casein
Gelatin	
Fibrin	
Fibrinogen	
Matrix proteins including:	
Collagen	
Elastin	

the enzyme. Thus matrix proteins like collagen and elastin
are used with enzymes such as helminth ES proteinases
which are involved in tissue invasion, haemoglobin, a widely;
used proteinase substrate, is also the natural substrate for
proteinases from intraerythrocytic protozoa and blood-
feeding helminths, while complement components, immunoglobulins and other serum proteins are likely targets for
a number of parasite enzymes.

Approaches to detecting and assaying activity using protein substrates fall into two broad categories, those which
measure gross changes, usually of the amount of peptides
and amino acids released, and those in which the hydrolysis
products are separated and analysed. The first approach is
more suitable for routine assays, especially of enzymes with
broad specificity, the second can be used for enzymes of high
specificity or when detailed information on substrate specificity is required. Note that there are few continuous assay
systems with proteins as substrates, although a decrease in

the turbidity of a powdered milk solution can be followed over a period of time (Bouvier et al. 1989).

For the first approach, the standard procedure is to incubate a soluble protein substrate with sample for a suitable length of time and then stop the reaction, usually by the addition of trichloroacetic acid (TCA). The undigested, acid-insoluble material is removed by centrifugation and the amount of acid-soluble product is determined. Alternatively, the reaction can be stopped with 6 M guanidine HCl and proteolysis products separated from substrate by ultrafiltration using a filter with a 10-kDa cutoff. Some substrate proteins are insoluble, and so it is possible to separate the undigested substrate and products without stopping the reaction with TCA. For broad specificity enzymes an inexpensive protein such as haemoglobin is appropriate and activity can be determined from the amount of product released (direct spectrophotometric measurement at 280 nm or after reaction with a reagent such as ninhydrin or Coomassie blue). Increased sensitivity can be achieved by using labelled substrates. Proteins with a chromophore attached are available, the products being measured spectrophotometrically. Selected proteins may be labelled with FITC, to produce a fluorescent substrate, or with biotin. Much use has been made of radiolabelled substrates: some are commercially available, e.g. ^{14}C-methylated haemoglobin, ^{14}C-methylated caseins and collagen-(1-^{14}C-acetylated) (all supplied, for example, by Sigma), others may be custom-synthesised by radioiodination (e.g. Robertson et al. 1989), acetylation with [^{14}C] acetic anhydride (e.g. Muñoz et al. 1984) or reduction with NaB[^{3}H]$_4$ (e.g. Hotez et al. 1985). Radiolabelled proteins may also be prepared biosynthetically, for example substrates for enzymes which hydrolyse extracellular matrix proteins have been prepared by incubating tissue culture cells with [^{3}H]-proline (McKerrow et al. 1985a) or haemoglobin by labelling erythrocytes with [3, 4, 5-^{3}H]-leucine (Goldberg et al. 1990).

Three examples of assays are given below. All require a spectrophotometer (UV for the haemoglobin assay), a centrifuge (bench or microfuge) and a water bath.

Haemoglobin

The following procedure can be used with different proteins as substrate, the pH may be adjusted as required. The assay conditions given are those used for recombinant plasmepsin I, a *P. falciparum* aspartic proteinase (Dame et al. 1994).

– Substrate, 2.5% (w/v) bovine haemoglobin solution (dialysed) – Buffer, 1 M sodium citrate, pH 4.7 or 3.1 – 4% (w/v) TCA	**Reagents**

1. Haemoglobin solution (0.3 ml) is preincubated with 0.075 ml buffer at 37 °C for 15 min to denature the substrate. **Procedure, Haemoglobin**

2. The reaction is started by adding proteinase sample (0.075 ml) and incubation continued for an appropriate time.

3. TCA (0.3 ml) is added and the sample centrifuged (2000 *g*, 5 min).

4. The proteolysis products in the supernatant are measured spectrophotometrically at 280 nm using blanks in which TCA is added prior to incubation with sample.

Azocasein

Azocasein is an example of a protein substrate labelled with a chromophore. The assay conditions are adapted from those of Coombs (1982). The pH may be varied, but note that azocasein precipitates at low pH.

– Substrate, azocasein (100 mg/ml in water) – Buffer, e.g. 0.1 M sodium acetate/acetic acid, pH 5.5 – 0.1 M DTT – 5% (w/v) TCA	**Reagents**

1. An assay mix consisting of 0.1 ml azocasein, 0.60 ml buffer and 0.01 ml DTT (needed for cysteine proteinases but otherwise optional) is set up. **Procedure, Azocasein**

2. The reaction is started by the addition of 0.05 ml sample and incubated for a suitable time, e.g. 30 min at 37 °C.

3. The reaction is stopped by addition of 0.75 ml TCA and the precipitate is removed by centrifugation (13 400 g, 5 min).

4. The absorbance at 366 nm of the supernatant is measured. Control and blank absorbances should be deducted. With 1 cm cuvettes, an increase of 1 absorbance unit represents the hydrolysis of 0.4 mg azocasein.

Hide Powder Azure

Hide powder azure is a non-soluble chromophoric substrate which is in the same physical state at all pHs. It can be dispensed as a suspension or weighed into individual assay tubes. This protocol is based on that described by North and Whyte (1984). It can be also used with other non-soluble substrates such as azocoll.

Reagents – Substrate, Hide powder azure suspension (10 mg/ml) in water. Batches of the substrate vary, and it may be necessary to grind the powder using a pestle and mortar or to use sonication (8 × 15 s) to obtain an even suspension.
– Buffer, e.g. 0.1 M acetic acid/sodium acetate, pH 5.5
– 0.1 M DTT
– 50% (w/v) TCA

Procedure, Hide powder azure

1. Hide powder azure (0.5 ml) is dispensed into tubes using an automatic pipette fitted with a tip cut to widen the nozzle. Buffer (0.5 ml) and 0.01 ml DTT (needed for cysteine proteinases but otherwise optional) are added.

2. The reaction is started by the addition of 0.1 ml sample. Assay tubes may be shaken gently.

3. After a suitable incubation period, e.g. 30 min at 37 °C (this can be judged by letting the substrate settle at the bottom of the tube and assessing the amount of blue colour in solution), 0.2 ml of TCA may be added (this is optional) and the assay mix spun in a microcentrifuge at 13 400 g for 5 min. Precipitation with TCA does give firmer pellets.

4. The amount of product in the supernatant is determined spectrophotometrically at 595 nm. Control and blank absorbances should be deducted. With a 1-cm cuvette, an increase of 1 absorbance unit represents the hydrolysis of 3.4 mg Hide powder azure.

Radiolabelled Substrates

Assays with radiolabelled substrates can be carried out essentially as above. Product release is determined by adding aliquots of supernatant to scintillation cocktail and determining the amount of radioactivity released in a liquid scintillation counter. As an example, aspartic proteinase activity in *P. falciparum* was assayed in a reaction mixture containing 150 mM sodium acetate buffer, pH 5, 60 000 cpm (0.00625 mM) $[^{14}C]$methylated globin and 0.01 ml sample in a final volume of 0.04 ml. The reaction is stopped by addition of TCA.

Microtitre Plate Assays with Labelled Proteins

Microtitre plate assays have been developed to handle a large number of samples simultaneously using very sensitive assay methods. A technique involving radiolabelled gelatin (Robertson et al. 1989), has been used to assay proteinases from a number of helminths. In outline, lysine residues in the gelatin substrate are radioiodinated with Bolton-Hunter reagent and this is then coated onto the wells of polyvinyl chloride microtitre plates (30 000 cpm/well). Samples are incubated in the wells and the radioactivity released determined by gamma counting without the need for scintillation fluid. Test volumes are minimised and the number of samples which can be handled maximised. Similar procedures have been used for assays with other substrates, e.g. ^{125}I-fibrin for *Ancylostoma caninum* metalloproteinase (Hotez et al. 1985), $[^{14}C]$haemoglobin for *S. mansoni* cysteine proteinases (Götz and Klinkert 1993), $[^{3}H]$-proline-labelled extracellular matrix used with cultured helminth larvae (Robertson et al. 1989). A method involving biotinylated gelatin has been described by Koritsas and Atkinson (1994) and used with *Globodera pallida*. Wells are coated with substrate prepared by reacting N-hydroxysuccinimidobiotin with gelatin. After incubation with sample, undigested substrate is measured by addition of an avidin-alkaline phosphatase conjugate, using p-nitrophenol phosphate as substrate.

Analysis of Proteolytic Products

Enzymes with narrow specificity will not generate sufficient small peptides to be measured as above and may only be detected by analysing the products. This may also be necessary if the amount of substrate is limited. Large protein fragments generated by limited proteolysis can be analysed by SDS-PAGE. Proteolysis can be judged from the loss of the substrate band and/or appearance of smaller product band(s). If sufficient substrate is available, bands may be detected by standard staining techniques, e.g. with Coomassie blue. Increased sensitivity can be achieved by using labelled proteins (radiolabelled protein detected by autoradiography/fluorography or biotinylated-protein detected with streptavidin/alkaline phosphatase, after blotting) or by using immunological detection methods. If smaller peptides are formed, or with peptides as substrates, the products may be separated by HPLC or TLC. Fragments separated by SDS-PAGE or HPLC may be subjected to sequence analysis to establish the position of the scissile bond.

Assaying Protein Processing Enzymes Using SDS-PAGE

Enzymes involved in the processing of endogenous proteins are likely to have very narrow specificity, as their activity in vivo may be restricted to the hydrolysis of one or a few bonds in a one or a small set of target proteins. It may be possible to assay them initially only by using the natural substrate and analysing the products. Cell preparations or recombinant protein may provide the only suitable source of substrate, although once the cleavage site is known, it should be possible to design peptide substrates based on the sequence surrounding the scissile bond. Processing enzymes of *P. falciparum* have been assayed as follows, for example. The processing of the merozoite surface antigen-1 (MSA1 or MSP-1) to fragment gp41 (MSP1$_{42}$) has been assayed using as substrate a recombinant protein (M10-11) containing a region of MSA1 known to be processed in the native molecule and a C-terminal [His]$_6$ peptide (Cooper and Bujard 1992).

Substrate can be incubated with parasite culture or schizont subfraction in wells of a multiwell tissue culture dish, and the contents of the wells then extracted into electrophoresis sample buffer and subjected to SDS-PAGE. Immunoblots are prepared and incubated with rabbit antisera to fragments of MSA1. Cleavage can be detected by the appearance of additional bands (39, 36 and 32 kDa) recognised by the antisera. Secondary processing of gp41 (MSP1$_{42}$) to fragments of 33 and 19 kDa by a calcium-dependent membrane-bound serine proteinase has also been assayed in merozoite samples (containing both the enzyme and its substrate) by immunoblotting using appropriate antiserum and by analysing immunoprecipitates of [^{35}S]-methionine-labelled merozoites (Blackman et al. 1993).

HPLC Analysis of Small Peptides

Information on the specificity of some proteinases can be obtained using chromogenic and fluorogenic substrates (see below), but a more detailed picture of the requirements in both P and P' positions can be gained by analysing the cleavage of peptides by purified enzymes. The substrates can be natural peptides of known sequence, such as oxidised insulin A and B chains and glucagon, fragments of natural polypeptides, for example those obtained from cleavage of cytochrome c with cyanogen bromide, or synthetic peptides. The following example is from an analysis of the specificity of leishmanolysin (*Leishmania* metalloproteinase). Substrate (0.25 mM) is incubated with purified enzyme (molar ratio of enzyme to substrate of 1:3 $\times$ 10^4 in a Tris-buffered saline. Samples (0.02 ml) removed at timed intervals are mixed with 0.01 ml 0.1 M HCl, heated for 2 min at 95 °C to inactivate the enzyme, and the hydrolysis products resolved by HPLC [Waters liquid chromatograph equipped with a Waters μ-Bondapak C$_{18}$ reverse-phase column equilibrated with 0.1% (v/v) trifluoroacetic acid in distilled water]. Full details of the separation and analysis of the products can be found in Bouvier et al. (1990). Once an appropriate substrate has been found, this may be used to assay the enzyme, either utilising HPLC or by modifying the substrate so that the products can be detected more directly (see below and Sect. 5.3.3).

TLC Analysis of a Radiolabelled Peptide

One of the leishmanolysin peptide substrates shown to have a single cleavage site was a nonapeptide (bovine cytochrome c residues 94–102) which has the sequence Leu-Ile-Ala-Tyr*-Leu-Lys-Lys-Ala-Thr (the scissile bond is indicated by the asterisk). The presence of a tyrosine residue allows the substrate to be radioiodinated and its hydrolysis analysed by TLC [cellulose plates using butanol/pyridine/acetic acid/water (20:15:3:12, v/v/v/v)]. The separated substrate and product peptides can be detected by spraying with ninhydrin and by fluorography. This substrate has been used to assay *Leishmania major* and *Herpetomonas samuelpessoai* metalloproteinases (Schneider and Glaser 1993a).

5.3.3
Synthetic Fluorogenic and Chromogenic Substrates

The use of fluorogenic and chromogenic derivatives of amino acids and small peptides has a number of advantages. Detection of activity depends on the hydrolysis of a single bond which can provide a rapid and sensitive assay and simplifies kinetic analysis. In some circumstances, substrates can be chosen to assay individual enzymes even when others are present. For proteinases without stringent requirements for P' residues, relatively simple derivatives which have a chromophore or fluorophore attached to the C-terminal carboxyl group can be used. The peptide or amino acid may be blocked at the N-terminus to prevent aminopeptidase activity. A variety of different types of substrate is available, exploiting both the esterase and amidase activity of proteinases. Commercial suppliers of substrates include Sigma, Bachem, Novabiochem, Peptide Institute and Enzyme Systems Products. Two types in particular, the fluorogenic amidomethylcoumarins and chromogenic nitroanilides, are widely used, although mostly for serine and cysteine proteinases. Some are also metalloproteinase substrates, but metalloproteinases and aspartic proteinases can have more complex requirements, and appropriate substrates may need to be custom-designed and synthesised (see below).

Table 3. Fluorogenic amidomethylcoumarins and amidotrifluoromethylcoumarins used for assaying parasite proteinases

Substrate	Proteinase type	Parasite
Amidomethylcoumarins		
Z-Phe-Arg-	Cysteine	Very widely used
H-Pro-Phe-Arg-	Cysteine	*Leishmania mexicana mexicana*
	Proteasomes	*Trypansoma brucei*
Z-Arg-Arg-	Cysteine	*Entamoeba histolytica, Trypanosoma cruzi Schistosoma mansoni*
	Serine	*T. brucei*
Bz-Phe-Val-Arg-	Cysteine	*L.m. mexicana*
Z-Val-Leu-Arg-	Cysteine	*Naegleria fowleri Dirofilaria immitis*
Z-Leu-Val-Arg-	Cysteine	*Plasmodium falciparum*
Boc-Val-Leu-Lys-	Cysteine	*Trichomonas vaginalis, Tritrichomonas foetus Paragonimus westermani*
Bz-Arg-	Serine	Trypanosomatids *Anisakis simplex*
Z-Gly-Gly-Arg-	Serine	*T. brucei, T. cruzi, P. falciparum Haplometra cylindracea*
Boc-Ile-Glu-Gly-Arg-	Serine	*P. falciparum*
Boc-Leu-Ser-Thr-Arg-	Serine	*P. falciparum*
Suc-Leu-Leu-Val-Tyr-	Proteasomes	*T. brucei*
	Metallo-	*Strongyloides stercoralis*
MeO-Ala-Ala-Pro-Val-	Metallo-	*Ancyclostoma caninum*
H-Gly-Pro-	Dipeptidylpeptidase	*Fasciola hepatica*
Ala-, Leu-, Phe-	Metallo-aminopeptidase	Widely used
Amidotrifluoromethylcoumarins		
Z-Phe-Arg-	Cysteine	*S. mansoni, Clonorchis sinensis*
Suc-Leu-Leu-Val-Tyr	Serine	*P. falciparum*

Amidomethylcoumarins

Peptidyl and amino acyl derivatives of 7-amino-4-methyl-coumarin (NHMec, also abbreviated as AMC and MCA) are frequently the substrates of choice when a serine or cysteine proteinase is being studied. 7-Amino-4-trifluoromethyl coumarins derivatives (AFCs, Enzyme Systems Products) are also available. Table 3 lists derivatives which have been used as substrates for parasite peptidases in sensitive fluorometric assays and for analysing specificity, either in conventional assays or in combination with SDS-PAGE. When multiple proteinases are present, a preliminary analysis of their specificity using SDS-PAGE is recommended (see Sect. 5.5.1). This would indicate the extent to which there are overlaps in substrate preference between different enzymes, and therefore how specific an assay based on one substrate is likely to be.

Serine proteinases with trypsin-like specificity hydrolyse substrates with a basic residue at the P_1 position. For some enzymes this may be the only requirement, and Z-Arg-NHMec will be appropriate. The inclusion of additional residues, usually small or hydrophilic amino acids, at the P_2 and P_3 positions may provide a better substrate for some enzymes. Amidomethylcoumarins are also available for serine proteinases with chymotrypsin-like specificity, e.g. Suc-Leu-Leu-Val-Tyr-AFC, but they may be less appropriate for elastase-like enzymes, which prefer leaving groups less bulky than aminomethylcoumarin (see Cohen et al. 1991).

Cysteine proteinases share with trypsin-like serine proteinases the ability to hydrolyse substrates with basic residues in the P_1 position, but other P_1 residues, e.g. Gly, Tyr, can be accommodated, and specificity is determined more by the nature of the P_2 and P_3 residues. The preferred substrate is often Z-Phe-Arg-NHMec, which is hydrolysed by enzymes whose specificities are similar to those of either mammalian cathepsin L or cathepsin B. Z-Arg-Arg-NHMec is a cathepsin B substrate that is not hydrolysed by cathepsin L and is preferred to Z-Phe-Arg-NHMec by some parasite cysteine proteinases (N.B. this does not necessarily correlate with a cathepsin B-like structure and relates to the presence of an aspartic acid or glutamic acid residue in the S_2 subsite of the enzyme). Derivatives with bulky residues at both the P_2 and P_3 positions are good substrates for enzymes with cathepsin L-like specificity and Z-Val-Leu-Arg-NHMec, Boc-Val-Leu-Lys-NHMec and Z-Leu-Val-Arg-NHMec have all been used. Other substrates may be useful for assaying enzymes with more unusual specificities, for example Bz-Phe-Val-Arg-NHMec for the *Leishmania mexicana* group D cysteine proteinases (Robertson and Coombs 1993).

The details below are based on those for the assay of *S. mansoni* cysteine proteinase (Sm31; Götz and Klinkert 1993). The same general procedure can be used with other substrates and buffers. The substrates have different solubilities, and to standardise conditions, DMSO should be used as the solvent for all substrates. Note that higher concentrations of substrate (up to 0.05 mM) have been used in some assays.

Assay

- Water bath
- Spectrofluorometer
- Buffer, 0.2 M acetate, pH 5.5, containing 2 mM DTT
- Substrate, Z-Arg-Arg-NHMec as 5 mM stock solution in DMSO

Equipment and reagents

1. Incubate buffer and sample (combined volume of 0.5 ml) and start reaction by adding substrate at a final concentration of 0.01 mM (0.001 ml of stock solution).

2. Activity is measured fluorometrically with the excitation wavelength at 360 nm and emission wavelength at 433 nm. Use a calibration curve for 0.1 to 100 μM 7-amino-4-methylcoumarin and express activity as μmol released per min per mg protein.

Procedure, aminomethylcoumarins

Assays may also be carried out in microtitre plates, but for quantitative fluorometric measurements an appropriate plate reader, e.g. Lab-systems Fluoroskan II will be required. Semi-quantitative estimations can be made simply by viewing the plate on a UV transilluminator: this can provide a rapid indication (within minutes) of peak fractions during purification steps, for example.

If pure enzyme is available, specificity requirements can be analysed in more detail by conducting a kinetic analysis involving a number of substrates. Values for k_{cat} and K_m can be determined by non-linear regression analysis using a programme such as ENZFITTER (Elsevier-Biosoft). K_m values for the best substrates are within the micromolar range.

Nitroanilides

A similar selection of amino acyl and peptidyl derivatives of nitroanilides (Nan) is available for colorimetric assays. The most widely used are Bz-Pro-Phe-Arg-Nan and Z-Arg-Arg-Nan for cysteine proteinases with cathepsin L-like and cathepsin B-like specificities, respectively, and Z-Arg-Nan for serine proteinases with trypsin-like specificity. MeO-Suc-Ala-Ala-Pro-Leu-Nan and MeO-Suc-Ala-Ala-Pro-Leu-Nan are suitable substrates for some other serine proteinases like the *S. mansoni* cercarial elastase.

Assay Nitroanilides can be substituted for amidomethylcoumarins in the assay described above. The substrates are used at a higher concentration, 0.05–0.1 mM, with the rate of product formation is determined spectrophotometrically at 405 nm. Assays can also be set up in microtitre plate wells. The following method has been used with flagellate protozoa to detect cysteine proteinases and putative serine proteinases.

Equipment and reagents
- 96-well microtitre plates
- Microtitre plate reader fitted with 405-nm filter
- Substrates (1 mM stock solutions): Z-Arg-Arg-Nan, Bz-Pro-Phe-Arg-Nan and Z-Arg-Nan can all be prepared in water. Some substrates may need to be dissolved in DMSO.
- Buffer, 0.1 M sodium phosphate, pH 6
- 0.1 M DTT

Procedure, Nitroanilides

1. A substrate/buffer solution is prepared containing 4.0 ml buffer, 0.8 ml substrate, 3.2 ml water and 0.08 ml DTT (can be replaced by water omitted for assays of serine or metalloproteinases). This is sufficient for 30 assays.

2. Samples (0.015 ml) are placed in wells and the reaction started by addition of 0.15 ml substrate/buffer solution.

3. Readings are taken at intervals and the activity determined from the initial rate of increase of absorbance at 405 nm. A change in optical density of 0.1 corresponds to 39.3 nmol nitroaniline released.

Other Substrates

Methoxy-β-naphthyl-amides Peptide and amino acid derivatives of 4-methoxy-β-naphthylamine (MeONap) have been used as substrates for a number of parasite proteinases, but have largely been superceded by nitroanilides and amidomethylcoumarins. The product can be detected fluorometrically (excitation, 340 nm;

emission, 425 nm). By using a coupling step which yields a coloured product the substrates can be useful in tests which require a simple positive or negative result, for example the Enzymeba test for *E. histolytica*, which uses Z-Arg-Arg-MeONap to detect the activity of the parasite cysteine proteinase (Luaces et al. 1992), or to give background colour against which proteinase inhibitors can be revealed (see Sect. 5.8). Z-Arg-Arg-MeONap, Arg-Arg-MeONap and Z-Val-Leu-Arg-MeONap have also been used for histochemical studies (e.g. Richer et al. 1993; Scholze and Tannich 1994). The substrate is added with 5-nitro-2-salicylaldehyde, which forms an insoluble fluorescent adduct with the naphthylamine derivative. The cells can be viewed under a fluorescence microscope (excitation, 360–430 nm; emission, 550–600 nm).

These fluorogenic substrates have been used for analyzing *Plasmodium* proteinases and were selected because of interesting spectroscopic properties, high sensitivity and very good coupling yield of aminoethyl carbazole (AEC) with amino acids. GlcA-Val-Leu-Gly-Lys(or Arg)-AEC is the most specific substrate for the *P. falciparum* Pf68 serine proteinase and related enzymes in other species. The assay is discontinuous and involves an ethyl acetate extraction which also stops the reaction: aminoethyl carbazole is determined fluorometrically (excitation, 370 nm; emission, 430 nm). The assay for Pf68 is conducted at pH 7.4. Further details can be found in Mayer et al. (1991).

Amino-9-ethyl-carbazole derivatives

Amino acyl esters of nitrophenol (NPEs) are in less frequent use than other substrates. Some can give high background activity, especially when reducing agents are included in assays of cysteine proteinases. However, they may be useful for the analysis of proteinases for which other types of substrate may not be appropriate, e.g. Z-Ala-NPE to measure the elastase-like activity of a *Porocephalus* metalloproteinase (Jones et al. 1991). Nitrophenol release is measured spectrophotometrically at 405 nm.

Amino acyl nitrophenol derivatives and other esters

Peptidyl thiolesters have proved to be good substrates for some serine proteinases, in particular for the *S. mansoni* cercarial elastase (Cohen et al. 1991) and a metalloproteinase produced by *Strongyloides stercoralis* (McKerrow et al. 1990). Suc-Ala-Ala-Pro-Phe-Sbzl is among the best of these. Assays are conducted in the presence of 4,4′-dithiopyridine and thiopyridone production followed at 324 nm.

Substrate for a metalloproteinase, *Leishmania* leishmanolysin

Commercially available synthetic substrates have not been widely used with parasite metalloproteinases. Indeed, studies of leishmanolysin have shown no activity on any of the nitroanilides and amidomethylcoumarins tested. A dansylated tryptophan-containing heptapeptide, dansyl-Ala-Tyr*Leu-Lys-Lys-Trp-Val-NH$_2$, whose structure is similar to that of the radiolabelled substrate (see above) has been developed as a fluorogenic substrate for this enzyme. Tryptophan fluorescence is quenched internally in the substrate but not in the product as proteolysis splits the dansylated portion of the peptide from the tryptophan-containing fragment. Assays are carried out at pH 7.5 (in TBS), and the product detected using excitation and emission wavelengths of 290 and 360 nm, respectively. Further details can be found in Bouvier et al. (1993).

Substrates for an aspartic proteinase, *Plasmodium* plasmepsin II

The only synthetic aspartic proteinase substrates to be used with parasite enzymes have a chromophoric reporter residue, *p*-nitrophenylalanine, at the P$_1$′ position, and additional P′ residues. Recombinant plasmepsin II has been assayed using a substrate based on the sequence in α-globin surrounding the Phe^{33}*Leu^{34} bond initially cleaved by the enzyme. The peptide is Ala-Leu-Glu-Arg-Thr-Phe*Nph-Ser-Phe-Pro-Thr (a similar non-chromophoric peptide is hydrolysed by recombinant plasmepsin I; Dame et al. 1994). A second substrate with a better K$_m$, Lys-Pro-Ile-Val-Phe*Nph-Arg-Leu, is also hydrolysed by plasmepsin II. This is a substrate for mammalian cathepsins D and E, suggesting that the plasmepsin is not a highly specific enzyme. Assays are carried out at pH 4.7 and activity determined spectrophotometrically from the decrease in absorbance at 300 nm. Further details may be found in Hill et al. (1994).

5.3.4
Exopeptidase Substrates

Methods for assaying exopeptidases are not considered here in detail, but Table 4 provides a summary of some of those

Table 4. Summary of exopeptidase assay methods

Substrate	Enzyme	Comment	Reference
Amino acyl nitroanilides and amidomethylcoumarins (not N-blocked)	Aminopeptidases	Assays as for endopeptidases	Vander Jagt et al. (1984) Curley et al. (1994)
Dipeptidyl nitroanilides and amidomethylcoumarins (not N-blocked)	Dipeptidylpeptidases	Assays as for endopeptidases	Carmona et al. (1994)
N-blocked dipeptides	Carboxypeptidases	Amino acid release measured with ninhydrin or fluorometrically	Doi et al. (1981) McLaughlin and Müller (1979)
Small peptides	All exopeptidases	Product analysis by HPLC or TLC: radiolabelled substrates can be used	Schneider and Glaser (1993b)
		Amino acid release measured with amino acid oxidase/peroxidase/tolidine	Knowles et al. (1989)

Table 5. Low molecular proteinase inhibitors

Inhibitor	Stock solution Solvent	concentration (mM)	Working concentration (mM)	Target enzymes	Comment
Pepstatin[a]	Methanol or DMSO	1.0	0.05–0.10	Aspartic	Very specific
E-64[a]	Water	1.0	0.005–0.02	Cysteine	Very specific irreversible inhibitor. Other derivatives: E-64c (Ep459) possibly more potent; E-64d membrane-permeant
Peptidyl diazomethanes	Acetonitrile or DMSO	10	0.05	Cysteine (serine)	Irreversible inhibitors. Specificity determined by amino acid residues. Most, including Z-Phe-Phe-CHN2 and Z-Phe-Ala-CHN2, are specific for cysteine proteinases
Antipain	Water	10	0.01–0.1	Cysteine/serine	Inhibits enzymes with trypsin-like specificity
Chymostatin	DMSO	10	0.01–0.1	Cysteine/serine	Inhibits enzymes with chymotrypsin-like specificity
Leupeptin	Water	10	0.01–0.1	Cysteine/serine	Inhibits enzymes with trypsin-like specificity
TPCK	Methanol or ethanol	10	0.01–0.1	Cysteine/serine	Irreversible inhibitor of enzymes with chymotrypsin-like specificity
TLCK	Water	10	0.01–1.0	Cysteine/serine	Irreversible inhibitor of enzymes with trypsin-like specificity. Unstable above pH 6

DFP	Dry pro pan-2-ol	200–500	1.0	Serine	Highly toxic irreversible inhibitor
PMSF[a]	Methanol or ethanol	200	1.0	Serine	Irreversible inhibitor. Unstable in aqueous solution: should be freshly prepared
3,4-DCI	DMF or DMSO	10	0.1	Serine	Irreversible inhibitor. Unstable in aqueous solution. Alternative to PMSF
EDTA	Water	500	2.0	Metallo-	Chelating agent. May inactivate other proteinases stabilized by divalent cations
1,10-Phenanthroline[a]	Methanol	200	1.0–2.0	Metallo-	Chelating agent. May inhibit cysteine proteinases. Absorbs in UV and can interfere with assay readings
Phosphoramidon	Water	1.0	0.1	Metallo-	Not effective against most mammalian enzymes
Bestatin	Methanol	1.0	0.1	Metallo-aminopeptidases	

[a]Recommended diagnostic inhibitors.

available. Some assays, with an appropriate change of substrate, are identical to those used for endopeptidases.

5.4
Proteinase Inhibitors

Agents which inhibit proteinase activity are important in parasite work for a number of reasons. First, the inclusion of appropriate inhibitors can limit unwanted proteolysis during the preparation of proteins from parasite samples (see North 1989). Second, use of inhibitors provides a means of determining proteinase class and can also give information on specificity. Third, the effects of inhibitors on parasite processes provides evidence on the role of particular proteolytic enzymes. Fourth, the possibility that proteinase inhibitors might be valuable anti-parasitic agents is under active investigation.

Table 5 gives details of the low molecular weight inhibitors which are most useful for determining proteinase class and preventing unwanted proteolysis. All are commercially available. Details of stock solutions and the most appropriate working concentrations are provided, with further information available in Beynon and Bond (1989). High molecular weight proteinaceous proteinase inhibitors are not listed: some, such as cystatin, a cysteine proteinase inhibitor, are specific enough to be of value in diagnosing proteinase class (see below), and some are useful as ligands for affinity chromatography (see Sect. 5.6). Some inhibitors which are not specific for proteinases are also in use. For example, cysteine proteinases are inactivated by thiol reagents such as iodoacetic acid, N-ethylmaleimide, mercuric chloride and organomercurials. These may be useful for stopping reactions (iodoacetic acid) or for the reversible inhibition of activity (mercuric chloride).

5.4.1
Standard Procedures for Enzyme Inhibition

Determining proteinase class Proteinase class can be determined by measuring the effect of inhibitors on activity in standard assays (Sect. 5.3) or using substrate-SDS-PAGE (Sect. 5.5). With standard assays, a sample should be preincubated with inhibitor under conditions in which the enzymes are active. Preincubation periods of up to 1 h have been used, but most inhibitors at the recommended concentrations should be effective within 5–10 min. Indeed, some, e.g. PMSF, are unstable in aqueous solutions and would be inactivated during prolonged preincubation. Appropriate controls must be set up to take account of the effects on enzyme activity of solvent and/or prolonged incubation.

If the aim of using inhibitors is to establish the class of proteinase responsible for activity present, it should not be necessary to test a large number of these. For most proteinases the catalytic mechanism and hence class should be deducible from the effects of a single inhibitor of each class. Those recommended are pepstatin, E-64, 1,10-phenanthroline and PMSF. Pepstatin and E-64 are highly specific for aspartic and cysteine proteinases, respectively, and inhibit most enzymes in these classes. 1,10-Phenanthroline and PMSF are metalloproteinase and serine proteinase inhibitors, respectively. Although both can have effects on cysteine proteinases, any ambiguities can be resolved by checking the effect of reducing agents (which reverse the effect on cysteine proteinases) or by extending the number of inhibitors used. Other inhibitors can be used to determine features of the specificity of the proteinases, for example the chloromethanes TLCK and TPCK can differentiate trypsin-like from chymotrypsin-like activities among serine proteinases (note that these agents also inhibit cysteine proteinases), and peptidyl diazomethanes can be used to distinguish between cysteine proteinases. Quantitative comparisons between inhibitors can be made by determining the concentration required for 50% inhibition (IC_{50}) or by determining inhibition constants (Salvesen and Nagase 1989).

Active site titration

Appropriate inhibitors can be used as active site titrants to determine the molar quantity of active enzyme present in preparations, E-64 for example to titrate cysteine proteinases. Working solutions of E-64 (1–10 μM) are prepared from a 1.0 mM stock. Add 0.025 ml of these to 0.075 ml activated enzyme solution (enzyme concentration should be about 3 μM) and incubate for 30 min at 37 °C. Samples from each are assayed to determine the residual activity and this is plotted as a function of inhibitor concentration. This should give a linear plot which intercepts the abscissa at the concentration of active enzyme (see Salvesen and Nagase 1989).

5.4.2
The Selection and Design of Specific Inhibitors

One aim of work with parasite proteinases is to find inhibitors which selectively inactivate parasite enzymes. Many of the compounds listed in Table 5 were first discovered or designed as agents which were effective against mammalian enzymes, and they thus lack the required specificity. Compounds with increased specificity and potency have therefore been sought. The aim is to find inhibitors with inactivation constants in the nanomolar and sub-nanomolar range. It is, however, important to note that some compounds which are good inhibitors of both host and parasite enzymes have proved to have a far greater effect on parasites than on animal cells. There are some excellent inhibitors of parasite cysteine and serine proteinases among the peptidyl diazomethanes, fluoromethanes, chloromethanes and acyloxymethanes. Examples include Z-Phe-Ala-CH$_2$F, which inhibits cysteine proteinases from many species, and Phe-Gly-Ala-Leu-CH$_2$Cl, which is very effective against the *S. mansoni* cercarial elastase. Other serine proteinase inhibitors examined include alkylamines and amino alcohols (Mayer et al. 1991). Aspartic proteinases from *P. falciparum* are inhibited by non-hydrolysable peptide-like agents (Francis et al. 1994; Hill et al. 1994) of the type which have proved useful for the inhibition of other aspartic proteinases such as renin and the HIV proteinase. No specific inhibitors of metalloproteinases have yet been reported.

The availability of sequence information for proteinases related to well-characterised enzymes whose three-dimensional structures have already been solved (cysteine proteinases of the C1 family, serine proteinases of the S1 family) has allowed molecular modelling to be used to predict the features required in an inhibitor for tight binding to enzyme. This has led to the refinement of known inhibitor types (e.g. Cohen et al. 1991) and has opened up the possibility of using computer programmes (DOCK 3.0) to screen a chemical directory for compounds which could also bind tightly and have potential as inhibitors (Ring et al. 1993).

When testing the effect of inhibitors on parasites, it is important to bear in bind the permeability of the compounds. In general, those requiring non-aqueous solvents (see Table 5) will penetrate cells, whereas the others may only get into cells slowly by endocytosis and fail to penetrate many compartments. If irreversible inhibitors are used, the effectiveness of an inhibitor can be checked by determining the activity remaining in samples prepared from parasites after inhibitor treatment, provided the cells have been thoroughly washed to remove all remaining traces of inhibitor.

5.5
Electrophoretic Methods

Parasite samples, whether derived from intracellular or extracellular material, may contain a number of proteinases, often of differing class and specificity. Although it is sometimes possible to distinguish these through suitable choice of assay conditions (inclusion of an inhibitor of a particular class of enzyme, use of a selective substrate), this will not always be the case, and the complexity of the proteolytic system may not necessarily be obvious from standard assays. It is therefore advantageous to use an approach which can give some indication of this complexity. This may be especially important if the amount of material is limited, making purification of individual enzymes from parasites an unrealistic goal. Two different methods can be used to detect proteinases using SDS-PAGE. The first takes advantage of the fact that many parasite proteolytic enzymes are still active following electrophoresis: this approach will be referred to as substrate-SDS-PAGE. In the second, enzymes are tagged prior to electrophoresis by incubation with labelled inhibitors.

5.5.1
Substrate-SDS-PAGE – General Procedure

Substrate-SDS-PAGE is now routinely used as the initial step in the analysis of proteinases from most parasites. It provides a simple method not only for revealing the multiplicity of proteinases but also for obtaining information on the classes of the proteinases present and, in some cases, their specificity. It is also useful for examining subcellular localization, for comparing enzymes in different isolates or clinical samples, and in parasites grown under different conditions, at different stages of their life cycle or undergoing in vitro differentiation.

Proteinases detected by substrate-SDS-PAGE must be stable under the conditions of the electrophoresis run. Thus the detection of monomeric, single-chain, SDS-resistant and alkali-stable proteinases is favoured. Fortunately, many of the parasite enzymes have these properties, and enzymes of all four classes have been detected. It is, however, also possible to use native gels to detect more complex proteinases, as in the case of purified proteasomes from *Trypanosoma brucei* (Hua et al. 1995). Standard conditions for SDS-PAGE are used, except that samples are not boiled and inclusion of a reducing agent is optional, indeed the presence of a reducing agent may lead to loss of activity of some enzymes. In the standard method, adapted from that of Heussen and Dowdle (1980), the substrate, gelatin, is incorporated into the gel. Fibrinogen has also been utilised, mostly for leishmanolysin and African trypanosome cysteine proteinases. Other proteins, e.g. immunoglobulin, albumin, haemoglobin and casein, have been used in comparative studies (e.g. Williams and Coombs 1995).

An alternative to incorporating protein into the gels is to add fluorogenic substrates, amidomethylcoumarins, after electrophoresis has been completed. Activity can be detected more rapidly, and by using different substrates it is possible to gain information on the specificity of a number of enzymes simultaneously. Enzymes must possess relatively high activity and as yet only cysteine and serine proteinases and some exopeptidases have been detected. Other substrates, including methoxynaphthylamine derivatives, e.g. Z-Arg-Arg-MeONap (Werries et al. 1991) have been used; however, these require a coupling step and activity cannot be monitored continuously as it can with the amidomethylcoumarins.

Proteinases may also be detected using overlay techniques but these have no apparent advantages over the direct visualisation of activity in the gels.

Basic Protocol

The following can be used for detection of activity on either gelatin or amidomethylcoumarins (North et al. 1990b; North 1994).

Equipment and reagents

- Mini gel system e.g. Bio-Rad mini Protean II
- Power Pack
- Plastic dishes or petri dishes
- Rocker
- UV transilluminator
- Polaroid camera fitted with a Wratten gelatin filter number 2E (Kodak)
- Acrylamide solution: 30% (w/v) acrylamide, 0.8% N,N'-methylenebisacrylamide
- Resolving gel buffer: 3 M Tris HCl, pH 8.8
- Stacking gel buffer: 0.5 M Tris HCl, pH 6.8
- 10% (w/v) SDS
- 1.5% (w/v) ammonium persulphate (freshly prepared)
- TEMED
- Gelatin: 2% (w/v) aqueous solution, stored at 4 °C and heated gently prior to use
- 0.05% bromophenol blue
- Water-saturated butanol
- Electrophoresis sample buffer: 1 ml stacking gel buffer, 0.8 ml glycerol, 1.6 ml 10% (w/v) SDS, 0.4 ml mercaptoethanol (may be omitted), 0.2 ml 0.05% bromophenol blue, 4 ml water
- Reservoir buffer (10× final concentration): 0.25 M Tris, 1.92 M glycine, 1% (w/v) SDS
- 2.5% (v/v) Triton X-100
- Gel incubation buffer. In general, enzymes detected by substrate-SDS-PAGE are active over a wide range of pH. Most will have some activity in phosphate buffer pH 6 or acetate buffer pH 5.5 and these would be suitable for initial runs.
- Peptidyl or amino acyl amidomethylcoumarin (5 mM stock solution in DMSO) (see Sect. 5.3.3 for appropriate choices)
- Stain: 0.1% Coomassie brilliant blue R-250 in 40% (v/v) methanol/10% (v/v) acetic acid
- Destain: 10% (v/v) acetic acid

Procedure, basic protocol

1. Separating gels are prepared by mixing 3.3 ml acrylamide solution (for a final concentration of 10%), 1.25 ml resolving gel buffer, 0.1 ml SDS, 1.0 ml gelatin solution (if required), 3.85 ml water (N.B. 4.85 ml for non-gelatin gels), 0.5 ml ammonium persulphate and finally 0.005 ml TEMED. This is sufficient for two 0.75-mm-thick gels poured to a standard height of 4 cm. Gels are allowed to set under water-saturated butanol.

2. Stacking gel is prepared by mixing 0.625 ml acrylamide solution, 1.25 ml stacking gel buffer, 0.05 ml SDS, 2.625 ml water, 0.2 ml bromophenol blue, 0.25 ml ammonium persulphate, and finally 0.004 ml TEMED. If wells are required (see below) a 10-well comb is used.

3. When a number of different samples are to be analysed on the same gel, these are run in individual lanes. If pH dependence, inhibitor sensitivity or substrate specificity of the enzymes in a single sample are being investigated, this is layered along the full length of a stacking gel lacking wells. Samples are prepared by mixing with an equal volume of electrophoresis buffer. A maximum of 20 μg protein per lane (samples loaded in wells) or 400 μg per gel (no wells) is recommended for gelatin gels. More protein can be loaded if activity is to be detected with fluorogenic substrates.

4. Electrophoresis is carried out at 16 mA per gel at room temperature until the dye front reaches the bottom of the gel (running time of approximately 40 min).

5. Following electrophoresis the gels are transferred to 2.5% (v/v) Triton X-100 in plastic dishes and placed on a rocker for 30 min at room temperature. Single sample gels can be cut into vertical strips for incubation under different conditions to establish the effect of pH, reducing agents, cations etc. or with different substrates. See 6.–7. for gelatin gels, 8.–9. for gels incubated with fluorogenic substrate.

6. Gelatin gels are incubated in buffer (10 ml per gel) at the required temperature for an appropriate period (1–48 h) depending on the level of activity. Most samples require overnight incubation, but with preparations of parasites

containing highly active enzymes the analysis can be completed within a few hours.

7. The gels are stained for 30 min and then destained. Activity is revealed as clear bands in the stained background.

8. For detecting activity with amidomethylcoumarins, gels are washed twice (with rocking) for 5 min in 10 ml incubation buffer and then incubated without rocking in 10 ml buffer containing the appropriate substrate (0.1 ml: final concentration of 0.05 mM). Different substrates may be compared by running single sample gels.

9. Activity is visualised by placing the gels on the transilluminator. This can usually be done within 5 min. Activity band patterns should be recorded immediately using a Polaroid camera (with 665 film (positive/negative) use an exposure time of 10–20 s at f11, with 667 film (positive only) an exposure time of 1 s at f11). Gels can be incubated with substrate for up to 1 h, after which time bands become too diffuse and the background too bright. If gelatin has been included in the gel, incubation can be continued and the gel stained and destained as above so that fluorescent bands can be matched to gelatinase bands.

The simplest way to determine the classes of proteinase present in samples is to use a gel method in combination with diagnostic inhibitors. Samples can be incubated with irreversible inhibitors prior to electrophoresis: this has the advantage of requiring only small quantities of inhibitor. Alternatively, gels or gel strips from a single sample gel can be incubated with inhibitor after the Triton X-100 wash. This will be essential with reversible inhibitors which could dissociate from the enzyme during electrophoresis. In some studies, inhibitor treatment is carried out both before and after electrophoresis to ensure effective inactivation of all possible proteinases.

Testing proteinase inhibitors

- The acrylamide concentration may be varied. For most samples 7.5% or 10% will be appropriate.

- The gel apparatus can be cooled during electrophoresis but this is not essential. In our experience short electrophoresis runs at room temperature give reproducible results

Technical hints on substrate-SDS-PAGE

and the quality is not improved by reducing current or voltage and running at lower temperature.

- The inclusion of a reducing agent in the electrophoresis buffer may eliminate problems of dimerisation but could also destabilise the proteinase. It is not possible to predict how this will affect every enzyme, and samples should be run initially with and without reducing agent to check this.

- It is important not to load too much sample on gelatin gels as the presence of endogenous protein bands, visible through the background of stained gelatin, can make it difficult to discern the clear protein-digesting bands. This certainly detracts from the appearance of the gels.

- The mobility of proteinases in substrate-SDS-PAGE may not correspond to that observed under totally denaturing conditions. There have been some reports of incorporated protein affecting the mobility, although comparisons of the same enzymes separated on gelatin and non-gelatin gels (detected with fluorogenic substrates) suggests that this is not often a problem. Because samples are not boiled and a reducing agent may be absent, some intermolecular interactions may occur. Indeed there are reported instances of single bands on denaturing SDS-PAGE yielding multiple bands on gelatin gels (e.g. Smith et al. 1993). Conclusions about molecular weight based solely on mobility in substrate-SDS-PAGE gels must therefore be treated with caution, although it is reassuring that in many cases the mobility of proteinases is close to that predicted from their molecular weight.

- While it is recommended that gels are always washed in Triton X-100 after electrophoresis to remove SDS, in practice activity may still be detected without a Triton wash. With amidomethylcoumarins a 10-min incubation in gel buffer may provide a sufficient wash, although we have noted that the activity towards some substrates is affected more than others if a Triton wash is omitted.

- Problems of autoproteolysis can be dealt with by treating samples with reversible proteinase inhibitors before electrophoresis.

- Substrate-SDS-PAGE has rarely been used to obtain quantitative data, but it is possible to scan gelatin gels. Checks should be made to confirm that the extent of substrate digestion, determined from trough size, is proportional to the activity loaded.

5.5.2
Substrate-SDS-PAGE – Additional Techniques

A number of useful adaptations have been made to the general method. The following provides a summary of some of these.

Substrate-SDS-PAGE can be used in combination with iso-electric focussing to produce two dimensional gels, e.g. for *Trichomonas vaginalis* cysteine proteinases (Neale and Alderete 1990). This allows greater resolution of enzymes in complex proteolytic systems and provides information on the isoelectric points of individual enzymes.   **Two dimensional gels**

Substrate-SDS-PAGE has been used to detect acid-activatable forms of cysteine proteinases. Following electrophoresis, the gel is incubated in 10% (v/v) acetic acid for 30 s prior to incubation in Triton X-100. The *L. m. mexicana* group H cysteine proteinases detected in this way appear to be zymogens activated by proteolytic cleavage (Robertson and Coombs 1992). However, in another protist, *Dictyostelium discoideum*, acid activation is likely to be the result of conformational change (North et al. 1995). Inclusion of this simple additional step may reveal novel proteinases in other organisms.   **Acid activation**

Substrate-SDS-PAGE can be used for the detection and analysis of anti-proteinase antibodies. For example, immunoprecipitated proteinases can be dissociated from antibody-antigen complexes by incubation in electrophoresis sample buffer and detected on gelatin gels. This can provide information on the specificity of prepared antisera or on anti-proteinase antibodies present in the serum of patients. The technique has been used with trichomonad cysteine proteinases (see Alderete et al. 1991a,b). The interaction of anti-   **Interaction of proteinases with antibodies**

bodies with proteinases has also been analysed by pre-incubating proteinase-containing samples with immunoglobulins and analysing the mixes directly on gels. Antibody binding leads to inhibition of enzyme activity or retardation of non-inhibitory antibody-enzyme complexes (e.g. Britton et al. 1992; Smith et al. 1994).

Interactions of proteinases with other host proteins

One of the concerns about substrate-SDS-PAGE is the possibility that the conditions do not prevent interactions between proteinases and other proteins, and that these might be responsible for some of the complex band patterns observed. However, with one group of parasites it has been possible to use substrate-SDS-PAGE to investigate interactions which may have some physiological significance. Cysteine proteinases of *T. brucei* and some other trypanosomatids interact with a serum protein, probably kininogen. This results in the appearance of multiple bands of proteinase activity. Not only does this affect the mobility of the enzyme but, unexpectedly, it enhances the activity (Lonsdale-Eccles et al. 1995).

5.5.3
Affinity Labelling

Affinity labels are derivatives of irreversible inhibitors of specific types of proteinase. With affinity labelling it is possible to detect proteinases after separation by SDS-PAGE under totally denaturing conditions. As labelling is carried out prior to electrophoresis, enzymes which are inactivated under conditions used for substrate-SDS-PAGE can be detected. The conditions used for denaturing SDS-PAGE also allow more accurate estimations of molecular weights based on electrophoretic mobility and should eliminate multiple bands due to interactions between protein molecules. Samples are incubated with the affinity label under conditions for optimal activity and any unbound label removed by dialysis. The specificity should be checked by confirming that the labelling is blocked by prior incubation of samples with other inhibitors which inactivate the same type of proteinase.

Affinity labels can be used with whole cells as well as cell lysates. Comparisons of the proteinases labelled can indicate whether an enzyme is active inside the cell (it will not become labelled if it is not) or if enzymes are differentially located (if inhibitors with different membrane permeabilities are used).

The use of radioalabelled inhibitors for investigating the active site of enzymes is a long established procedure and has been used with a number of different types of proteinase. Parasite serine and cysteine proteinases have both been labelled in this way. Following incubation with inhibitor (for 1–4 h) samples are dialysed to remove excess probe. They are denatured by heating at 100 °C with electrophoresis sample buffer and subjected to SDS-PAGE. Labelled proteins are detected by autoradiography/fluorography.

Radiolabelling methods

Serine proteinases can be labelled with [³H]DFP, which is commercially available (Amersham), and has been used, for example, to label the cercarial elastase of *S. mansoni* (McKerrow et al. 1985b) and the *P. chabaudi* p68 proteinase (Braun-Breton et al. 1992).

Radioiodinated tyrosine-containing peptidyl diazomethanes have been utilised as affinity labels for cysteine proteinases: Z-Tyr-Ala-CHN$_2$ with *L. m. mexicana* (Alfieri et al. 1995), *Schistosoma haematobium* (Rege et al. 1992) and *S. mansoni* (Götz et al. 1993), Z-Tyr-Phe-CHN$_2$ with *T. cruzi* (Meirelles et al. 1992) and Z-Leu-Tyr-CHN$_2$ for *Entamoeba histolytica* (De Meester et al. 1990). The method for preparation of iodinated inhibitor by the Iodogen method is as follows (see Götz and Klinkert 1993). Inhibitor [25 μl of a 1 mM solution in 25% (v/v) DMSO] is added to an iodogen (1,3,4,6-tetrachloro-3,6-diphenylglycouril)-coated glass tube (Pierce) in the presence of 50 mM sodium phosphate buffer, pH 7.5 and 10 μl of Na^{125}I (200 μCi) and incubated for 10 min at 0 °C. Sodium phosphate buffer (0.455 ml) is then added and the reaction stopped by removing the mixture from the tube. A tritiated fluoromethylketone, Z-Ala-[³H]Phe-CH2F has also been used with *T. cruzi* (Ashall et al. 1990).

Other types of label are now being developed. Biotinylated diazomethanes and chloromethanes which bind to cysteine and serine proteinases and can be detected using streptavidin/alkaline phosphatase system have been produced (Biosyn Diagnostics Ltd., Belfast, Northern Ireland, UK). The peptides are similar to those used for radiolabelling except that

Other labelling methods

there is no requirement for a tyrosine residue (the biotin is coupled to the N-terminus of the peptide). They have been used successfully with *Fasciola hepatica* cysteine proteinases (biotin-Phe-Ala-CHN$_2$ and biotin-Phe-Cys(Sbzl)-CHN$_2$; McGinty et al. 1993) and *Haplometra cylindracea* (biotin-Phe-Ala-CHN$_2$; Hawthorne et al. 1994). Serine proteinase-specific biootinylated amino acid phosphanates have also been tested with *H. cylindracea* (Hawthorne et al. 1994). These reagents should be commercially available from Biosyn Ltd, soon.

5.6
Purification Methods

As it is not possible to predict whether or not the exact protocol for the purification of one parasite proteinase will be suitable for another enzyme, the steps used in the purification of a representative selection of enzymes of all four classes from protozoa (Table 6) and helminths (Table 7) are summarised. For full details the cited references should be consulted. Most of the procedures are based on combinations of standard techniques: ion exchange chromatography, gel filtration, chromatofocusing and affinity chromatography. Improvements in purifications have often been due to the application of HPLC (FPLC) technology and the inclusion of affinity chromatography steps which have helped improve yields from small amounts of starting material. Some general points can be noted.

- Many of the helmith ES proteinases can be purified by one- or two-step procedure from culture fluids because the starting material contains relatively few proteins.

- Affinity techniques specific to proteinases use immobilised substrates and inhibitors. The success depends on strong but reversible binding. Some are highly selective (pep-statin-Sepharose for aspartic proteinases) others will bind a number of different types of proteinase (bacitracin-Se-pharose). Other possible ligands include proteinaceous proteinase inhibitors such as antitrypsin and soybean trypsin inhibitor.

Table 6. Purification schemes for proteinases from parasitic protozoa

Species/Enzyme	Stage	Properties Substrate, pH (M_r, pI)	Purification steps[a]	Comment	Reference
Cysteine proteinases					
Leishmania mexicana Groups A,B,C	Amastigotes	Bz-Pro-Phe-Arg-Nan, 6.0 (20–35 000; -)	Concanavalin-A Sepharose CL43 (7) Superose 12 (4) MonoQ (B and C only) (1)	All these proteinases are products of the same gene set (lcmpb) A separated from B and C by ConA affinity chromatography	Robertson and Coombs (1990)
Group D	Amastigotes	Bz-Phe-Val-Arg-Nan, 6.0 (31 000/33 000; -)	Superose 12 (4) Alkyl Superose (8) Superdex 75 (4) MonoQ (1)	Ampiphilic Two enzymes in preparation No activity with gelatin	Robertson and Coombs (1993)
Trypanosoma cruzi cruzipain	Epimastigotes	Azocasein, 5.0 (58–50 000; -)	Ammonium sulphate precipitation (9) DEAE-Sephacel (1) Sephadex G200 (4) MonoQ (1) Superose 6 (4)	Contains C-terminal fragment as product of autoproteolysis	Cazzulo et al. (1989)
Trypanosoma rangeli	Epimastigotes	Bz-Pro-Phe-Arg-Nan, 5.5 (51 000; -)	Concanavalin-A Sepharose (7) Cystatin-Sepharose (6)		Labriola and Cazzulo (1995)

Table 6 (*Contd.*)

Species/enzyme	Stage	Properties Substrate, pH (M_r, pI)	Purification steps[a]	Comment	Reference
Trypanosoma congolense trypanopain-Tc	Bloodstream forms lysosomes	Z-Phe-Arg-NHMec, 6.0 (31–32 000; 7.4)	Thiopropyl Sepharose 6B (7) Sephacryl S-200 (4)	Lysosomes prepared by Percoll density gradient centrifugation Sample reduced with 2.5 mM DTT before application to thiopropyl Sepharose Enzyme also purified using cystatin	Mbawa et al. (1992)
Tritrichomonas foetus	Axenic cells	Z-Arg-Arg-Nan, 6.0 (25 000; -)	Bacitracin-Sepharose (6)	Suitable for N-terminal determinations	Irvine et al. (1993)
Entamoeba histolytica histolysain	Trophozoites	Z-Phe-Arg-NHMec, 9.5 (26–29 000; -)	Sepharose-aminohexanoyl-Phe-Phe semicarbazone (6)	Z-Arg-Arg-NHMec a superior substrate	Luaces and Barrett (1988)
Amoebapain	Trophozoites	Azocasein, 5.0 (22-27 000; 4.9)	Sephadex G-100 (4) Organo-mercury-Sepharose-6B (6) DEAE-cellulose (1) Sephadex G-75 (4)		Scholze and Tannich (1994)
Serine proteinases					
Trypanosoma brucei	Bloodstream forms	Bz-Arg-NHMec, 8.1, (80 000; 5.0–5.2)	DE52 cellulose (1) Sephacryl 200 (4) Preparative isoelectric focusing (3) Immobilized benzamidine (6)	E-64 added to buffer to inactive cysteine proteinases Lacks general proteolytic activity	Kornblatt et al. (1992)

Trypanosoma cruzi Tc 120	Epimastigotes	Z-Gly-Gly-Arg-NHMec, 8.0 (120 000; 5.0)	Ammonium sulphate precipitation (9) DEAE-Sepharose (1) MonoQ (1) MonoP (3)	No activity detected on protein substrates	Santana et al. (1992)
Eimeria tenella	Sporulated oocysts	Azocasein, 8.0 (20 000; 8.6)	Bacitracin-Sepharose (6) BioSil SEC-250 (4)		Michalski et al. (1994)
Plasmodium falciparum Pf68	Schizonts	GlcA-Val-Leu-Gly-Lys(Arg)-AEC, 7.5 (68 000; 4.4)	MonoP 5/20 (3) Superose 12 (4)	M_r of native forms is 105 000	Grellier et al. (1989)

Metalloproteinases

Leishmania major Leishmanolysin (gp63)	Promastigotes Ttriton X-114 detergent phase	Powdered milk, 8.5 (63 000; -)	Fractogel TSK DEAE-650 (1) MonoQ (1)	Membrane-bound Buffer contains a non-ionic detergent, 2.2 mM lauryldodecylamine N-oxide	Bouvier et al. (1989)

Table 6 (*Contd.*)

Species/enzyme	Stage	Properties Substrate, pH (M_r, pI)	Purification steps[a]	Comment	Reference
Aspartic proteinases					
Plasmodium falciparum Plasmepsin I and II	Late trophozoites Digestive vacuoles	3H-haemoglobin, 5.0 (40 000; -)	Hydroxyapatite chromatography (8) DEAE-Sephacel (1) Waters 200SW (4)	Vacuoles prepared by sorbitol lysis and differential centrifugation A second aspartic proteinase has been obtained from the hydroxyapatite step and further purified by DEAE chromatography	Goldberg et al. (1991) Francis et al. (1994)

[a]Purification Techniques: (1) anion exchange chromatography; (2) cation exchange chromatography; (3) chromatofocusing; (4) gel filtration; (5) substrate affinity chromatography; (6) inhibitor affinity chromatography; (7) other affinity chromatography; (8) other chromatographic methods; (9) selective precipitation.

Table 7. Purification schemes for proteinases from parasitic helminths

Species/enzyme	Stage	Properties substrate, pH (M_r, pI)	Purification steps[a]	Comment	Reference
Cysteine proteinases					
Spirometra mansoni	Larval ES products or Plerocercoids	Z-Phe-Arg-AFC, 5.5 (28 000; -)	(DEAE)-Trisacryl M (1) Thiolpropyl-Sepahrose (7)		Song and Chappel (1993)
Globodera pallida	Immature females	Biotin-gelatin, 6.0 (62 000; -)	MonoQ (1) p-Chloromercuribenzoic acid (PMBA)-agarose (6)		Koritsas and Atkinson (1994)
Fasciola hepatica L1	Mature parasite culture medium	Z-Phe-Arg-NHMec, 4.5 (27 000; -)	Sephacryl S-200 (4) QAE 400 Sephadex (1)	Unbound at pH 7	Smith et al. (1993)
L2	Mature parasite culture medium	Tos-Gly-Pro-Arg-NHMec, 7.0 (29 500; -)	Sephacryl S-200 (4) QAE400 Sephadex (1)	Bound at pH 7	Dowd et al. (1994)
Schistosoma mansoni Sm31 (cathepsin B)	Disrupted adult worms	Azocoll, 5.0 (30 000; 5.7–6.0)	Ultrogel ACA-54 (4) MonoP (3)		Lindquist et al. (1986)
Sm32	Disrupted adult worms	Haemoglobin, 4.9 (32 000; -)	Ultrogel ACA-54 (4) Dialysis and precipitation (9)		Chappel and Dresden (1987)
Serine proteinase					
Spirometra mansoni	Spargana	Haemoglobin, 7.5 (36 000; -)	Gelatin-Sepharose 4B (5) (DEAE)-Trisacryl M (1)		Kong et al. (1994)

Table 7 (*Contd.*)

Species/enzyme	Stage	Properties substrate, pH (M_r, pI)	Purification steps[a]	Comment	Reference
Schistosoma mansoni cercarial elastase	Crude larval secretion	3H-Elastin, 9.0 (27 000; 8.0)	Ultrogel AcA54 (4) MonoP (3)	Assayed in the presence of 2 mM CaCl2	McKerrow et al. (1985b)
Metalloproteinase					
Porocephalus crotali	Nymph frontal glands	Azocoll, 7.6 (48 000; 8.7)	Ammonium sulphate precipitation (9) MonoQ (1)	Frontal glands purified by Percoll density gradient centrifugation	Jones et al. (1991)
Ancylostoma caninum	Homogenized worms	1251-fibrin, 9.0 (37 000; -)	CM-Sepharose 6L-6B (2) Sephadex G-50 (4) Phenyl-Sepharose (8)		Hotez et al. (1985)
Haemonchus contortus	Exsheathing fluids	Azocoll, 8.0 (44 000; -)	SP5PW (HPLC) (2) I-125/I-60 (HPLC) (4)		Gamble et al. (1989)
Trichuris suis	Adult culture fluid	Azocoll, 7.0 (45 000; 8.0)	SynChropak CM-300 (HPLC) (2)		Hill et al. (1993)
Aspartic protenases					
Dirofilaria immitis	Adult worms lyophilized	Haemoglobin, 3.1 (42 000; 5.8–6.4)	Sephadex G-75 (4) Pepstatin-agarose (6)		Sato et al. (1993)

[a]Purification techniques (1)–(9) as in Table 6.

- Other affinity techniques may be useful for separating closely related proteinases. For example, immobilised concanavalin A can be used to separate different similar enzymes with different degrees of glycosylation.

- Schemes for recombinant enzymes are not included in Tables 6 and 7. However, active forms of a number of parasite enzymes have now been recovered from *E. coli* and insect cells expressing cloned parasite genes, and these will be an increasing importantly source. This will overcome problems resulting from the lack of availability of adequate starting material or from the presence of multiple proteinases with very similar properties.

5.7
Endogenous Inhibitors

Endogenous proteinase inhibitors can be detected by adding sample to a standard assay system, normally for a readily available and well-characterized proteinase. Proteinase inhibitors are usually heat-stable, and so endogenous proteinases can be inactivated by heating the sample without affecting the inhibitor. However, appropriate controls must always be included in assays to measure for residual proteinase activity in any sample.

Examples of systems used for assaying inhibitors include the following. The microtitre plate method with radiolabelled gelatin (see Sect. 5.3) has been used with trypsin, chymotrypsin and elastase to detected a serine proteinase inhibitor secreted by *Echinocoocus granulosus* (Shepherd et al. 1991). An elastase assay with Suc-Ala-Ala-Ala-Nan as substrate has been used to measure a serine proteinase inhibitor from *S. mansoni* (Ghendler et al. 1994). Various cysteine proteinases (cathepsin B and enzymes from *E. histolytica, C. elegans*) have been used with Z-Arg-Arg-NHMec and Z-Phe-Arg-NHMec to assay the recombinant onchocystatin from *Onchocerca volvulus* (Lustigman et al. 1992).

Ghendler et al. (1994) have detected the Smpi56 *S. mansoni* serine proteinase inhibitor by incubating biosynthetically labelled samples with biotinylated elastase and then precipitating with streptavidin-agarose to which the biotin binds. The precipitates can be analysed by SDS-PAGE.

Irvine et al. (1992) employed an SDS-PAGE method to detect endogenous cysteine proteinase inhibitors in protozoa. In outline, samples are run on 20% acrylamide gels in which papain is incorporated into gel (in the original method proteinase was added after electrophoresis). Gels are washed in Triton X-100 for 30 min and then incubated with 0.2 M MES, pH 5.5 with 5 mM DTT and 2.5 mM Z-Phe-Arg-MeO-Nap (a substrate for papain). After incubation at 37 °C for 30 min, papain activity can be detected on a UV transilluminator. Inhibitors showed as dark areas against a bright fluorescent background.

The purification of endogenous inhibitors can be achieved using affinity chromatography with immobilised proteinases, although care should be taken that binding of inhibitor to proteinase does not inactivate the inhibitor.

5.8
Final Comments

The application of molecular biology approaches to the study of parasite proteinases is providing new routes to studying the properties and function of the enyzmes. Genes for a number of parasite proteinases of all four classes have now been cloned. The sequences indicate the family of the enzymes and allow predictions of the three-dimensional structure. Recombinant enzymes produced in heterologous hosts can be used to obtain direct structural information and for detailed studies of enzymic properties. Recombinant mutant forms of the enzymes can be produced to aid the analysis of structure/function relationships. As parasite transformation systems are developed, so it will become possible to analyse cellular function through the production of null mutants or strains which overexpress particular proteinases. This is exemplified by the recent studies on *L. m. mexicana* cysteine proteinases (Souza et al. 1994).

In most studies of parasite proteinases, the first steps have inevitably involved those enzymes which are most easily detected by conventional assay methods. This has favoured the analysis of high activity, broad specificity enzymes. While it is certain that many of these have important roles in parasite biology and represent possible targets for anti-parasitic agents, other enzymes involved in more subtle proteolytic processes are likely to be of equal importance. The protein-processing enzymes of *Plasmodium* represent one such group. If the spectrum of proteolytic enzymes under investigation is to be widened, then it is likely that more sensitive and specific assays will have to be developed, and it is possible that more unusual proteolytic enzymes might be encountered.

Abbreviations

AEC	aminoethyl carbazole
AMC	alternative abbreviation for NHMec
AFC	7-amino-4-trifluoromethyl coumarin
MCA	alternative abbreviation for NHMec
MeONap	4-methoxy-β-naphthylamine
Nan	4-nitroaniline
NHMec	7-amino-4-methylcoumarin
NPE	nitrophenyl ester
Nph	nitrophenylalanine
Sbzl	benzyl thioester
TBS	Tris buffered saline
TCA	trichloroacetic acid
Boc	N-tert-butoxycarbonyl
Bz	N-benzoyl
CH_2Cl	chloromethane (chloromethyl ketone)
CH_2F	fluoromethane (fluoromethyl ketone)
CHN_2	diazomethane
DEAE	diethylaminoethyl
DFP	di-isopropylfluorophosphate
DMSO	dimethylsulphoxide
DTT	dithiothreitol
EDTA	ethelenediaminetetraacetic acid
ES	excretory/secretory
FITC	fluorescein isothiocyanate
GlcA	N-gluconyl
MeO	methoxy
PBS	phosphate buffered saline
PMSF	phenylmethane sulphonyl fluoride

SDS	sodium dodecyl sulphate
Suc	N-succinyl
TPCK	1-chloro-4-phenyl-3-tosylamido-L-butan-2-one
TLCK	tosyl lysine chloromethane (chloromethyl ketone)
Tos	tosyl
Z	N-benzyloxycarbonyl

References

Alderete JF, Newton E, Dennis C, Neale KA (1991a) Antibody in sera of patients infected with *Trichomonas vaginalis* is to trichomonad proteinases. Genitourin Med 67:331–334

Alderete JF, Newton E, Dennis C, Neale KA (1991b) The vagina of women infected with *Trichomonas vaginalis* has numerous proteinases and antibody to trichomonad proteinases. Gentiourin Med 67:469–474

Alfieri SC, Balanco JMF, Pral EMF (1995) A radioiodinated peptidyl diazomethane detects similar cysteine proteinases in amastigotes and promastigotes of *Leishmania (L) mexicana* and *L (L) amazonensis*. Parasitol Res 81:240–244

Ashall F, Angliker H, Shaw E (1990) Lysis of trypanosomes by peptidyl fluoromethyl ketones. Biochem Biophys Res Commun 170:923–929

Barrett AJ (1994) Proteolytic enzymes: serine and cysteine peptidases. Methods Enzymol 244: pp 1–15

Barrett AJ (1995) Proteolytic enzymes: aspartic and metallo peptidases. Methods Enzymol 248: pp 105–120

Barrett AJ, Rawlings ND (1995) Families and clans of serine peptidases. Arch Biochem Biophys 318:247–250

Beynon RJ, Bond JS (1989) Proteolytic enzymes: a practical approach. IRL Press Oxford

Blackman MJ, Chappel JA, Shai S, Holder AA (1993) A conserved parasite serine protease processes the *Plasmodium falciparum* merozoite surface protein-1. Mol Biochem Parasitol 62:103–114

Bordier C (1981) Phase separation of integral membrane proteins in Triton X-114 solution. J Biol Chem 256:1604–1607

Bouvier J, Bordier C, Vogel H, Reichelt R, Etges R (1989) Characterization of the promastigote surface protease of *Leishmania* as a membrane-bound zinc endopeptidase. Mol Biochem Parasitol 37:235–245

Bouvier J, Schneider P, Etges R, Bordier C (1990) Peptide substrate specificity of the membrane-bound metalloprotease of *Leishmania*. Biochemistry 29:10113–10119

Bouvier J, Schneider P, Malcolm B (1993) A fluorescent peptide substrate for the surface metalloprotease of *Leishmania*. Exp Parasitol 76:146–155

Braun-Breton C, Blisnick T, Jouin H, Barale JC, Rabilloud T, Langsley G, Pereira da Silva LH (1992) *Plasmodium chabaudi* p68 serine protease activity required for merozoite entry into mouse erythrocytes. Proc Natl Acad Sci USA 89:9647–9651

Britton C, Knox DP, Canto GJ, Urquart GM, Kennedy MW (1992) The secreted and somatic proteinases of the bovine lungworm *Dictyocaulus vivparus* and their inhibition by antibody from infected and vaccinated animals. Parasitology 105:325–333

Carmona C, McGonigle S, Dowd AJ, Smith AM, Coughlan S, McGowran E, Dalton JP (1994) A dipeptidylpeptidase secreted by *Fasciola hepatica.* Parasitology 109:113–118

Cazzulo JJ, Couso R, Raimondi A, Wernstedt C, Hellman U (1989) Further characterization and partial amino acid sequence of a cysteine proteinase from *Trypanosoma cruzi.* Mol Biochem Parasitol 33:33–42

Chappell CL, Dresden MH (1987) Purification of cysteine proteinases from adult *Schistosoma mansoni.* Arch Biochem Biophys 256:560–568

Cohen FE, Gregoret LM, Amiri P, Aldape K, Railey J, McKerrow JH (1991) Arresting tissue invasion of a parasite by protease inhibitors chosen with the aid of computer modeling. Biochem 30:11221–11229

Coombs GH (1982) Proteinases of *Leishmania mexicana* and other flagellate protozoa. Parasitology 84:149–155

Cooper JA, Bujard H (1992) Membrane-associated proteases process *Plasmodium falciparum* merozoite surface antigen-1 (MSA1) to fragment gp41. Mol Biochem Parasitol 56:151–160

Curley GP, O'Donovan SM, McNally J, Mullally M, O'Hara H, Troy A, O'Callaghan S, Dalton JP (1994) Aminopeptidases from *Plasmodium falciparum, Plasmodium chabaudi chabaudi* and *Plasmodium berghei.* J Eur Microbiol 41:119–123

Dame JB, Reddy GR, Yowell CA, Dunn BM, Kay J, Berry C (1994) Sequence, expression and modeled structure of an aspartic proteinase from the human malaria parasite *Plasmodium falciparum.* Mol Biochem Parasitol 64:177–190

De Meester F, Shaw E, Scholze H, Stolarsky T, Mirelman D (1990) Specific labeling of cysteine proteinases in pathogenic and nonpathogenic *Entamoeba histolytica.* Infect Immun 58:1396–1401

Doi E, Shibata D, Matoba T (1981) Modified colorimetric ninhydrin methods for peptidase assay. Anal Biochem 118:173–184

Dowd AJ, Smith AM, McGonigle S, Dalton JP (1994) Purification and characterisation of a second cathepsin L proteinase secreted by the parasitic trematode *Fasciola hepatica.* Eur J Biochem 223:91–98

Francis SE, Gluzman IY, Oksman A, Knickerbocker A, Mueller R, Bryant ML, Sherman DR, Russell DG, Goldberg DE (1994) Molecular characterization and inhibition of a *Plasmodium falciparum* aspartic hemoglobinase. EMBO J 13:306–317

Gamble HR, Purcell JP, Fetterer RH (1989) Purification of a 44 kilodalton protease which mediates the ecdysis of infective *Haemonchus contortus.* Mol Biochem Parasitol 33:49–58

Ghendler Y, Arnon R, Fishelson Z (1994) *Schistosoma mansoni:* isolation and characterization of Smpi56, a novel protease inhibitor. Exp Parasitol 78:121–131

Gluzman IY, Francis SE, Oksman A, Smith CE, Duffin KL, Goldberg DE (1994) Order and specificity of the *Plasmodium falciparum* hemoglobin degradation pathway. J Clin Invest 93:1602–1608

Goldberg DE, Slater AF, Cerami A, Henderson GB (1990) Hemoglobin degradation in the malaria parasite *Plasmodium falciparum*: an ordered process in a unique organelle. Proc Natl Acad Sci USA 87:2931–2935

Goldberg DE, Slater AF, Beavis R, Chait B, Cerami A, Henderson GB (1991) Hemoglobin degradation in the human malaria pathogen *Plasmodium falciparum*: a catabolic pathway inititated by a specific aspartic protease. J Exp Med 173:961–969

Götz B, Klinkert MQ (1993) Expression and partial characterization of a cathepsin B-like enzyme (Sm31) and a proposed "haemoglobinase" (Sm32) from *Schistosoma mansoni*. Biochem J 290:801–806

Grellier P, Picard I, Bernard F, Mayer R, Heidrich H-G, Monsigny M, Schrével J (1989) Purification and identification of a neutral endopeptidase in *Plasmodium falciparum* schizonts and merozoites. Parasitol Res 75:455–460

Hawley JH, Peanasky RJ (1992) *Ascaris suum*: are trypsin inhibitors involved in species specificity of ascarid nematodes? Exp Parasitol 75:112–118

Hawthorne SJ, Halton DW, Walker B (1994) Identification and characterization of the cysteine and serine proteinases of the trematode, *Haplometra cylindracea* and determination of their haemoglobinase activity. Parasitology 108:595–601

Heussen C, Dowdle EB (1980) Electrophoretic analysis of plasminogen activators in polyacrylamide gels containing sodium dodecyl sulfate and copolymerised substrates. Anal Biochem 102:196–202

Hill DE, Gamble HR, Rhoads ML, Fetterer RH, Urban JF (1993) *Trichuris suis*: a zinc metalloprotease from culture fluids of adult parasites. Exp Parasitol 77:170–178

Hill J, Tyas L, Phylip LH, Kay J, Dunn BM, Berry C (1994) High level expression and characterisation of Plasmepsin II, an aspartic proteinase from *Plasmodium falciparum*. FEBS Lett 352:155–158

Hotez PJ, LeTrang N, McKerrow JH, Cerami A (1985) Isolation and characterization of a proteolytic enzyme from the adult hookworm *Ancylostoma caninum*. J Biol Chem 260:7343–7348

Hua SB, Li XQ, Coffino P, Wang CC (1995) Rat antizyme inhibits the activity but does not promote the degradation of mouse ornithine decarboxylase in *Trypanosoma brucei*. J Biol Chem 270:10264–10271

Irvine JW, Coombs GH, North MJ (1992) Cystatin-like cysteine proteinase inhibitors of parasitic protozoa. FEMS Microbiol Lett 96:67–72

Irvine JW, Coombs GH, North MJ (1993) Purification of cysteine proteinases from trichomonads using bacitracin Sepharose. FEMS Microbiol Lett 110:113–120

Jones DAC, Riley J, Kerby NW, Knox DP (1991) Isolation and preliminary characterization of a 48-kilodalton metalloproteinase from the excretory/secretory components of the frontal glands of *Porocephalus pentastomids*. Mol Biochem Parasitol 46:61–72

Knowles G, Abebe G, Black SJ (1989) Detection of parasite peptidase in the plasma of heifers infected with *Trypanosoma congolense*. Mol Biochem Parasitol 34:25–34

Kong Y, Chung YB, Cho SY, Choi SH, Kang SY (1994) Characterization of three neutral proteases of *Spirometra mansoni* Plerocercoid. Parasitology 108:359–368

Koritsas VM, Atkinson HJ (1994) Proteinases of females of the phytoparasite *Globodera pallida* (potato cyst nematode). Parasitology 109:357–365

Kornblatt MJ, Mbawa ZR, Lonsdale-Eccles JD (1992) Characterization of an endopeptidase of *Trypanosoma brucei brucei*. Arch Biochem Biophys 293:25–31

Labriola C, Cazzulo JJ (1995) Purification and partial characterization of a cysteine proteinase from *Trypanosoma rangeli*. FEMS Microbiol Letts 129:143–148

Lindquist RN, Senft AW, Petitt M, McKerrow J (1986) *Schistosoma mansoni*: purification and characterization of the major acidic proteinase from adult worms. Exp Parasitol 61:398–404

Lonsdale-Eccles JD, Mpimbaza GWN, Nkhungulu ZRM, Olobo J, Smith L, Tosomba OM, Grab DJ (1995) Trypanosomatid cysteine protease activity may be enhanced by a kininogen-like moiety from host serum. Biochem J 305:549–556

Luaces AL, Barrett AJ (1988) Affinity purification and biochemical characterization of histolysin, the major cysteine proteinase of *Entamoeba histolytica*. Biochem J 250:903–909

Luaces AL, Picó T, Barrett AJ (1992) The ENZYMEBA test: detection of intestinal *Entamoeba histolytica* infection by immuno-enzymatic detection of histolysin. Parasitology 105:203–205

Lustigman S, Brotman B, Huima T, Prince AM, McKerrow JH (1992) Molecular cloning and characterization of onchocystatin, a cysteine proteinase inhibitor of *Onchocerca volvulus*. J Biol Chem 267:17339–17346

Mayer R, Picard I, Lawton P, Grellier P, Barrault C, Monsigny M, Schrével J (1991) Peptide derivatives specific for a *Plasmodium falciparum* proteinase inhibit the human erythrocyte invasion by merozoites. J Med Chem 34:3029–3035

Mbawa ZR, Gumm ID, Shaw E, Lonsdale-Eccles JD (1992) Characterisation of a cysteine protease from bloodstream forms of *Trypanosoma congolense*. Eur J Biochem 204:371–379

McGinty A, Moore M, Hamilton DW, Walker B (1993) Characterization of the cysteine proteinases of the common liver fluke *Fasciola hepatica* using novel, active-site directed affinity labels. Parasitology 106:487–493

McKerrow JH (1989) Parasite proteases. Exp Parasitol 68:111–115

McKerrow JH, Jones P, Sage H, Pino-Heiss S (1985a) Proteinases from invasive larvae of the trematode parasite *Schistosoma mansoni* degrade connective-tissue and basement-membrane macromolecules. Biochem J 231:47–51

McKerrow JH, Pino-Heiss S, Lindquist R, Werb Z (1985b) Purification and characterization of an elastinolytic proteinase secreted by cercariae of *Schistosoma mansoni*. J Biol Chem 260:3703–3707

McKerrow JH, Brindley P, Brown M, Gam AA, Staunton C, Neva FA (1990) *Strongyloides stercoralis*: identification of a protease that facilitates penetration of skin by the infective larvae. Exp Parasitol 70:134–143

McKerrow JH, Sun E, Rosenthal PJ, Bouvier J (1993) The proteases and pathogenicity of parasitic protozoa. Annu Rev Microbiol 47:821–853

McLaughlin J, Müller M (1979) Purification and characterization of a low molecular weight thiol proteinase from the flagellate protozoon *Tritrichomonas foetus*. J Biol Chem 254:1526–1533

Meirelles MNL, Juliano L, Carmna E, Silva SG, Costa EM, Murta ACM, Scharfstein J (1992) Inhibitors of the major cysteinyl proteinase (GP57/51) impair host cell invasion and arrest the intracellular development of *Trypanosoma cruzi*. Mol Biochem Parasitol 52:175–184

Michalski WP, Crooks JK, Prowse SJ (1994) Purification and characterisation of a serine-type protease from *Eimeria tenella* oocysts. Int J Parasitol 24:189–195

Muñoz ML, Rojkind M, Calderón J, Tanimoto M, Arias-Negrete S, Martínez-Palomo A (1984) *Entamoeba histolytica*: collagenolytic activity and virulence. J Protozool 31:468–470

Neale KA, Alderete JF (1990) Analysis of the proteinases of representative *Trichomonas vaginalis* isolates. Infect Immun 58:157–162

North MJ (1989) Prevention of unwanted proteolysis. In: Beynon R, Bond JS (ed) Proteolysis: a practical approach. IRL Press, Oxford, pp 105–124

North MJ (1991) Proteinases of parasitic protozoa: an overview. In: Coombs GH, North MJ (ed) Biochemical protozoology. Taylor & Francis, London pp 180–185

North MJ (1994) Cysteine endopeptidases of parasitic protozoa. In: Barrett AJ (ed) Proteolytic enzymes: serine and cysteine peptidases. Academic Press San Diego pp 523–539

North MJ, Lockwood BC (1995) Amino acid and protein metabolism. In: Müller M, Marr JJ (ed) Biochemistry and molecular biology of parasitic organisms. Academic Press, New York pp 67–88

North MJ, Whyte A (1984) Purification and characterization of two acid proteinases from *Dictyostelium discoideum*. J Gen Microbiol 130:123–134

North MJ, Mottram JC, Coombs GH (1990a) Cysteine proteinases of parasitic protozoa. Parasitol Today 6:270–274

North MJ, Robertson CD Coombs GH (1990b) The specificity of trichomonad cysteine proteinases analysed using fluorogenic substrates and specific inhibitors. Mol Biochem Parasitol 39: 183–194

North MJ, Nicol K, Sands TW, Cotter DA (1995) Acid activatable cysteine proteinases in the cellular slime mold *Dictyostelium discoideum*. J Biol Chem (submitted)

Rawlings ND, Barrett AJ (1993) Evolutionary families of peptidases. Biochem J 290:205–218

Rege AA, Wang W, Dresden MH (1992) Cysteine proteinases from *Schistosoma haematobium* adult worms. J Parasitol 78:16–23

Richer JK, Hunt WG, Sakanari JA, Grieve RB (1993) *Dirofilaria immitis*: effect of fluoromethyl ketone cysteine protease inhibitors on the third stage to fourth stage molt. Exp Parasitol 76:221–231

Ring CS, Sun E, McKerrow JH, Lee GK, Rosenthal PJ, Kuntz ID, Cohen FE
    (1993) Structure-based inhibitor design using model built structures.
    Proc Natl Acad Sci USA 90:3583–3587
Robertson BD, Bianco AT, McKerrow JH, Maizels RM (1989) *Toxocara
    canis*: proteolytic enzymes secreted by the infective larvae in vitro. Exp
    Parasitol 69:30–36
Robertson CD, Coombs GH (1990) Characterisation of three groups of
    cysteine proteinases in the amastigotes of *Leishmania mexicana mex-
    icana*. Mol Biochem Parasitol 42:269–276
Robertson CD, Coombs GH (1992) Stage-specific proteinases of *Leishma-
    nia mexicana mexicana* promastigotes. FEMS Microbiol Lett 94:127–132
Robertson CD, Coombs GH (1993) Cathepsin B-like cysteine proteases of
    Leishmania mexicana. Mol Biochem Parasitol 62:271–280
Salvesen G, Nagase H (1989) Inhibition of proteolytic enzymes. In: Beynon
    R, Bond JS (ed) Proteolysis: a practical approach. IRL Press, Oxford pp
    83–104
Santana JM, Grellier P, Rodier MH, Schrével J, Teixeira A (1992) Pur-
    ification and characterization of a new 120 kDa alkaline proteinase of
    Trypanosoma cruzi. Biochem Biophys Res Commun 187:1466–1473
Sato K, Nagai Y, Suzuki M (1993) Purification and partial characterization
    of an acid proteinase from *Dirofilaria immitis*. Mol Biochem Parasitol
    58:293–299
Schechter I, Berger A (1967) On the size of the active site in proteases.
    Biochem Biophys Res Commun 27:157–162
Schneider P, Glaser TA (1993a) Characterization of a surface metallopro-
    teinase from *Herpetomonas samuelpessoai* and comparison with the
    Leishmania major promastigote surface protease. Mol Biochem Para-
    sitol 58:277–282
Schneider P, Glaser TA (1993b) Characterization of two soluble me-
    talloexopeptidases in the protozoan parasite *Leishmania major*. Mol
    Biochem Parasitol 62:223–232
Scholze H, Tannich E (1994) Cysteine endopeptidases of *Entamoeba his-
    tolytica*. In: Barrett AJ (ed) Proteolytic enzymes: serine and cysteine
    peptidases. Academic Press San Diego, pp 512 523
Shepherd JC, Aitken A, McManus DP (1991) A protein secreted in vivo by
    Echinococcus granulosus inhibits elastase activity and neutrophil che-
    motaxis. Mol Biochem Parasitol 44:81–90
Smith AM, Dowd AJ, McGonigle S, Keegan PS, Brennan G, Trudgett A,
    Dalton JP (1993) Purification of a cathepsin L-like proteinase secreted
    by adult *Fasciola hepatica*. Mol Biochem Parasitol 62:1–8
Smith AM, Carmona C, Dowd AJ, McGonigle S, Acosta D, Dalton JP (1994)
    Neutralization of the activity of a *Fasciola hepatica* cathepsin L pro-
    teinase by anti-cathepsin L antibodies. Parasite Immunol 16:325–328
Song CY, Chappell CL (1993) Purification and partial characterization of
    cysteine proteinase from *Spirometra mansoni* plerocercoids. J Parasitol
    79:517–524
Souza AE, Bates PA, Coombs GH, Mottram JC (1994) Null mutants for the
    lmcpa cysteine proteinase gene in *Leishmania mexicana*. Mol Biochem
    Parasitol 63:213–220

Vander Jagt DL, Baack BR, Hunsaker LA (1984) Purification and characterization of an aminopeptidase from *Plasmodium falciparum*. Mol Biochem Parasitol 10:45–54

Werries E, Franz A, Hippe H, Acil Y (1991) Purification and substrate specificity of two cysteine proteinases of *Giardia lamblia*. J Protozool 38:378–383

Williams AG, Coombs GH (1995) Multiple protease activities in *Giardia intestinalis* trophozoites. Int J Parasitol 25:771–778

Chapter 6

Neurobiology of Helminth Parasites

Aaron G. Maule, David W. Halton, Timothy A. Day,
Ralph A. Pax and Chris Shaw

6.1
Introduction

Neurobiology embraces the study of nerves and the muscles
they innervate, and in helminth parasites it is a subject which
is very much in its infancy. This may seem somewhat sur-
prising, considering the immense potential that neuromuscular systems in parasites provide as targets for
chemotherapeutic intervention (Geary et al. 1992a). However,
it has to be appreciated that most parasitic organisms are
small and their study is often technically challenging, and
that when separated from their hosts in vitro, they present
unique problems in studies of their neuromuscular physiology. Nevertheless, significant advances have been made in the
last 5 years from immunochemical investigations of parasites
which have laid the foundations for a better understanding of
helminth neurochemistry (Halton et al. 1994), and which,
through improved means of physiological testing, analysis
and screening, may help promote the discovery of more effective anthelmintics.

6.2
Immunocytochemical Localisation of Neuroactive Substances
in Helminth Parasites

A number of candidate neurotransmitters and neuromodu-
lators have been identified in helminth parasites. These include: acetylcholine (ACh) [demonstrated indirectly by the
detection of cholinesterase (ChE), or choline acetyltransfer-
ase (ChAT) activities], biogenic amines [e.g. serotonin = 5-
hydroxytryptamine (5-HT), dopamine, noradrenalin and, in

nematodes, γ-aminobutyric acid (GABA)], together with an increasing number of neuropeptides [e.g. neuropeptide F (NPF), FMRFamide-related peptides (FaRPs), substance P]. The single most successful means of demonstrating these neuroactive substances has been by immunocytochemistry (ICC), that is, by using labelled antibodies directed against either the transmitter itself or against a specific enzyme (e.g. ChAT) in its biosynthetic pathway. The label may be fluorescent (e.g. fluorescein isothiocyanate, FITC) for fluorescence microscopy, a chromogen-generating enzyme (e.g. peroxidase) for histochemistry, radioactive for autoradiography, or metal (e.g. gold spheres) for electron microscopy. The practice of using antibodies as probes for localising tissue constituents in situ was developed by Coons et al. (1955) into the indirect immunofluorescence technique that is used most frequently today in ICC studies. It is the technique of choice for localising neuroactive substances in helminth parasites, for light microscopy. Essentially, it involves the addition of unlabelled primary antibody to the tissue and the detection of any binding by a labelled secondary antibody directed against the immunoglobulins (IgG) of the species which donated the first antibody (Fig. 1). This indirect procedure has many advantages, not least the economy of using a single fluorescent secondary antiserum to immunostain any number of primary antibodies to different antigens, provided the same donor species is used throughout. It is also more sensitive than direct binding because of the signal amplification, since anti-IgG sera are usually of very high titre and avidity (see later), and several (averaging a factor of 4–5) labelled anti-IgG molecules can bind to each primary antibody. Although there is always the possibility of non-specific staining in ICC, with suitable controls and the use of highly specific, well-characterised antisera whose staining can be totally abolished by the antigen (see Fig. 1), sources of spurious staining can be kept to a minimum.

6.2.1
Production of Antisera

While there are many commercially available antisera, the most frequently used in ICC are produced by immunising animals, such as rabbits (see Chapter 10.3), guinea pigs, goats

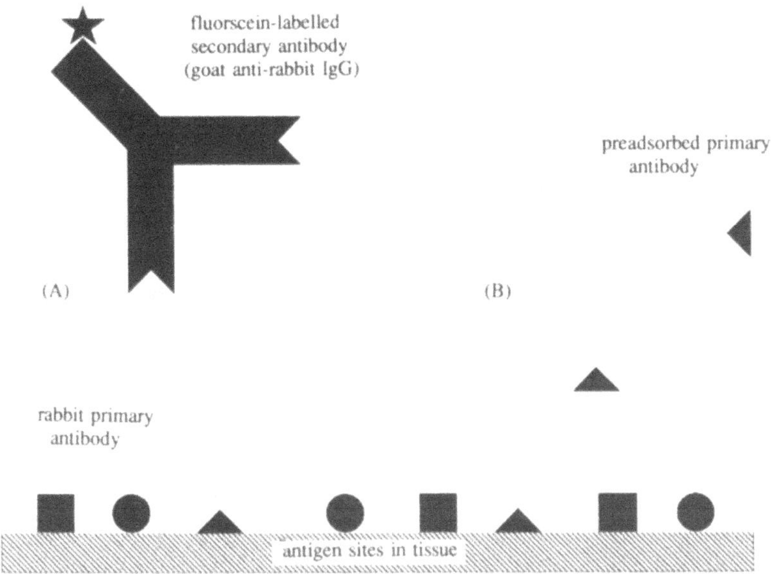

Fig. 1. A Indirect immunofluorescence technique in which the primary antibody (rabbit anti-antigen) binds to the antigen site (▲) in tissue, then serves as a γ-globulin antigen for the second, fluorescent (★) antibody. **B** Preadsorption of the primary antibody with appropriate antigen (▲) is an essential control for staining specificity

and sheep. This means that the donor serum is heterogenous and contains several populations of different antibodies directed to several portions of the immunising antigen; it also means that some of the antibodies may recognise and react with related peptides that share the same amino acid sequence (i.e. the antigenic determinant or epitope) as the required specific antigen; in other words, they are polyclonal. To minimise these cross-reactions, it is best to immunise with small fragments of the antigen and so increase the chances of producing region-specific antisera (see below). Alternatively, monoclonal antibodies to unshared portions of the molecule theoretically offer the advantage of specificity for a single epitope. However, because fixation often destroys some epitopes of a given antigen, many monoclonal antibodies may no longer react and so have often proved unsuitable for use in ICC. In general, synthetic antigens are preferred for raising polyclonal antisera, and in the case of bioactive peptides, which are relatively small molecules (generally 2–80 amino acids), they need to be chemically coupled or conjugated (e.g. by glutaraldehyde or by carbodiimide) to a larger, carrier protein to make them immunogenic. Typical carrier proteins are bovine serum albumin (BSA) and keyhole limpet haemocyanin (KLH). Clearly, it is important to know the nature of the carrier proteins used in immunisation, since antibodies to them could interfere with the staining, and this often necessitates their removal by the addition of the appropriate antigen (i.e. BSA or KLH) for preadsorption (see below). In general, polyclonal antisera are raised by initially injecting the animal of choice with the peptide/protein conjugate in complete adjuvant. After 4 weeks, the animals are boosted with the peptide/protein conjugate in incomplete adjuvant, and initial bleeds removed and assessed for immunogenicity after 10–14 days. Often, several booster injections are given at monthly intervals until such time as a high-titre-specific antiserum is generated which is suitable for the immunological studies to be undertaken.

Use of Heterologous Antisera

With the exception of the invertebrate FMRFamide family of peptides (or FaRPs) and NPF, much of the immunocytochemical mapping of parasite neuropeptides has relied on their cross-reactivity with antibodies directed to vertebrate peptides, rather than by immunostaining with antibodies to native invertebrate peptides. This is because relatively few authentic invertebrate peptide antisera have been available. Moreover, since many vertebrate peptide families are considered to be phylogenetically ancient and highly conserved, especially in their biologically active portions, homologues to them are likely to be distributed throughout the invertebrate phyla. Indeed, it was cross-reactivity with a vertebrate antiserum, namely that directed to the C-terminal hexapeptide amide of pancreatic polypeptide (PP), that led to the isolation and sequencing of NPF, the first known flatworm neuropeptide (Maule et al. 1991; see Table 1).

Region-Specific Antisera

Antisera are described as region-specific when they are known to have an absolute requirement for a particular recognition site or antigenic determinant. The identification of such antisera usually involves extensive preadsorption studies to establish its cross-reactivity with a series of similar and dissimilar antigens. The most extensive and complete preadsorption studies, carried out using specific antisera, enable the identification of antigens with a relatively high degree of confidence. However, even where supported by extensive controls, results with a single antibody cannot be taken as absolute evidence for the presence of an antigen. One solution to this is to use a number of region-specific antibodies, in that each can recognise a different site in the antigen under investigation such that the identification of an antigen does not rely on a single antibody/antigen interaction. Thus, by immunostaining tissues with antibodies to several different synthetic fragments of the peptide, e.g. to N-terminal, mid-portion, C-terminal amino sequences, its localisation as the genuine peptide is more likely.

6.2.2
Immunocytochemistry Protocol

The critical factors in ICC are twofold: namely, *the nature of the antiserum* insofar as its titre, avidity (stickiness) and specificity are concerned, and the *preservation of the antigen* (see Fixation, below). The nature of the antiserum is perhaps best assessed using a known positive immunocytochemical system. In this way, a new serum can be tested for the presence of antibody, evaluated for the quality of specific staining, and then titrated before use to determine the best working dilution. Where possible, a high titre (i.e. low concentration) of polyclonal antibody (>1:200) should be used. While this often means a longer staining time, there will be less chance of non-specific staining from contaminating antibody subpopulations, since these will have been diluted out; it is also economical to use. High-avidity antibodies are those that bind strongly to the specific antigen, and their use is essential in ICC if the antibodies are not to be washed off the tissue during staining protocols.

Specificity Controls

For the critical interpretation of ICC data, a number of controls are essential to ensure correct staining procedure as well as specificity of staining. These should include:

- immunostaining a known *positive control* preparation containing the antigen in question,

- *negative controls,* whereby the omission of primary or secondary antisera, or the use of a non-immune serum as the first layer, should give negative results; the possibility of false positives through ionic binding of antibodies to highly charged tissue components can be minimised by raising the salt (NaCl) content of the buffer-wash and/or the buffer used as antibody diluent to 0.5 M,

- *adsorption controls,* by which the immunoreactivity is quenched through liquid-preadsorption or solid-phase preadsorption of the antiserum with a range of concentrations of specific antigen (preferably synthetic), but which is not abolished by adsorption with similar quantities of related or unrelated substances.

Fixation

The sensitivity of ICC is limited by the compromise of retaining antigenicity for a desirable immunocytochemical reaction, and immobilising tissue constituents for an acceptable cytological structure. Cryostat sections or wholemount preparations of worms prefixed by immersion in buffered formaldehyde or para-benzoquinone have been found to be the most useful for tracing biogenic amine- or peptide-containing nerves in helminth parasites. A formaldehyde-based fixative, such as 4% paraformaldehyde (PFA), is the fixative of choice because it penetrates tissues rapidly and lightly cross-links peptides by forming methylene bridges via the N-terminal α-amino group and the ε-amino functions of lysyl residues. However, if either of these groups resides within the epitope to which the antiserum is raised, there may be a loss of antigenicity, and an alternative fixative, such as p-benzoquinone, would have to be used. Since most of the reactions with PFA are partially reversible, it is possible to wash most of the formaldehyde out of the tissue, and so minimise the degree of cross-linking.

- *0.2 M phosphate-buffered saline (PBS, pH 7.2)*

  In 200 ml distilled water, dissolve 85 g NaCl; 39 g NaH$_2$PO$_4 \cdot$2H$_2$O; 107 g Na$_2$HPO$_4 \cdot$ Adjust pH to 7.2 with concentrated ammonia solution. Make up to 5 l with distilled water.

- *4% Paraformaldehyde (PFA) in 0.1 M PBS*

  Under a ventilated hood, add 16 g solid PFA to 200 ml distilled water and warm to 55–60 °C until completely dissolved (60 min minimum). Add 5 M NaOH solution dropwise to the turbid suspension until it clears. When cool, make up to 400 ml with 0.2 M PBS. Filter before use. The solution can be stored at 4 °C for up to 2 weeks.

Preparation of 4% PFA in phosphate-buffered saline (PBS)

Problems of Antibody Penetration

Since prefixed tissues are not subjected to lipid solvents during processing, penetration of immunoreagents to sites occupied by neuroactive substances can present problems. For example, IgG antibody molecules require pores of 4–8 nm in diameter to pass through, and as labelled reagents, even larger pores, necessitating some permeabilisation of cell membranes. This can be achieved by adding detergents like Triton X-100, Tween 20 or saponin to the PBS used in the processing, i.e. as the *PBS-wash* and/or *antibody diluent*. The detergent not only facilitates antibody penetration but also helps reduce any background staining due to non-specific protein binding.

- *0.1 M PBS buffer-wash*

  As for 0.2 M PBS, except make up to 10 l with distilled water.

PBS antibody diluent

- *Antibody diluent*

  0.1 M PBS containing: 0.3–0.5% (v/v) Triton X-100 (to permeabilise membranes and so aid antibody penetration); 0.1% (w/v) BSA or KLH (to preadsorb primary antisera against the conjugating protein used in immunisation);

0.1% (w/v) sodium azide (as a preservative, but **note** it is highly toxic).

Cryostat-Section Immunostaining

Example 5-HT-immunoreactivity in *Fasciola hepatica*, using primary antisera raised in New Zealand white rabbits by immunisation with 5-HT hydrochloride (Sigma, Poole, UK) conjugated to BSA Fraction V via 0.8% depolymerised paraformaldehyde.

Procedure, immuno-staining

1. Wash freshly recovered worms in warm saline and fix whole specimens and/or slices for 24 h at 4 °C in 4% PFA in PBS.

2. Cryoprotect specimens by first transferring them to 5% sucrose in PBS at 4 °C overnight, then to 30% sucrose in PBS at 4 °C until submerged.

3. Mount specimens in Cryo-M-Bed on a stub and, either snap-freeze using liquid nitrogen, or freeze in cryostat (–20 °C) for 30 min-1 h.

4. Cut sections (4–20 μm in thickness) and collect onto subbed slides.

Subbed slides: In 1 l of distilled water heated to 80 °C, dissolve 5 g gelatin powder and 0.5 g chrome alum (chromic potassium sulphate). Filter if necessary and store cold (4 °C). Dip clean slides in this solution, remove and allow to dry in a hot oven (60 °C). Repeat at least twice.

Note: The use of detergents (such as Triton X-100) in the buffer-wash and/or antibody diluent, may reduce the effectiveness of the subbing and result in the removal of tissue sections from the slides. In such cases, it is recommended that detergents are removed from the buffers or that their concentrations are reduced.

5. Leave sections to air-dry on slides for 30 min at room temperature.

6. Wash slides in PBS for 10 min.

7. Remove excess PBS from around sections, using tissue paper.

8. Place slides horizontally in a plastic tray, lined with buffer-moistened filter paper, and cover sections with primary antiserum made up in PBS (antibody diluent) to a working dilution of 1:500. Cover the tray and incubate for 18 h at 4 °C.

9. Wash slides in PBS for 10 min and remove excess PBS with paper tissue.

10. Incubate sections for 1 h at 4 °C with secondary antiserum (swine anti-rabbit IgG) conjugated to fluorescein isothiocyanate (FITC) (Dako Immunoglobulins, Denmark), made up in PBS (antibody diluent) to a working dilution of 1:500. ·

11. Wash in PBS for 10 min.

12. Mount in PBS-glycerol (1:9) containing 2.5% antifade (1,4-diazobicyclo-[2.2.2] octane) and examine by fluorescence microscopy.

13. Store slides horizontally under tin foil at 4 °C.

Essential controls are: **Controls**

- omission of the primary antiserum,

- substitution of primary antiserum with non-immune rabbit serum (Dako), and

- liquid-phase pre-adsorption of the antiserum with 5-HT hydrochloride (1–50 μg/ml diluted antiserum).

Note: Fixation (and conjugation) alter the configuration of 5-HT, such that it is very difficult to preadsorb.

Whole-Mount Immunostaining

In whole-mount ICC studies of helminth endoparasites, the body wall of the worm can often present serious permeability problems beyond those resolved by treatment with detergents. For example, trematodes with thick teguments, such as *Fasciola*, are best sliced (in fixative) prior to treatment, and with adult and metacestode stages of cestodes, like *Diphyllobothrium*, consistently good immunostaining of nerves is generally only possible by actually removing the tegument before fixing, a procedure most easily done by placing the worms alive in distilled water for 2–8 h (Gustafsson 1991). The nematode cuticle is an absolute impediment to antibody penetration and, as in *Ascaris*, requires the worms to be cut open so as to expose the internal surfaces directly to the antiserum. Since there is often considerable autofluorescence induced by the muscle cells in *Ascaris*, it may also be necessary to remove the muscle bag cells from the cuticle and nervous tissue, using a 0.5% (w/v) solution of collagenase (after Johnson and Stretton 1987). Monogeneans, like *Diclidophora*, and most trematode larval stages immunostain quite readily following treatment with detergents, although in *Schistosoma*, the ciliated coat of the miracidium and the cercarial glycocalyx can present penetration problems.

Example Neuropeptide F (NPF)-immunoreactivity in *Moniezia expansa*, using primary antisera raised in New Zealand White rabbits by immunisation with the C-terminal decapeptide of synthetic NPF (30–39) coupled to BSA via glutaraldehyde.

Procedure, immuno -staining

1. Wash freshly collected worms thoroughly in warm saline, and then immerse in distilled water at 4 °C for 2 h to remove the tegument.

2. Flat-fix portions of the scolex region and strobila in 4% (w/v) PFA in 0.1 M PBS (pH 7.2) for 1 h at 4 °C between microscope slides, weighted as appropriate.

3. Free-fix (in 4% PFA) the flattened specimens for a further 3 h at 4 °C.

4. Transfer specimens to antibody diluent (containing Triton X-100 and BSA) at 4 °C for 48 h.

5. Incubate worms in primary antiserum at a working dilution of 1:800 for 48 h at 4 °C.

6. Wash in PBS at 4 °C for 24 h.

7. Immerse in FITC-labelled secondary antiserum (swine anti-rabbit IgG) for 24 h at 4 °C.

8. Wash in PBS at 4 °C for a further 24 h.

9. Mount on microscope slides in PBS/glycerol (1:9) containing antifade and view.

10. Store slides horizontally at 4 °C in slide tray covered with tinfoil.

Essential controls include

Controls

- omission of primary antiserum,

- use of non-immune rabbit serum as the primary antiserum, and

- preadsorption of the primary antiserum with 50–500 ng NPF/ml of diluted antiserum.

To maximise the use of immunostained whole-mount preparations of parasites, they are best examined by confocal scanning laser microscopy.

6.2.3
Confocal Scanning Laser Microscopy (CSLM)

The main advantage of confocal microscopy over conventional epi-fluorescence microscopy is that it permits visualisation of fluorescent molecules in a *single plane of focus*, thereby creating a vastly sharper image, without the problem of out-of-focus blur or fluorescence flare. This is done by confining illumination and detection to the same spot in the specimen; in other words, the illumination and detection apertures are confocal. Then, by scanning a laser beam as point probe over the specimen, an image is rapidly constructed, as in a TV raster, which is totally in focus and blur-free. Fluorescent images of planes at different depths in the specimen, i.e. collected as an extended-focus series from a scan in the Z-axis, can be recorded as a number of high quality optical sections and then projected in perfect register. In this way, CSLM can provide accurate spatial resolution of the specimen in three dimensions, making it a powerful tool for non-invasive serial sectioning. Using two confocal channels, it is possible to simultaneously image the emission from two different fluorescent dyes in the same specimen, and so detect two or more neuronal substances in colocalisation studies. Since all of the images in CSLM are digitised, they can be stored on computer for further manipulation, including quantitation and comparison of staining intensities between specimens.

6.2.4
Electron Microscope Immunogold Labelling

In common with light microscopic ICC, there are many different procedures for electron microscope (EM) immune detection, but the choice of treatment will be governed by the nature of the antigen investigated and the affinity of the antisera used. While in light microscopy, subcellular integrity can be sacrificed to a fair degree for good antigen preservation, this is not the case with electron microscopy. Here, retention of cell organelles is paramount if the fine-structural localisation of antigens is to be achieved, and this usually means inclusion of glutaraldehyde in fixation (in combination with formaldehyde) and the use of resin embedding, both at the expense of antigenicity. On balance, *post-embedding immunogold labelling* is the method of choice in helminth parasite studies, whereby the tissue is embedded and immunolabelling is carried out on sections of the parasite. Preferred embedding techniques include use of a standard non-polar epoxy resin, when examining heat-stable antigens, or a low-temperature polar resin for the more sensitive epitopes. The immunogold labelling employs electron-dense spheres as reporter molecules, which when coupled either to antibodies or to antibody detectors, enable precise subcellular localisation of neuronal immunoreactivities. Immunogold probes are small enough not to obscure the underlying tissue; they are available commercially in different sizes (from 1–100 nm diameter), allowing two or more antigens to be detected in the tissue; and they are easily quantitated. To save time and effort, it is advisable that when evaluating particular procedures and antisera for EM immunogold labelling, the various parameters (e.g. fixation and antibody binding) are established first at light microscope level.

Low Temperature Post-Embedding Immunogold Labelling

Example Localisation of FaRP-immunoreactivity in *Diclidophora merlangi*, using antisera raised in guinea pigs by immunisation with whole molecule synthetic GNFFRFamide conjugated to BSA via glutaraldehyde (GTA).

1. Cut transverse slices (1 mm in thickness) of fresh worms **Procedure** in cold fix, using a mixture of 4% paraformaldehyde and 1% double-distilled glutaraldehyde (Agar Scientific Ltd, Stansted, UK) in 0.1 M phosphate or cacodylate buffer (pH 7.2), containing 3% (w/v) sucrose. Leave for 1 h at 4 °C.

2. Wash slices in cold buffer containing 100 mM sucrose (and 100 mM NaCl for marine worms), 3×10 min changes.

3. Dehydrate through a graded ethanol series at –20 °C.

4. Infiltrate for 12 h at –20 °C in a 50:50 mixture of pure ethanol and Lowicryl K4M resin (Taab, Watford, UK).

5. Infiltrate for 12–18 h at –20 °C in pure resin.

6. Transfer to BEEM capsules containing fresh K4 M resin and polymerise under UV light (360-nm-long wavelength) for 30 h at –20 °C, and then allow resin to cure for 72 h at room temperature.

7. Cut sections (60–70 nm in thickness) and mount on bare 200-mesh nickel grids. Float the grids, section side down, on the following (steps **8–19**) at room temperature.

8. Tris-HCl buffer (20 mM, pH 8.2) containing 0.1% BSA and Tween 20 (1:20 dilution) (5×1 min washes).

9. Normal goat serum (1:20 dilution in Tris-HCl buffer) to block non-specific proteins (30 min). In trials, this treatment can be left out until labelling has been achieved.

10. Tris-HCl buffer (5×1 min).

11. Tris-HCl buffer containing 0.02 M glycine, to block free aldehyde groups in the tissue (5 min).

12. Primary antibody suitably diluted with 0.1% BSA/Tris-HCl ($20\text{-}\mu l$ drop, overnight).

13. BSA/Tris-HCl buffer (5×1 min).

14. Secondary antibody of 10 or 15 nm gold-conjugated goat anti-guinea pig IgG (Biocell, Cardiff, UK), $20\text{-}\mu l$ drop, 2 h.

15. BSA/Tris-HCl buffer (5 × 1 min).

16. Lightly fix with 2% double-distilled GTA (2 min).

17. Tris-HCl buffer (5 × 1 min).

18. Rinse in double-distilled water (7 × 1 min).

19. Double-stain with uranyl acetate (10 min) and lead citrate (16 min); wash in distilled water.

20. Examine by TEM.

Controls   Controls should include:

- incubation with non-immune rabbit serum (Dako Ltd, High Wycombe, UK);

- use of gold-labelled antiserum in the absence of primary antiserum; and

- liquid-phase preadsorption of the primary antiserum with GNFFRFamide standard (100–5000 ng/ml diluted antiserum).

As a positive control, a portion of *Moniezia expansa* should be processed as described.

Sequential Double-Immunogold Labelling (e.g. GNFFRFamide and NPF)

This requires goat anti-guinea pig and goat anti-rabbit secondary antisera conjugated with different sized gold probes, e.g. 10 and 15 nm. For example, immunostain for GNFFRFamide by proceeding to stage **14** as above, using 10 nm gold-conjugated anti-guinea pig IgG; wash grids in buffer and incubate in NPF antiserum as in **12** and proceed to stage **14**, using 15 nm gold-conjugated anti-rabbit antiserum, then continue to stage **20**.

6.3
Characterisation and Purification of Neuropeptides

Numerous methods are used to quantify regulatory peptides in tissue extracts and in tissue sections, including bioassay and immunochemical techniques. Parasite neuropeptides have been quantified most commonly using radioimmuno-assay (RIA) methodologies. However, the accuracy of these methods is affected by a number of factors, including the type of extraction procedure used, and the specificity of the antiserum employed, i.e. its degree of cross-reactivity with the native peptide.

A wide array of regulatory peptides have been isolated and sequenced from vertebrate and invertebrate tissues, the majority having been isolated from tissues containing relatively high concentrations of the peptide in question, e.g. endocrine/exocrine glands; neuropeptides are most commonly purified from isolated brain/nervous tissue extracts. To date, the full primary structures of 26 helminth neuropeptides, and the partial structure of one, have been determined by peptide sequencing (Table 1). In addition, the primary structures of other peptides have been deduced from gene sequences, but these remain to be isolated from tissue extracts (see Table 1).

Parasitic helminth neuropeptides have been isolated and sequenced from either whole-worm extracts or from extracts of worm regions rich in neural tissue, e.g. the anterior 1 cm ("head region") of *Ascaris suum*. The use of whole-worm or worm-head extracts in neuropeptide isolation procedures have been necessary because of the large quantities of parasite tissue required to enable peptide purification and sequencing, as well as by the difficulty in isolating nervous tissue per se from helminth parasites. However, increasingly, advances in peptide/protein purification technologies and in the sensitivity of sequencing systems have enabled the chemical and structural characteristics of neuropeptides to be determined from relatively low levels of purified messenger. Indeed, primary sequences can now be deduced from low picomole quantities of pure peptide.

6.3.1
Tissue Extraction

It is important to establish the extraction procedure which optimises peptide solubility and stability from a particular extract. It is also necessary to determine the quantity of starting material required to enable the purification of sequenceable amounts of peptide from crude tissue extracts.

Table 1. The primary structures[a] of helminth regulatory peptides

	Neuropeptide Y/neuropeptide F (NPF)-related peptides
Platyhelminthes	
Fasciola hepatica[1]	PSVQEVEKLLHVLDRNG-KV-AE ----------------------- NH$_2$
Moniezia expansa[2]	PDKDFIVNPSDLVLVDNKAALRDYLRQINEYFAIIGRPRF NH$_2$
Artioposthia triangulata[3]	KVVHLRPRSSFSSEDEYQIYLRNVSKYIQLYGRPRF NH$_2$

	FMRFamide-related peptides (FaRPs)
Platyhelminthes	
Moniezia expansa[4]	GNFFRF . NH$_2$
Artioposthia triangulata[5]	RYIRF . NH$_2$
Dugesia tigrina[6]	GYIRF . NH$_2$
Bdelloura candida[7]	GYIRF . NH$_2$
	YIRF . NH$_2$
Nematodes	
Panagrellus redivivus[8,9,10,11]	(PF1) SDPNFLRF . NH$_2$
	(PF2) SADPNFLRF . NH$_2$
	(AF2) KHEYLRF . NH$_2$
	(PF3/AF8) KSAYMRF . NH$_2$
	(PF4) KPNFIRF . NH$_2$
Caenorhabditis elegans[12,13,14]	(PF1) SDPNFLRF . NH$_2$
	(PF2) SADPNFLRF . NH$_2$
	SQPNFLRF . NH$_2$
	ASGDPNFLRF . NH$_2$
	AAADPNFLRF . NH$_2$
	[b]AGSDPNFLRF . NH$_2$

Ascaris suum[15,16,17]

[b] (K)PNFLRF . NH_2
[b] (K)PNFMRY . NH_2
(AF2) KHEYLRF . NH_2

(AF1) KNEFIRF . NH_2
(AF2) KHEYLRF . NH_2
(AF3) AVPGVLRF . NH_2
(AF4) GDVPGVLRF . NH_2
(AF5) SGKPTFIRF . NH_2
(AF6) FIRF . NH_2
(AF7) AGPRFIRF . NH_2
(PF3/AF8) KSAYMRF . NH_2
(AF9) GLGPRPLRF . NH_2
(AF10) GFGDEMSMPGVLRF . NH_2
(AF11) SDIGISEPNFLRF . NH_2
(AF12) FGDEMSMPGVLRF . NH_2

Haemonchus contortus[18]

(AF2) KHEYLRF . NH_2

Others

Nematodes
Ascaris suum[19]

TKQELE

[a]Amino acid sequences shown using single letter notation. [b]Putative peptides from the gene sequence.
References: 1. Magee et al. (1991); 2. Maule et al. (1991); 3. Curry et al. (1992); 4. Maule et al. (1993); 5. Maule et al. (1994a); 6. Johnston et al. (1995); 7. Johnston et al. (1996); 8. Geary et al. (1992); 9. Maule et al. (1994b) 10. Maule et al. (1994c); 11. Maule et al. (1995a); 12. Rosoff et al. (1992); 13. Rosoff et al. (1993); 14 Marks et al. (1995) 15. Cowden et al. (1989); 16. Cowden and Stretton (1993); 17. Cowden and Stretton (1995); 18. Keating et al. (1995); 19. Smart et al. (1992)

Small quantities of the tissue are usually extracted using a range of different methods and different extraction media, and thereby the extraction procedure which maximises peptide recovery is identified. Although many extraction media are available, those most widely used include: acidified ethanol (ethanol/0.7 M HCl; 3:1, v/v) or methanol, 2 M acetic acid, boiling 0.5 M acetic acid, boiling water, and acetone. Table 2 shows the recovery of neuropeptide F (NPF) from *Moniezia expansa* using different extraction media. Extracts should be stored at low temperatures (4 °C) to reduce the activity of endogenous peptidases.

Acidified ethanol was the optimal extraction medium in the purification of the neuropeptides, NPF and GNFFRFamide from *M. expansa*, and in the extraction of TKQELE (TE6) from *A. suum*. The procedures employed in the extraction of NPF and GNFFRFamide and of TE6 are summarised in the Acid ethanol extraction protocol. The FMRFamide-related peptides (FaRPs), KNEFIRFamide (AF1) and KHEYLRFamide (AF2) were extracted from *A. suum*, using acidified methanol (90% methanol in 1% acetic acid and 1 mM dithiothreitol). Acetone has been employed in the extraction of FaRPs from the free-living nematodes, *Panagrellus redivivus* and *Caenorhabditis elegans* (Acetone extraction Protocol). However, AF2 and the related FaRP, KSAYMRFamide, were extracted from *P. redivivus* using acidified ethanol. It seems likely that different profiles of neuropeptides can be obtained using different extraction methods. This solubilisation phenomenon was clearly demonstrated in *P. redivivus*, where in acetone extracts the

Table 2. Recovery of neuropeptide F (NPF) from *Moniezia expansa* using different extraction media

Extraction medium	NPF-immunoreactivity (ng/g wet weight *M. expansa*)
Acid ethanol	193
Acid methanol	144
2 M acetic acid	101
0.5 M boiling acetic acid	72
Boiling water	59
Acetone	21

most abundant FaRP was found to be SDPNFLRFamide (PF1), whereas in acidified ethanol extractions it was KHEYLRFamide (AF2). See Table 1 for further details and references.

1. Homogenise tissue in acidified ethanol (ethanol/0.7 M HCl; 3:1, v/v; 8 ml/g wet weight of tissue). Agitate extract for 24 h at 4 °C. **Acid ethanol extraction**

2. Centrifuge (3000 *g* for 30 min) to remove tissue debris.

3. Rotary evaporate supernatant to remove ethanol and add trifluoroacetic acid (TFA) to a final concentration of 0.1%.

4. Store at 4 °C for 18 h, centrifuge (5000 *g* for 1 h) and remove the supernatant.

1. Place tissue in acetone (1:4, vol/vol) and store for 3 weeks at –20 °C. **Acetone extraction**

2. Rotary evaporate to remove acetone.

3. Filter remaining aqueous residue prior to purification procedure and lyophilise.

6.3.2
Peptide Detection

Extracted peptides may be detected in tissue extracts or chromatographic fractions using specific bioassay or immunoassay systems. The detection system usually relies on the lyophilisation of a small aliquot (10–100 μl) of the sample and the reconstitution of this in the appropriate bioassay or immunoassay buffers. Numerous different radioimmunoassay/enzyme-linked immunosorbent assay (ELISA) detection systems exist. However, details of the range of different methods employed are beyond the scope of this chapter (see Chap. 10 for details). One of the most widely employed immunoassay techniques is *radioimmunoassay*. This relies on the availability of a relatively specific antibody which cross-reacts with the peptide of interest (native helminth neuropeptide) under competitive conditions, ie in the presence of the appropriate radiolabelled ligand. Undiluted antisera should be stored at –20 °C; however, the number of freeze-thaw events should be kept to a minimum. This can be achieved by freezing small aliquots of antisera, such that only the amount required is defrosted.

The radioimmunoassay scheme employed in the isolation of GNFFRFamide (*M. expansa*), RYIRFamide (*Artioposthia triangulata*), KHEYLRFamide, KSAYMRFamide and KPNFIRFamide (*P. redivivus*) is outlined in the Protocol below. Some radioimmunoassay systems have been developed, such that small sample aliquots may be diluted (usually 1:100) in assay buffer and assayed directly, thereby, negating the need for a lyophilisation stage. During the purification of peptides, it is desirable to keep the number of lyophilisation stages to a minimum as they often account for the greatest reductions in peptide recovery. Although most radioimmunoassays require the incubation of sample, radiolabelled tracer and antibody for 12–24 h at 4 °C, some proceed to equilibrium within 1 h at 37 °C. This approach greatly facilitates the rapid detection of the neuropeptides of interest and enables a series of purification steps to be carried out in a short period of time.

FMRFamide-Related Peptide (FaRP) Radioimmunoassay (RIA)

Reagents – Assay buffer 40 mM sodium phosphate (pH 7.2) containing 70 mM sodium chloride and 0.2% (w/v) RIA grade bovine serum albumin
– Antiserum (RIN 8755, Peninsula Laboratories, St. Helens, UK) used at a working dilution which gives 50% binding of the radiolabel
– Tracer, monoradioiodinated (100 Bq = 1.4 fmol per assay tube) YGGFMRFamide purified by reverse-phase HPLC].
– Standard, 2–500 pg FMRFamide/assay tube
– Sample, 10 μl sample diluted (1:100) with assay buffer
– Separation medium, 40 mM sodium phosphate (pH 7.2) containing 0.5% (w/v) dextran-coated charcoal and 10% (v/v) heat-inactivated horse serum

Procedure, RIA 1. Total assay volume (300 μl) consisting of 100-μl sample or standard; 100-μl antiserum RIN 8755 (diluted 1:100 000 in assay buffer); 100 μl YGGFMRFamide tracer.

2. Incubate assay for 18 h at 4 °C or for 2 h at 37 °C.

3. Add 1 ml of separation medium to each assay tube and centrifuge (1200 *g* for 30 min). Antibody-peptide complex will remain in the supernatant and unbound peptide and tracer will reside in the pellet.

4. Separate supernatant and pellet and record results using a gamma counter. The non-specific binding of samples (i.e. in the absence of antibody) should be determined.

6.3.3
Primary Purification Steps

The use of whole-worm or worm-head extracts to enable neuropeptide isolation necessitates the employment of primary (high-capacity) purification steps prior to analytical HPLC procedures. Several primary purification methods are commonly used, including solid-phase extraction with Sep-Pak cartridges (usually reverse-phase, C18, C8, CN) and gel-permeation chromatography. Sep-Pak cartridges should be activated/conditioned with a low-polarity solvent, and washed in a high-polarity solvent prior to sample loading. It is essential that the sample is dissolved in a high-polarity solvent. The supernatant, obtained following acid-ethanol extraction (See protocol 6.3.1, step **4**), may be loaded directly onto Sep-Pak cartridges. Lyophilised extracts may be reconstituted in high polarity solvent, e.g. 0.1% trifluoroacetic acid in water, prior to Sep-Pak cartridge extraction (protocol see below).

Sephadex G50 (medium or fine) or the Sephacryl equivalent are suitable for the gel-permeation fractionation of most peptides with a molecular mass 1–30 kDa. This procedure will partition the peptide such that higher and lower molecular weight contaminants may be removed. It will also give an indication of the size of the biomolecule in question and help determine the types of HPLC column to be used during subsequent purification steps. Both of these methods were employed as primary purification steps in the isolation of NPF from extracts of *M. expansa* (see protocol below).

1. Activate/condition C18 Sep-Pak cartridges with 10× cartridge volume of 0.1% TFA in acetonitrile.   **Sep-Pak cartridge extraction**

2. Flush cartridge with 10× cartridge volume 0.1% TFA in water.

3. Pump sample (dissolved in 0.1% TFA in water) through Sep-Pak series (10–12 ml/h flow-rate ensures a high degree of adsorption) and repeat as necessary.

4. Elute unwanted components from Sep-Pak series with 0.1% TFA in water.

5. Elute peptides with less polar solvent (e.g. 0.1% TFA in acetonitrile) At this stage, further purification may be carried out using progressively less polar solvents, such that bound peptides may be partitioned. Discard used Sep-Paks.

6. Lyophilise eluted sample.

Gel-permea-tion chro-matography

1. Reconstitute lyophilised sample in 2 M acetic acid.

2. Centrifuge extract (6000 g for 1 h) and remove supernatant.

3. Load supernatant onto gel-permeation column and fractionate [general chromatography column (1.6 × 90 cm) packed with Sephadex G50 (fine) gel equilibrated with 2 M acetic acid].

4. Elute gel-permeation column (2 M acetic acid at a flow rate of 10–15 ml/h).

5. Collect fractions at 12–15 min intervals and, following the elution of the predetermined total column volume, store at 4 °C. Fractions should be collected into polypropylene tubes.

6. Subject fractions to peptide-detection system. The total (V_t) and void (V_o) volumes are usually predetermined, using potassium dichromate and Dextran blue, respectively.

6.3.4
Reverse-Phase HPLC

The increasing diversity, sophistication and resolution capacity of HPLC column matrices has greatly simplified the purification procedures required to isolate pure peptides from crude tissue extracts. The majority of reverse-phase HPLC columns employ silica as the stationary phase. These matrices vary considerably with respect to particle size, shape and density, as well as to particle surface chemistry. Semi-preparative and analytical HPLC columns suitable for regulatory peptide isolation most commonly possess carbon chain (C3, C4, C8, C18)-, diphenyl-derivatised or ion exchange (anion or cation) stationary phases. Peptide messenger purification is most successful when combinations of these column chemistries are employed.

The initial HPLC fractionation is normally performed using a semi-preparative column with a peptide loading capacity appropriate to the sample. If the sample was lyophilised following the primary purification steps (see protocol Sep-Pak cartridge extraction), it may be reconstituted in the aqueous mobile phase which will be employed in the HPLC fractionation. Conversely, if the sample was fractionated by gel-permeation chromatography, using an acidic mobile phase, e.g. 2 M acetic acid, the peptide-containing fractions may be pumped directly onto a semi-preparative reverse-phase HPLC column. A semi-preparative reverse-phase HPLC column (1×30 cm or 1×60 cm) with a C18-derivatised stationary phase is ideal for the partition of most peptides with a molecular mass < 4 kD. Once fractionated, aliquots of the eluate can be assessed for the presence of the peptide, using the appropriate bioassay or immunoassay systems (see Sect. 6.2.2 above). At this stage, it is not necessary to monitor the absorbance wavelengths of column eluates.

Identified immunoreactive fractions can then be subjected to a combination of analytical reverse-phase HPLC fractionations, as required. Makers recommendations should be consulted for mobile phase flow-rates, although these are normally 2–3 ml/min for semi-preparative, and 0.5–2 ml/min for analytical HPLC columns. Peptide losses, as a result of lyophilisation between purification steps, can be avoided by diluting the fractionated eluate in the appropriate mobile phase and pumping directly onto the reverse-phase columns. The dilution factor required will be dictated by the column chemistry to be employed; however, normally, an eluate to aqueous mobile phase ratio of 1:4 (v/v) is adequate. If ion-exchange chromatography is employed as one of final steps in the purification of a peptide, an additional reverse-phase HPLC run is recommended to facilitate the removal of buffer salts. The mobile phase employed in ion-exchange chromatography should incorporate 20–30% acetonitrile to avoid the loss of peptides through strong hydrophobic interactions.

The absorbance profile of analytical reverse-phase HPLC eluates should be monitored at several wavelengths. The most commonly employed absorbance wavelength is 214 nm, which monitors the abundance of peptide bonds. If the peptide is known to possess tyrosyl or tryptophanyl residues or, if its structure is unknown, it may be useful to monitor the eluate simultaneously at 280 nm. As purification proceeds, the symmetry of the absorbance peaks is indicative of peptide purity.

Polypropylene tubes should be used for the collection of column effluents. Fractions should be collected automatically, e.g. at 1 min intervals, during the initial semi-preparative and analytical purification steps. However, as the peptide is purified in subsequent analytical HPLC fractionations, it is helpful to collect individual peaks and shoulders of absorbance manually. This enables the removal of closely-eluting contaminants and greatly facilitates peptide purification by reducing the number of purification steps required. However, it is essential to determine the time-delay between peptide absorbance and elution, prior to the adoption of manual collection, by system-calibration using peptide standards.

Although a number of helminth neuropeptides have been purified using reverse-phase HPLC techniques, the procedures employed differ in detail. The following protocol gives a typical hypothetical helminth neuropeptide (X) purification scheme. An example of the chromatographic profiles generated during the purification of the *M. expansa* FaRP, GNFFRFamide, is presented in Fig. 2.

HPLC Purification of Helminth Neuropeptide X

Tissue extracted in acidified ethanol (see Sect. 6.3.1, Acid ethanol extraction) and subjected to Sep-Pak partition and gel-permeation chromatography using a 2-M acetic acid mobile-phase (see Sect. 6.3.3, Primary purification steps)

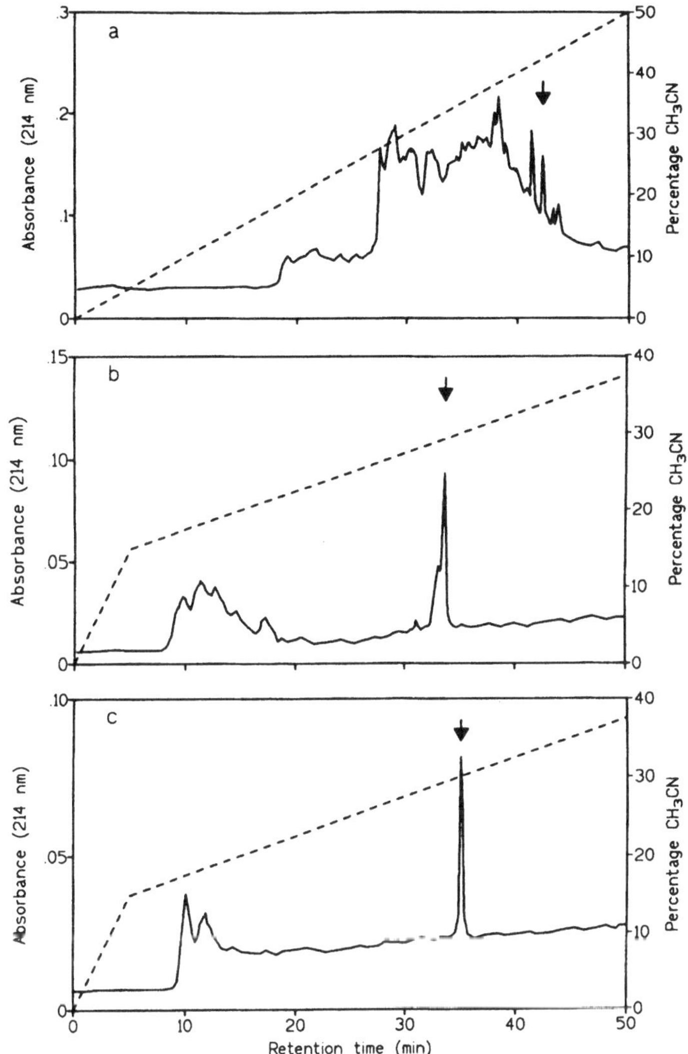

Fig. 2a–c. Absorbance profiles of the FMRFamide-immunoreactivity (IR) in an extract of *Moniezia expansa* following **a** semi-preparative (C18); **b** analytical (C18); **c** analytical (C8) reverse-phase HPLC steps. *Arrows* indicate the elution position of the FMRFamide-IR

- Aqueous mobile-phase (A), 0.1% trifluoroacetic acid (TFA)   **Reagents** in HPLC grade water
- Organic mobile-phase (B) 0.1% TFA in acetonitrile (far UV or supergradient grade)

– Phosphoric acid and heptafluorobutyric acid counter-ions may be employed as alternatives to TFA in (A) and (B)

Procedure, HPLC purification At each purification stage, the immunoreactive fractions were identified using a radioimmunoassay for peptide X (see protocol Sect. 6.3.2). Peptide X-containing fractions were diluted in aqueous mobile-phase (1:4, v/v) prior to each HPLC step.

1. Pump diluted peptide X-containing gel-permeation fractions onto a semi-preparative reverse-phase HPLC column (mobile-phase flow rate 3 ml/min) and elute using a linear gradient of 100% A:0% B to 30% A:70% B in 70 min. Collect 1 min fractions.

2. Dilute and pump immunoreactive-fractions onto:
   a) Diphenyl reverse-phase HPLC column (flow-rate 1 ml/min) and elute using a linear gradient of 100% A:0% B to 50% A:50% B in 100 min. Collect 1 -min fractions.
   b) C8 reverse-phase HPLC column (flow-rate 1 ml/min) and elute with a linear gradient of 100% A:0% B to 50% A:50% B in 100 min. Collect 1 min fractions.
   c) C18 reverse-phase HPLC column (flow-rate 1 ml/min) and elute with a linear gradient of 100% A:0% B to 40% A:60% B in 60 min. Manually collect peak fractions.
   d) C8 reverse-phase HPLC column (flow-rate 1 ml/min) and elute with a gradient of 100% A:0% B to 85% A:15% B in 5 min and to 55% A:45% B in a further 60 min. Manually collect peak fractions.

3. Store the purified peptide in a sealed polypropylene container prior to structural characterisations.

4. The peptide-containing fraction should either be lyophilised before being reconstituted in the appropriate buffers for structural characterisations or evaporated to a volume appropriate for direct mass-spectrometry, amino acid sequencing or amino acid analysis.

Occasionally, neuropeptides prove difficult to purify using the schemes outlined above. In such cases, the organic solvents, e.g. methanol, acetic acid, acetone, propanol and ethyl acetate, may be employed in the mobile phase. Alternatively, different counter ions, ie other than trifluoroacetic acid (TFA) may be used in the mobile-phase. Counter-ions, including phosphoric acid and heptafluoroacetic acid, reduce and strengthen, respectively, the peptide-binding properties, and often allow the separation of peptides which could not be separated by standard methods. Maker's recommended pH ranges should be noted, as silica-based stationary phase of HPLC columns may be damaged if the pH is neutral or alkaline. If basic counter ions, e.g. triethylamine, are required to effect purification then organic-polymer stationary-phase columns may be employed. Also, microbore HPLC systems greatly improve the ability to resolve extremely small samples.

6.3.5
Endoproteinase Digestion

Direct gas-phase Edman degradation results in the sequential loss of peptide during each sequencing cycle, such that the larger the peptide to be sequenced, the greater the quantity required to enable full structural determinations. The most widely studied invertebrate neuropeptide family are the FaRPs. However, their relatively small size (the majority are <10 amino acid residues) enables their full primary structure to be determined from relatively small quantities of purified peptide. Conversely, greater quantities of larger peptides, e.g. NPF (36–40 residues), are needed if Edman degradation is to prove successful. Large peptides (>30 residues) may be subjected to endoproteinase digestion such that smaller sequenceable fragments are generated. A range of HPLC-purified, sequencing grade endoproteinases are available commercially: eg Arg-C, Asp-N, Glu-C and Lys-C.

In order to determine the endoproteinases suitable for the digestion of a particular peptide, some primary structural information is required. In this respect, partial primary structures may be deduced by Edman degradation of the intact peptide (Maule et al. 1991). However, it may be possible to deduce the presence of potential cleavage sites in peptides of unknown structure by analysis of the conserved residues in related peptides.

Once the procedures have been designed, the peptide and endoproteinase are incubated in the buffer and in the ratio outlined in the maker's instructions. Following incubation, the digest is subjected to analytical HPLC fractionation and the absorbance peaks are collected manually. Subsequently, each fraction may be subjected to direct gas-phase sequencing and the primary structure of each fragment and, therefore, the intact peptide determined. This technique was employed to enable the full primary structural determination

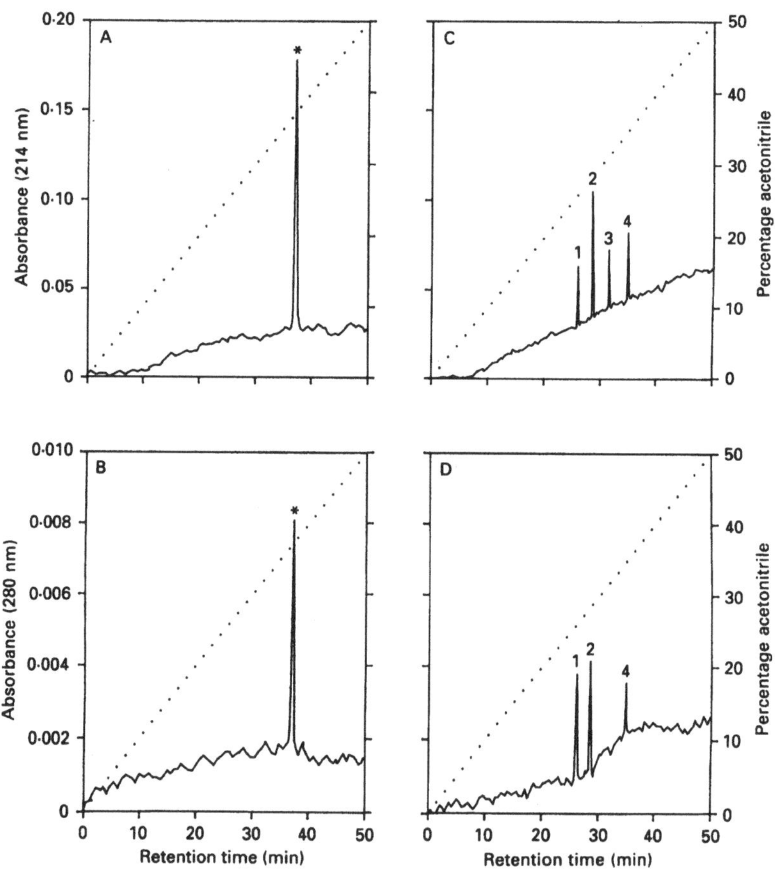

Fig. 3A–D. The absorbance profile of 1 nmol of neuropeptide F (NPF) at A 214 nm and B 280 nm is shown, following the final chromatographic run on a Supelcosil LC-308 analytical reverse-phase column eluted at 1.5 ml/min. The symmetry of both peaks of absorbance indicates a high degree of purity. PP-immunoreactivity is indicated (*). The absorbance profile of the four NPF fragments produced by endoproteinase Glu-C digestion at C 214 nm and D 280 nm are also shown

of NPF (*M. expansa*; Maule et al. 1991). The absorbance profiles of NPF (*M. expansa*) before and after endoproteinase Glu-C digestion are shown in Fig. 3.

Endoproteinase Glu-C Digestion of Neuropeptide F (NPF)
(*M. expansa*)

1. Lyophilise 1 nmol of purified NPF.

2. Incubate lyophilised peptide with 50 μl of 25 mM ammonium carbonate buffer (pH 7.8) containing 5 μg of endoproteinase Glu-C (Boehringer AG, Mannheim, Germany) for 18 h at 25 °C.

3. Fractionate digest, using reverse-phase HPLC, and collect peptide fragments manually.

4. Subject fragments to structural analyses.

Procedure, digestion

6.4
Physiological Action of Transmitters and Drug Assessment In Vitro

6.4.1
Recording of Motor Activity in Whole Animals or in Muscle Strips

Because of the anatomy of the parasitic worms, it is virtually impossible to obtain nerve-muscle preparations free of significant extraneous and interfering tissues on which controlled experiments can be performed. For this reason, assessment of the action of transmitters and drugs on the nerve and muscle physiology of parasitic helminths is, for the most part, carried out on whole animals or segments of the parasites.

In physiological experiments in which whole animals or segments of worms are used, conventional physiological equipment and experimental protocols published in the general physiological literature are often appropriate, and can be used with little or no modification. However, for the smaller parasitic forms, extensive modifications of conventional equipment are often needed, and methods and procedures must be specifically adapted for use with these animals.

The Basic Technique

Equipment and Reagents
- Tissue chamber or other suitable recording chamber
- Muscle transducer
- Chart recorder or other read-out device
- Incubation medium suitable for parasite being studied

Basic procedure

Depending on the size of the parasite under study, whole worms or sections of worms of appropriate length are used. For nematodes, characterised by cuticles that are only slightly permeable, lengthwise slitting of the preparation may be necessary to assure penetration of test substances to their sites of action. For some of the larger parasitic flatworms, longitudinal strips can be used. Attachment of the preparation to the transducer system can be accomplished by tying silk or other suitable thread to either end of the preparation, with one end attached to the transducer and the other to some fixed point in the recording chamber. The transducer may be any of a variety of commercially available types. Isotonic transducers are generally more appropriate, but their sensitivity must be matched to the force that the preparation is expected to generate. Temperature and gas phase should be appropriate for the animal being studied. Once the preparation is mounted in the tissue chamber, allow 5–15 min equilibration time to establish baseline tone and activity in the preparation. Agents to be tested may be added to the chamber directly or via an exchange of the medium in the tissue chamber. The activity of the preparation can be recorded on a chart recorder, but analogue-to-digital conversion of the signal and computer storage of the data may be more convenient, since it facilitates later analysis of the data.

Recording from small parasites

For the larger parasites, the attachment to the transducer system can be accomplished via thread. However, many of the parasitic helminths are of such small size and delicate structure that even the finest available suture thread is unsuitable, in which case other mechanisms for interfacing the parasite with the transducer are required. The following describes a system which has been used for studies on *Schistosoma mansoni* and smaller nematode species, but it is

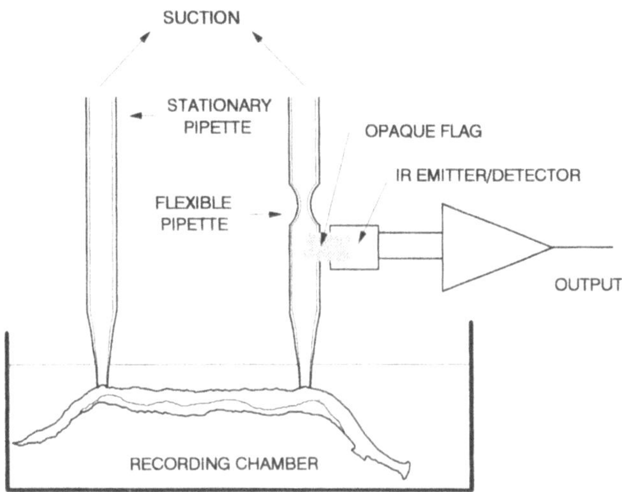

Fig. 4. Schematic representation of the apparatus used for the recording of motor activity in small helminths

readily adaptable to other small parasites or to strips of tissue from larger parasites. In this system, two thin flexible pipettes are attached to the surface of the parasite via suction, and, as the muscles contract, changes in the distance between the two pipettes are monitored (Fig. 4).

Pipette construction. Pieces of small gauge polyethylene tubing, about 7 cm in length with a bore of 0.25 to 0.50 mm, are used for construction of the pipettes. By using heat from a low-wattage soldering iron or other comparable heat source, the tip of each pipette is drawn out to a fine tip with an inside diameter of 40 to 80 μm.

The pipettes are mounted in micromanipulators and connected via plastic tubing to 5-ml syringes. The parasite to be studied is placed in a small recording chamber with appropriate incubation medium. With the aid of a dissecting microscope, one of the pipettes is manoeuvered into a position such that it is at a right angle to the parasite surface and is just touching it. Suction is applied to the pipette via the syringe, causing the worm surface to be pulled against the pipette tip. By maintaining the suction on the pipette, the parasite remains firmly attached to the pipette. After the first pipette is attached in this way, the other pipette is manoeuvered into a position on the parasite at a set distance away from the first pipette. Suction is then applied to this pipette in the same way. With this attachment, any shortening or lengthening of muscles between the pipettes is translated into changes in position of the flexible pipette relative to the non-flexible pipette. The amount of stretch and load imposed on the muscle can be controlled by adjusting the flexibility of the pipettes. Usually, one of the pipettes is kept as rigid as possible while the other is made as flexible as possible. This can be done by placing a constriction in the pipette 2–3 cm above its tip by heating and drawing it out at that point. By controlling the degree of constriction, flexibilities of varying amounts can be obtained.

Recording. Because of the delicate nature of the muscles, the apparatus used for detecting the changes in distance between the pipettes should not impose a significant load on the movements of the pipettes. The movements expected are also minute. Photooptic devices have proven to be best suited for this system. They impose no load on the pipettes and they can be designed to detect even the slightest movement of the pipettes. A small opaque flag is attached to the flexible pipette near its tip. A light source and a photodetector are arranged so that the flag interrupts the light path by varying amounts as the pipette is moved by the movements of the parasite. Molded IR emitter-detector pairs with appropriate electronic circuitry may be used as the detectors for the pipette movement. The varying voltage output can then be displayed on an appropriate read-out device.

The suction pipettes attached to the parasite surface can also serve as devices for delivering electrical stimulation to the parasite or for recording of the electrical potentials generated by the parasites in the vicinity of the pipettes. To do this, a silver chloride-coated wire is inserted into the pipette to near its tip. A column of incubation medium is drawn up into the pipette so that it comes in contact with the silver wire. Attachment of the pipette to the parasite is then accomplished as already described. A second silver wire placed in the recording chamber, serves as a reference electrode. The electrode can be attached to a physiological stimulator for delivery of the electrical stimulation or it may be connected to a physiological amplifier for measurement of the electrical activity associated with parasite movements.

6.4.2
Action of Transmitters and Drugs at the Cellular Level

Assessment of the physiological action of transmitters and drugs can also be carried out at the cellular level. This can be achieved either with the use of conventional microelectrode techniques, or, if a suitable preparation of cells from the parasites can be obtained, by direct visual assessment or use of patch-clamp techniques.

Intracellular Microelectrode Recordings

- Microelectrode amplifier
- Microelectrode puller
- Micromanipulators
- Oscilloscope or other suitable read-out device
- Faraday cage
- Recording chamber
- Incubation medium
- Microelectrode capillary glass with microfilament

Equipment and Reagents

For microelectrode studies on the muscle of the larger nematode parasites, e.g. *Ascaris*, a short segment is removed from the parasite, cut lengthwise along the lateral line and the internal viscera removed. This body-wall segment is then securely pinned, muscle-side up, in the recording chamber and covered with the appropriate physiological saline.

Procedure, recording

Microelectrodes are pulled from capillary glass, filled with 3 M KCl or other appropriate salt solution, and connected via silver-chlorided silver wire to the input of the microelectrode amplifier. Electrodes with tip diameters in the micron range, and resistances in the 10–50 Mohm range, are generally suitable for studies on the large muscle cells of nematodes, such as *Ascaris*. A second, larger electrode, also filled with 3 M KCl or other suitable saline solution, is placed in the bath to serve as a reference electrode. For these studies, electrical shielding is essential and the whole system is usually shielded by doing all recordings within a Faraday cage. Micromanipulators are essential for insertion of microelectrodes into the cells.

Cell sizes in parasitic flatworms are generally quite small. The muscle cells are located below the syncytial tegumental layer and hence are accessible to microelectrodes only by penetration through the tegument. For such microelectrode studies, it is imperative that the animals be immobilised. However, since there are no skeletal elements to give effective anchoring points for immobilisation, and any mechanical immobilization is generally unsatisfactory, paralyzing agents, such as carbachol or Na-pentobarbital, must be used. This introduces the question of how these agents in themselves might be affecting the functioning of the muscles and interfering with the very processes under study. Altogether, the small size of the muscle cells, their location below the tegument, and the need to use immobilising agents makes the use of microelectrodes not a particularly useful approach for the study of flatworm muscle.

The Dispersed Cell Preparation

To circumvent these difficulties with respect to flatworms, dispersed muscle fibres can be prepared from the parasites. In this way the effects of various agents can be assayed for their direct effects on muscle function without interference from other tissues. The methods presented here have been developed specifically for *S. mansoni*, but should be readily adaptable for use with other flatworm parasites.

Equipment   – Incubator
– Shaker table
– Incubation media

• Medium A. Dulbecco's Modified Eagle's Medium (DMEM, w/o Na_2HPO_4 or $NaHCO_3$, Catalog No. 3656, Sigma, St. Louis, MO, USA), reduced to 67% of its usual concentration and supplemented with: 2.2 mM $CaCl_2$, 2.7 mM $MgSO_4$, 0.04 mM Na_2HPO_4, 61.1 mM glucose, 1.0 mM dithiothreitol (DTT), 1 μM serotonin, 10 mg/ml pen-strep and 15 mM 4-(2-hydroxyethyl)-1-piperazine ethanesulphonic acid (HEPES), pH 7.4; the final osmolality of this medium is about 290 mosm, which is similar to that of the host blood.

- Medium B. This is identical to Medium A except there is added 1 mM ethyleneglycol-bis-(β-aminoethyl ether) N, N, N′, N′-tetraacetic acid (EGTA), 1 mM ethylenediamine tetraacetic acid (EDTA) and 0.1% bovine serum albumin.

- Medium C. This medium consists of only the inorganic components of the supplemented DMEM (Medium A), plus 5 mM Hepes, 80 mM glucose and 1 μM serotonin as well as any drug to be tested.

The source of the cells used is 20–25 pairs of adult *S. mansoni* recovered 45–60 days postinfection from the portal and mesenteric veins of female Swiss Webster mice. Once isolated, the parasites are rinsed several times then stored at 37 °C in Medium A until ready for use. Throughout the dispersion procedure the parasites are kept at 37 °C. Each step in the process is done as aseptically as possible, with all media passed through 0.45-μm syringe filters before use. **Procedure, cell preparation**

The worms are placed onto a glass microscope slide into a drop of Medium B. They are then coarsely chopped with a razor blade to obtain worm pieces, averaging about 1 mm in length. The pieces are rinsed several times to remove the debris produced during the chopping process and then placed in a scintillation vial with several ml of Medium B, and incubated on a shaker table for 15 min.

After this, the medium is replaced with 1 ml of Medium B to which is added 1 mg/ml papain (EC 3.4.22.2, Boehringer-Manheim). The pieces are incubated in the enzyme at 37 °C on a shaker table for a total of 45 min. After 15 min and 30 min, the bathing medium is decanted and replaced with fresh medium containing papain. After 45 min, the enzyme medium is removed and the worm pieces washed with three exchanges of Medium A and incubated on the shaker table for a further 10 min in 10 ml of enzyme-free medium.

The preparation is then aliquoted at room temperature into four 12-mm glass test tubes. At this point, the worm pieces are still largely intact. The pieces are then broken up by forcing them back and forth through the orifice of a Pasteur pipette some 75–100 times. This final disruption step is critical for obtaining a viable dispersion of cells. The pipetting must be gentle and should continue only until the medium appears cloudy and small pieces of the worms are still present. Successful completion of this step yields a mixture of isolated cells as well as pieces of worms broken up to varying degrees.

The volume in each of the test tubes is then adjusted to 6 ml with Medium A and, depending on the use to which the cells will be put, the contents of each tube are dispensed into two plastic 30 mm petri dishes or other suitable containers. The suspension is allowed to settle for a minimum of 30 min at room temperature, after which the medium is exchanged. This removes, for the most part, any remaining worm pieces in the preparation, as well as cells that have failed to adhere to the bottom of the dishes. The preparation is then stored at 18–20 °C until ready for study.

Visual Assays of Muscle Effectors on Dispersed Cells

Equipment
- Inverted microscope
- Perfusion pipettes
- Micromanipulator
- Microelectrode puller
- Video camera
- VCR recorder
- TV monitor

Procedure, Visualisation
The isolated schistosome cells are quite small, and any visual observations on them must be done at high power under an inverted microscope. In plates isolated by the above procedure, there is a mixture of cell morphologies, some partially digested pieces and debris. Within this mixture, some cells have the appearance of muscle fibres and they seem to fall into several morphological types. Occasionally, active flame cells are present.

For visual observations of the effects of possible transmitters, drugs etc, on the isolated muscles, a microperfusion system, by which agents can be directly perfused onto individual isolated muscles in the dispersed cell preparation, is very effective. For this microperfusion, micropipettes which have been pulled on a microelectrode puller are used. The tips of the pipettes, when pulled, are too small for practical use, and need to be broken back to give tip diameters of 10–20 μm. The pipettes are connected to small diameter polyethylene tubing and a syringe. The pipettes are filled by immersing the tip of the pipette into a filtered solution of the substance to be perfused and suction applied via the syringe attached to the pipette. Normally, the filling solution will be identical to that bathing the cells but with the addition of the agent to be tested. This medium should be filtered before attempting to fill the pipettes to prevent clogging of the tip of the pipette. The plates, stored at 18–20 °C, are warmed to 34–36 °C about 10 min prior to and during the time observations are made.

The end point used in these studies is visual observation of contractions of individual muscle fibres in response to microperfusion of test substances. Alternatively, test substances can be added to the plates to assess how they affect responsiveness of the fibres to perfused substances. A video camera and monitor are useful for these observations since all observations can be videotaped for later analysis. For each plate, 20–30 muscle fibres are randomly selected and each micro perfused with the test medium. **Alternative procedure**

Patch-Clamp Recording from Dispersed Muscle Cells

The dispersed muscles also give a preparation suitable for application of patch-clamp recording methodologies.

- Patch-clamp amplifier
- Micromanipulators
  Microelectrode puller
- Patch-clamp electrodes
- Analogue-to-digital converter hardware
- Computer
- Data acquisition and analysis software
- Electrode puller
- Inverted microscope
- Recording chambers
- Faraday cage

Equipment

For patch-clamp recordings, the recording chamber used is a 24 × 45 mm cover glass on which is placed a 1/8th-inch-thick square of Teflon with a 1.5-cm circular hole. The Teflon is adhered to the glass coverslip by coating the bottom of the Teflon with a thin layer of petroleum jelly. The dispersed cells **Procedure, patch-clamp recording** are plated onto these chambers instead of petri dishes. The cells are allowed to settle for 30 min and then the medium is exchanged to remove debris and unattached cells from the dish. The cells are viewed through an inverted microscope at high power.

Patch-clamp electrodes are pulled from capillary glass to resistances of 5–15 Mohm. These resistances are somewhat higher than are commonly used, but they are most suitable for the small schistosome muscle fibres. The filling solution for the pipettes depends on the particular type of study being performed. A pipette filled with 130 mM KCl serves as a reference electrode in the bath.

For patch-clamp recording, the electrode is brought into contact with the membrane of an isolated muscle cell, and gentle suction is applied. This, if successful, results in the formation of a tight seal of the electrode to the cell membrane. If the membrane under the pipette is ruptured at this point, it is possible to perform either current-clamp or voltage-clamp experiments on the whole cell. Alternatively, the pipette can be withdrawn to obtain an isolated patch of membrane.

In the whole-cell configuration, whole-cell currents can be measured in response to applied voltage pulses or, alternatively, changes in membrane potential in response to applied currents can be measured. In this configuration, test substances can be applied to the cell via the microperfusion system described above, so as to assess the effects of various potential transmitters, modulators or other drugs on the membrane potential or on membrane currents. When working with isolated patches of membrane, characterisation of individual ion channels in the patch is possible and the effects of exposure of the patch to exogenous agents can be determined. The voltage or current data obtained from these experiments are digitized and stored to computer for further analysis. Using software specifically developed for analysis of such data, the data can later be analysed.

References

Coons AH, Leduc EH, Connelly JM (1955) Studies on antibody production. I. A method for the histochemical demonstration of specific antibody and its application to a study of the hyperimmune rabbit. J Exp Med 102:49–60

Cowden C, Stretton AOW (1993) AF2, an *Ascaris* neuropeptide: isolation sequence and bioactivity. Peptides 14:423–430

Cowden C, Stretton AOW (1995) Eight novel FMRFamide-like neuropeptides isolated from the nematode *Ascaris suum*. Peptides 16(3):491–500

Cowden C, Stretton AOW, Davis AE (1989) AF1, a sequenced bioactive neuropeptide isolated from the nematode *Ascaris suum*. Neuron 2:1465–1473

Curry WJ, Shaw C, Johnston CF, Thim L, Buchanan KD (1992) Neuropeptide F: primary structure from the turbellarian, *Artioposthia triangulata*. Comp Biochem Physiol 101C:269–274

Geary TG, Klein RD, Vanover L, Bowman JW, Thompson DP (1992a) The nervous system of helminths as targets for drugs. J Parasitol 78:215–230

Geary TG, Price DA, Bowman JW, Winterrowd CA, Mackenzie CD, Garrison RD, Williams JF, Friedman AR (1992b) Two FMRFamide-like peptides from the free-living nematode *Panagrellus redivivus*. Peptides 13:209–214

Gustafsson MKS (1991) Skin the tapeworms before you stain their nervous system! A new method for whole-mount immunocytochemistry. Parasitol Res 77:509–516

Halton DW, Shaw C, Maule AG, Smart D (1994) Regulatory peptides in helminth parasites. Adv Parasitol 34:163–227

Johnson CD, Stretton AOW (1987) GABA-immunoreactivity in inhibitory motor neurons of the nematode *Ascaris*. J Neurosci 7:223–235

Johnston RN, Shaw C, Halton DW, Verhaert P, Baguna J (1995) GYIRFamide: novel FMRFamide-immunoreactive peptide (FaRP) from the triclad turbellarian, *Dugesia tigrina*. Biochem Biophys Res Commun 209(2):689–697

Johnston RN, Shaw C, Halton DW, Verhaert P, Blair KL, Brennan GP, Price DA, Anderson PAV (1996) Isolation, localization and bioactivity of the FMRFamide-related neuropeptides GYIRFamide and YIRFamide from the marine turbellarin, *Bdelloura candida*. J Neurochem (in press)

Keating C, Thorndyke MC, Holden-Dye L, Williams RG, Walker RJ (1994) The FMRFamide-like neuropeptide AF2 is present in the parasitic nematode *Haemonchus contortus*. Parasitology 111:515–521

Magee RM, Shaw C, Fairweather I, Thim L, Johnston CF, Halton DW (1991) Isolation and partial sequencing of a pancreatic polypeptide-like neuropeptide from the liver fluke, *Fasciola hepatica*. Comp Biochem Physiol 100C:507–511

Marks NJ, Shaw C, Maule AG, Davis JP, Halton DW, Verhaert P, Geary TG, Thompson DP (1995) Isolation of AF2 (KHEYLRFamide) from *Caenorhabditis elegans*: evidence for the presence of more than one FMRFamide related peptide-encoding gene. Biochem Biophys Res Commun 217(3):845–851

Maule AG, Shaw C, Halton DW, Thim L, Johnston CF, Fairweather I, Buchanan KD (1991) Neuropeptide F: a novel parasitic flatworm regulatory peptide from *Moniezia expansa* (Cestoda: Cyclophyllidea). Parasitology 102:309–316

Maule AG, Shaw C, Halton DW, Thim L (1993) GNFFRFamide: a novel FMRFamide-immunoreactive peptide isolated from the sheep tapeworm, *Moniezia expansa*. Biochem Biophys Res Commun 193:1054–1060

Maule AG, Shaw C, Halton DW, Curry WJ, Thim L (1994a) RYIRFamide: a turbellarian FMRFamide-related peptide (FaRP). Regul Pept 50:37–43

Maule AG, Shaw C, Bowman JW, Halton DW, Thompson DP, Geary TG, Thim L (1994b) KSAYMRFamide: a novel FMRFamide-related heptapeptide from the free-living nematode, *Panagrellus redivivus*, which is myoactive in the parasitic nematode, *Ascaris suum*. Biochem Biophys Res Commun 200:973–980

Maule AG, Shaw C, Bowman JW, Halton DW, Thompson DP, Geary TG, Thim L (1994c) The FMRFamide-like neuropeptide AF2 (*Ascaris suum*) is present in the free-living nematode, *Panagrellus redivivus* (Nematoda, Rhabditida). Parasitology 109:351–356

Maule AG, Shaw C, Bowman JW, Halton DW, Thompson DP, Thim L, Kubiak TM, Martin RA, Geary TG (1995) Isolation and preliminary biological characterization of KPNFIREamide, a novel FRMFamide-related peptide from the free-living nematode, *Panagrellus redivivus*. Peptides 16(1):87–93

Rosoff ML, Burglin TR, Li C (1992) Alternatively spliced transcripts of the *flp-1* gene encode distinct FMRFamide-like peptides in *Caenorhabditis elegans*. J Neurosci 12:2356–2361

Rosoff ML, Doble KE, Price DA, Li C (1993) The *flp-1* propeptide is processed into multiple highly similar FMRFamide-like peptides in *Caenorhabditis elegans*. Peptides 14:331–338

Smart D, Shaw C, Curry WJ, Johnston CF, Thim L, Halton DW, Buchanan KD (1992) The primary structure of TE-6: a novel neuropeptide from the nematode *Ascaris suum*. Biochem Biophys Res Commun 187:1323–1329

Chapter 7

Electron Microscopy in Parasitology

ANDREW HEMPHILL and SIMON L. CROFT

7.1
Introduction

Electron microscopy has had a major influence on the development and direction of cell biology and parasitology for many years. However, during the past decade, most of the spectacular advances in our understanding of the functions of cells have come through the techniques of molecular biology, immunology and biochemistry. It has become clear that a cell is not just a bag containing a complex mixture of chemicals and molecules, but an intricate machine dependent upon a high level of structural, spatial and temporal organization.

A new range of microscopical techniques have been developed that are able to match these advances in the molecular fields, and reveal more about the relationships between structure and function and the nature of cellular interactions. Microscopy has now become more analytical and less descriptive, quantitative and three-dimensional, more dynamic with greater emphasis on live cells, and an expanded role for fluorescent studies. The new or improved techniques available for these studies range from immunocytochemistry using immunogold labelling techniques or fluorescent probes, cryopreservation and cryosectioning, in situ hybridization, fluorescent markers for subcellular localization, microanalytical methods for elemental distribution, confocal microscopy with laser technology and scanning tunneling microscopy which can image macro-molecules. Many of these new techniques are already making an impact on parasitology. However, not all these techniques involve electron microscopy, and certainly not all are possible in the normal electron microscope laboratory; indeed some require highly specialized skills, and equipment, and should therefore be first learned and acquired in a specialized laboratory.

This chapter will concentrate on the basic electron microscopy techniques; techniques which can be used by most parasitologists in order to obtain information on the ultrastructural aspects of parasites or their interactions with their hosts. In addition information is provided on the most commonly used immunolocalization methods based on colloidal goldlabelling of antibodies. These techniques are mostly applicable in a normal laboratory using widely available standard transmission and scanning electron microscopes.

It is simply impossible to do justice to a discipline on which tomes (whole volume series) are written on the plethora of specialist techniques in electron microscopy. In many cases, where specialized techniques are involved, we will refer to a number of excellent books, reviews, or original scientific articles for more detailed information.

7.2
Preparation for Transmission Electron Microscopy (TEM)

In TEM the electrons are emitted at high voltage(60–100 kV) from an electron source, and pass through the specimen under investigation. Heavy metal stains are introduced into the specimen prior to examination (see Sect. 7.3) and deflect electrons to give the contrast necessary for the visualization of fine structures. As the electron beam passes through and interacts with the specimen, only thin sections can be investigated by TEM.

There exists a wide range of techniques, which have been applied in order to visualize biological material by TEM. Whole-mount preparations have been used in many studies on whole cells, membrane- and organelle fractions, liposomes, viruses, macromolecules and even single filaments and protein molecules. For detailed information on the whole-mount preparation technique see Harris and Horne (1990). Other related, but clearly more sophisticated, methods include the shadowing techniques (Souto-Padron et al. 1984; Slayter 1990; Hemphill et al. 1991a, b), the freeze-fracture techniques (Pinto da Silva 1984; Tetley and Vickerman 1985; Vickerman et al. 1988; Fujikawa 1990) and the wet cleaving technique (Sherwin and Gull 1989; De Priester 1991). They all have been applied to parasites, also in conjunction with immunogold labelling (for further references see Sect. 7.5.3).

In this section we will focus on the most popular EM technique: thin-section TEM will be discussed. In brief, tissues or cells are fixed, dehydrated and embedded in resin which is polymerized to provide a stable matrix. Thin sections (ideally 60–80 nm) are then cut with an ultramicrotome, mounted onto a grid and treated with a heavy metal stain. Examples are shown in Fig. 1a and 2a.

7.2.1
Fixation: General Background

The aim of fixation is to prepare cells or tissues to withstand the rigours of the preparative techniques as well as irradiation by the electron beam, and to preserve a "near-life image" of the specimen.

Paraformaldehyde (pFA) and/or glutaraldehyde (GA) are the two most commonly used fixatives. They act by chemical cross-linking of proteins. Nucleic acids are probably retained because they are coupled to proteins such as histones and ribosomal components. Lipids, phospholipids and carbohydrates remain virtually unfixed by aldehydes. Therefore, aldehyde fixation is normally applied in conjunction with osmium tetroxide (OsO_4) as a secondary fixative.

No one fixation protocol is ideal for all problems. Therefore, procedures have described the use of the two aldehydes alone or together containing various additives. For example, the addition of $CaCl_2$ has been shown to enhance the preservation of microfilaments (Hopwood and Milne 1990), and NaN_3 is used to enhance mitochondrial preservation (Minassian and Huang 1979). There are an enormous number of different fixation protocols for various problems, and often it is a matter of trial and error to find out which method is most suitable for a particular problem. Excellent reviews on fixation for TEM are provided by Griffiths (1993) and Hopwood and Milne (1990).

It is also important to choose the right buffer. Some will react with the fixative, thereby reducing the effective concentration of both reagents (e.g. amine-containing buffers will react with aldehydes). Other buffers will extract material from the tissues (e.g. phosphate buffer), which in some cases might be an advantage rather than a disadvantage, depending on the structures to be visualized. The most commonly used buffers are sodium cacodylate, phosphate and imidazole. It is not advisable to alter the buffer when changing from primary to secondary fixation, since this might cause precipitation. The temperature is also an important factor in fixation, since it will affect the rate of penetration of a fixative, its rate of reaction and the rate at which substances are extracted.

For fixation of tissue, the most commonly used methods are immersion and perfusion, and to a lesser degree, injection of the fixative. During immersion fixation, blocks of tissue (1 mm^3) are cut using a razor blade, and the tissue pieces are rapidly placed into the fixative. This method is generally used for surgical specimens. However, there is a problem of diffusion of the fixative into the tissue, and gradients are set up. This can be overcome using perfusion fixation: with animal tissue (e.g. mice and rats), a cannula is placed in the left ventricle and a cut made in the inferior cava. The vasculature is washed out with heparinized saline to remove the blood, and the fixative solution is then perfused. This gives much more even fixation. For special purposes, it is also possible to stabilize/fix tissue by non-chemical means, for example microwaves (Kok et al. 1987; Login and Dvorak 1988).

Monolayers of cells grown on plastic or glass containers are generally fixed by replacing the medium with fixative. Shorter fixation times (up to 30 min) are usually sufficient in this case. Single cells or organelles are best centrifuged into a pellet, the supernatant is removed, and the fixative is added by slowly running it down the side of the centrifuge tube in order to keep the pellet intact. For large pellets it is advisable to resuspend the cells carefully by gentle shaking.

Fig. 1a–d. *Cryptosporidium parvum* in mouse ileum, immunogold labeled for specific antigens, *arrowed, Bar* 500 nm. **a** Standard glutaraldehyde/osmium tetroxide fixation, TAAB resin embedded, uranyl acetate and lead citrate-stained. **b** Specimen fixed in 2% paraformaldehyde/0.1% glutaraldehyde, soaked in 2.3 M sucrose, rapidly quenched in liquid N_2 and cryosectioned. Labelled with 5-nm gold particles, silver enhanced.

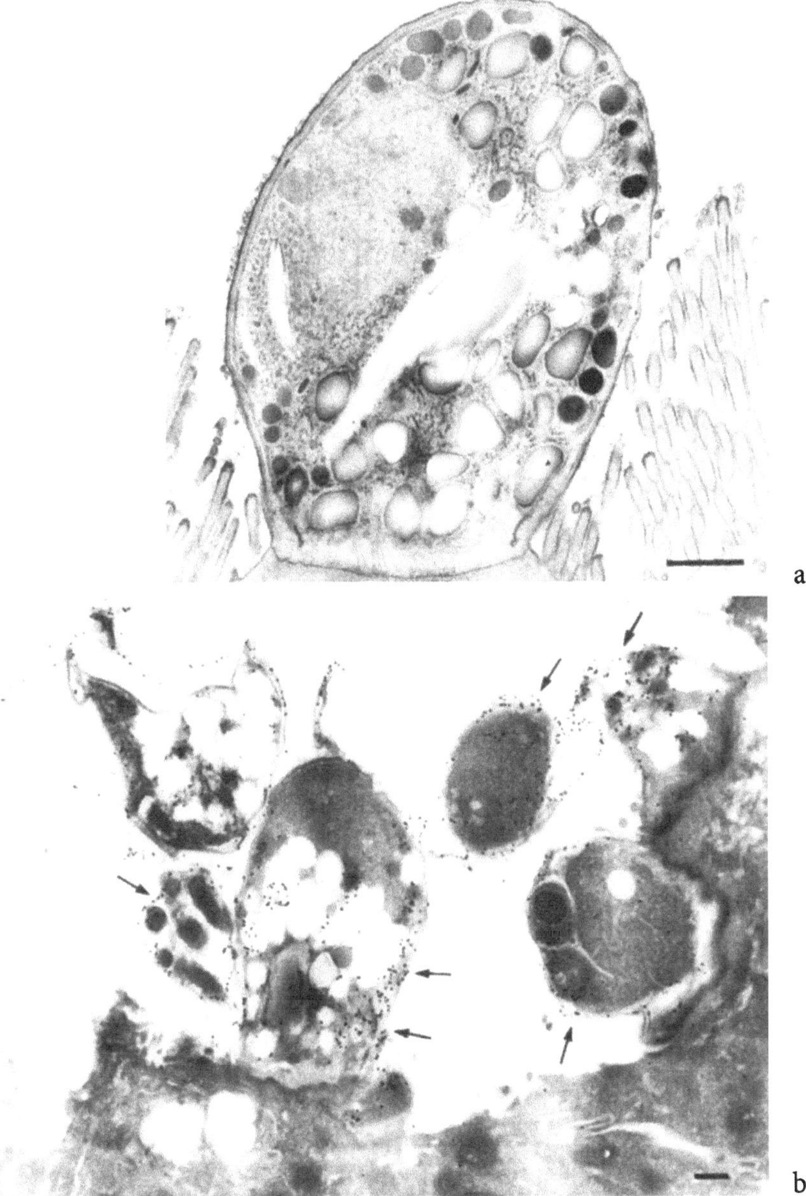

a

b

Fig. 1a,b

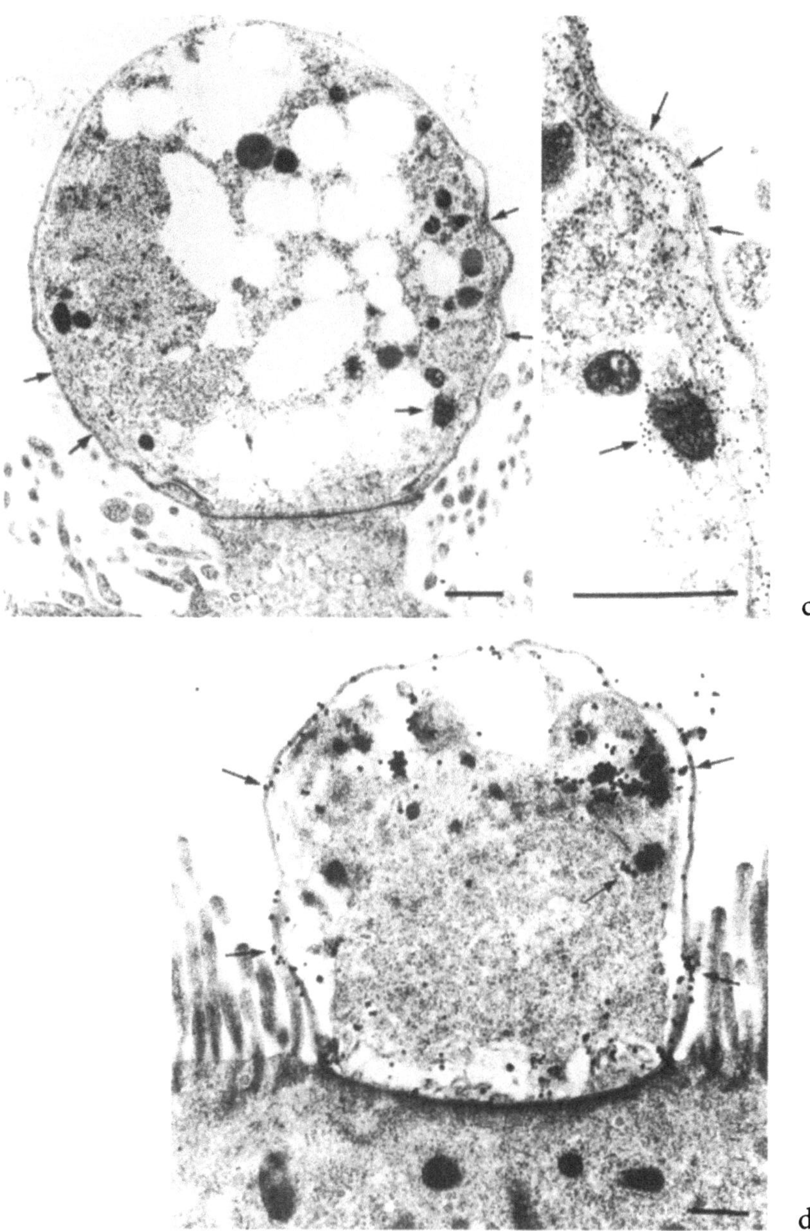

Fig. 1. c Fixed in 2% paraformaldehyde/0.1% glutaraldehyde, embedded in LR Gold resin at −20 °C Labelled with 10-nm gold particles. d Prepared and labelled as c, but then silver enhanced for 6–8 min. (Micrographs courtesy of Maria V. McCrossan and Dr. Vincent McDonald, London School of Hygiene and Tropical Medicine)

7.2.2
Standard Fixation Methods

The following buffers and fixatives are commonly used in
TEM

- 0.2 M phosphate buffer pH 7.3   **Solutions**

  0.2 M Na_2HPO_4: 28.39 g in 1000 ml H_2O dist. (Sol A)
  0.2 M NaH_2PO_4 $2H_2O$: 31.21 g in 1000 ml H_2O dist. (Sol B)
  Mix 39 ml of Sol A with 11 ml of Sol B to achieve pH 7.3

- 0.2 M sodium cacodylate buffer pH 7.3

  4.28 g sodium cacodylate in 80 ml H_2O dist.
  Adjust to pH 7.3 with 0.1 N HCl and fill up to 100 ml with
  H_2O dist.

Note. Sodium cacodylate is toxic and should be handled with
care.

- 3% GA in 0.1 M phosphate- or cacodylate buffer

0.2 M phosphate- or cacodylate buffer pH 7.3	25 ml
25% GA (glutaraldehyde)	6 ml
H_2O dist.	19 ml

Note. This fixation solution is made up fresh from a commercially available 25% stock solution. The stock should be
stored at 4 °C and have a pH of about 5. It should be discarded if the pH drops below 3.5!

- 10% pFA in H_2O dist.

  10 g pFA in 100 ml H_2O dist.
  Heat up to 65 °C in a fume cupboard, and add 6–10 drops 1
  N NaOH. Wait until the solution is clear.
  Make small aliquots and freeze them at –20 °C.

Note. Discard the aliquots after thawing!

- 1% buffered OsO_4

  Prepare 2% OsO_4 solution by breaking the ampoules
  containing 1 g of OsO_4 and dropping the content into a
  clean glass bottle.

Add 50 ml H_2O dist.

Allow 24 h to dissolve, store in a fume cupboard.

Mix 5 ml of the same 0.2 M buffer pH 7.3 which was used for primary fixation and 5 ml of the OsO_4 solution.

Note. 2–4% aequous OsO_4 solution in glass vials can also be bought commercially from various companies. Always handle in a fume cupboard, and wear gloves, since this substance is very toxic.

Standard Fixation Protocols

The following protocols represent only the most commonly used fixation procedures:

Glutaraldehyde fixation

1. Fix blocks of tissue not exceeding 1 mm^3 in 3%GA in 0.1 M phosphate or cacodylate buffer pH 7.3 for at least 2 h at 4 °C. Cell suspensions or monolayers require only 30-min fixation.

2. Wash in phosphate- or cacodylate buffer overnight at 4 °C.

3. Postfix in 1% OsO_4 in same buffer as primary fixative for 2 h at 4 °C.

Note. This fixation procedure is known to provide good preservation for routine TEM.

PFA/GA – fixation

1. 10% pFA in H_2O dist.	20 ml
2. 0.2 M phosphate- or cacodylate buffer	50 ml
3. 25% GA stock	10 ml
4. H_2O dist.	20 ml

5. Prepare small aliquots. Freeze and thaw only once.

6. Fixation is carried out for 4 h at 4 °C.

7. Postfix tissue in OsO_4 as above

Note. This fixative contains 2% pFA and 2.5% GA. pFA penetrates tissues more rapidly than GA and is thought to initially stabilize structures, which are then fixed more permanently with GA (Karnovsky 1965).

1. 25% GA   6 ml

2. 0.2 M sodium cacodylate   25 ml

3. Tannic acid   0.5 g

4. Mix the components and adjust the pH to 7.3 with 2 N HCl

5. Fill up to 50 ml with H_2O dist.

6. Fix for at least 6 h before postfixation with 1% OsO_4.

Tannic acid-GA fixation

Note. Tannic acid reacts with both proteins and lipids, holding them firmly in the tissue. Enhanced preservation of carbohydrates is achieved by this fixation procedure (Simionescu and Simionescu 1976). It is also used to demonstrate damaged cells, which appear particularly electron-dense.

1. Primary fixation in 3% GA in 0.1 M sodium cacodylate pH 7.3.

2. Secondary fixation in 1% OsO_4, 0.05 M $KMnO_4$ in 0.1 M sodium cacodylate, 2–24 h, 4 °C.

OsO_4-$KMnO_4$ fixation

Note. Carbohydrates are difficult to retain in electron microscopy, since they do not react with aldehydes. This fixation procedure increases contrast of the outer membrane surface. It is also recommended for the preservation of phospholipids (Elbers 1966). Other procedures for increased surface glycocalix and/or lipid preservation include Cuprolinic blue, Ruthenium red, and imidazole-buffered OsO_4 fixation (see Sect. 7.3).

Prior to embedding, it is essential to dehydrate the sample carefully, since none of the commercial embedding media are soluble in water. Ethanol, methanol and acetone are commonly used as dehydrating agents, with propylene oxide (1,2-epoxy propane) being used as an intermediate solvent with tissues which are difficult to infiltrate, e.g. insect or bone tissue.

Dehydration

If ethanol is used for dehydration, specimens should be extensively washed first as OsO_4 precipitation can occur. Time and temperature are important factors. If dehydration is too slow, there is a risk of chemical leaching, if dehydration is carried out too rapidly, osmotic shock can result, and precipitation of the buffer might occur. It is therefore sensible to dehydrate samples using a gradient of increasing concentration of dehydration reagents. Generally, it is advisable to perform several short changes (5–10 min each) rather than one long incubation. The samples should be agitated using a rotor. If necessary, dehydration can be carried out at 4 °C in order to reduce the rate of extraction. Block staining in 1% (w/v) aqueous uranyl acetate for 1 h at 4 °C prior to dehydration, or at the 30% methanol stage, will reduce the rate of extraction (for protocol look up Sect. 7.3).

A standard dehydration time schedule for tissue blocks is given below:

1.	H$_2$O dist.	2 × 30 min
2.	30% methanol	2 × 10 min
3.	60% methanol	10 min
4.	80% methanol	10 min
5.	90% methanol	10 min
6.	100% methanol	2 × 10 min
7.	Propylene oxide	2 × 10 min

Tissue blocks can be processed by simply changing the solvents in the vials. Small specimens or cell pellets can be processed in microfuge tubes or embedded in agarose prior to dehydration:

1. Prepare a 2% agarose solution in H$_2$O dist.

2. Heat until the agarose has just melted.

3. Form the specimen into a pellet, e.g. by centrifugation.

4. Remove as much of the supernatant as possible.

5. Add a small amount of agar to the pellet, mix very gently and centrifuge immediately to reform the pellet.

6. Allow to cool and harden.

7. Remove the hardened agar from the tube and cut out the area of agar containing the pellet with a razor blade.

8. Process as for a tissue sample.

Generally, epoxy resins are mostly used for embedding conventional TEM specimens. There are principally three main types of epoxy-embedding media: araldite, epon and spurr. All resins polymerize uniformly into a three-dimensional cross-linked structure with little shrinkage. Araldite has a high viscosity which prolongs the infiltration time, while epon and especially spurr resin exhibit lower viscosities. Because of the higher viscosities of epon and araldite, it may be necessary to leave these resins to infiltrate from one night (soft tissue) to several days (e.g. skin), depending on the rigidity of the specimen. However, a rapid processing procedure has been developed, achieving complete processing of tissue within 3 h (Hayat 1981).

Resins can be made up conveniently according to the instructions provided by the manufacturers. They are usually stored at –20 °C for several months. The final hardness of the resin can be adjusted to suit the consistency of the specimen by varying the ratio of the hardener to resin (see Smith and Croft 1990). For routine purposes, it is advisable to use a hard resin mix, which would be most stable in the electron beam. Polymerization of epoxy resins is usually carried out at 60 °C for 16–24 h. However, the standard resins will also polymerize at 50 °C (48 h), 70 °C (8 h) or 100 °C (1–2 h).

Other resin types include melamine, polyester, urea-aldehyde, gelatin and glutaraldehyde-carbohydrazide (Smith and Croft 1990). They have generally not found widespread acceptance, since they mostly exhibit poor characteristics such as instability in the electron beam, shrinkage of tissue, and/or extreme hydrophilicity which makes them difficult to section. There are different embedding procedures depending on the specimen:

Tissue blocks and cell pellets can be embedded either in capsules (gelatin or Beem polyethylene capsules) with pointed or flat ends. Flat moulds are also commercially available or can be made from polyethylene or aluminium foil. It is important that moulds are predried in an oven, since traces of water will interfere with the polymerization process.

Monolayers can be processed in several ways. If the orientation of the cells is not important, layers can be scraped off their substrate after fixation, prior to embedding, and handled as a pellet. Cells grown on glass slides can be embedded in situ by covering them with resin, and after infiltration, positioning inverted capsules full of resin over selected areas. After polymerization, excess resin is trimmed away, and the glass coverslip is released from the resin by carefully dipping the back of the glass into liquid nitrogen. The monolayer samples can be reembedded in flat moulds for sectioning perpendicularly to the plane of the culture. For a recent applications see Hemphill et al. (1994) and Hemphill and Ross (1995). Monolayers can also be grown on polyester or plastic substrates such as Mellinex or Thermonox cover slips which can easily be peeled away from the polymerized resin, and the block is reembedded as mentioned above. As an alternative, it is possible to section through these cover slips (see, for example Vickerman et al. 1988), or cells can be grown on other sectionable substrates such as Millipore filters or cling film. It is also possible to grow cells on resin sheets, although one should be aware of possible cytotoxic effects of resin components. In these cases, sheets should be incubated in full medium for at least 24 h prior to adding the cells.

Trimming and cutting of resin blocks Uniform embedding of the specimen is a prerequisite for good sectioning. The block should be trimmed with a sharp razor blade to give a trapezoid shaped surface. It is then positioned into the ultramicrotome holder so that the shorter of the parallel edges of the block face trapezium will contact the knife first. This reduces the shock and hence the compression during cutting of the section.

Sections are cut using glass, diamond or saphire knives. It is important to adjust the knife to its correct angle. Sections are generally cut onto a waterbath positioned on the back of the knife, containing distilled water with up to 5% acetone. After cutting, sections will have suffered 5–20% compression. They are flattened before being picked up from the water bath surface by carefully waving an artist's brush or strip of filter paper dipped in chloroform over the sections. The chloroform must not touch the water surface.

Sections should be picked up from below using a suitable grid held carefully by a pair of watchmaker's forceps, and dried immediately with a piece of filter paper. However, if coated grids are used, it might be an advantage to pick up the sections from above.

It is of prime importance to maintain a high level of cleanliness within the equipment and all the solutions used in order to avoid contamination of the sections with unwanted particles and dirt.

7.3
Cytochemistry and Staining Protocols in TEM

7.3.1
Introduction

Staining of biological materials gives both

- contrast and

- specificity and localization of intrinsic components.

Whereas in light microscopy, contrast is produced by the differential distribution of coloured stains, it depends in electron microscopy upon the deposition of electron-dense metal stains. In this section we will introduce

- standard stains used for contrast of biological material in TEM,

- some specific stains for chemical composition of cells, and

- stains for some enzymes in host-parasite interactions.

The methods for staining of antigen are considered in Sect. 7.5 on Immunolabelling. Further details on staining can be found in Lewis and Knight (1992).

Two basic points have to be considered before embarking on particular staining procedures. When examining multicellular parasites, or parasites in tissue or vectors, the orientation of the specimen and the identification of the infected area are important. Thick (1-μm) resin sections stained rapidly with methylene blue or toluidine blue will provide this information:

1. Transfer 1-μm sections to a clean glass slide

2. Place slide on a hot plate at 60 °C

3. Immediately cover section with toluidine blue stain (1% toluidine blue in 1% borax) for 1 min

4. Drain off stain, wash with H_2O dist. and examine on the light microscope

Toluidine blue staining

Most staining procedures are carried out after sectioning. The most convenient method is to mount sections onto copper or nickel grids (unless other grid types are specified). Staining is performed in a petri dish covered with dental wax containing moist filter paper. Place discrete drops of stain on the surface and float the grids, section side down, on the surface of the drop while raising the lid as little as possible to exclude dust. After staining, hold the grid with tweezers and gently rinse with H_2O dist. from a wash bottle.

7.3.2
Standard Stains

Fixatives and metal stains used in TEM preparation have chemical affinities. The primary aldehyde fixatives have no staining properties, but OsO_4 and $KMnO_4$ used as secondary fixatives have high affinities for phospholipids and will give good organelle membrane structure in protozoa.

Uranyl acetate (UA) staining

1. Sectionstaining. Prepare saturated solution in 70% or 50% ethanol (keep at 4 °C in the dark) and centrifuge or filter before use. Ethanolic solutions give better staining than aqueous or methanolic solutions. Stain for 15 min at room temperature and rinse with H_2O dist.

2. Blockstaining. During the dehydration stage of tissue processing, include 2% UA in the 30% methanol/ethanol stage for 30 min.

Lead citrate (LC) staining

To prepare LC, mix 1.33 g lead nitrate and 1.76 sodium citrate with 30 ml boiled (to remove CO_2) then cooled distilled water in a 50-ml volumetric flask. Shake for 1 min in every 5 over a 30-min period. Add 8 ml 1 N NaOH, make up to 50 ml with H_2O dist. mix and store this in the fridge (pH 12). Centrifuge prior to use.

For staining, place NaOH pellets in the staining dish to remove CO_2. Stain the grids for 15–30 min (epoxy resins), or 3–10 min (araldite or acrylic resins) at room temp. After staining, wash grids in 0.2 N NaOH followed by extensive rinsing in H_2O dist.

7.3.3
Special Cytochemical Stains

Proteins

There are no good specific stains for proteins although ethanolic phosphotungstic acid has been used to localize basic proteins. The best approach is to use antibodies raised to specific proteins and the immunocytochemical techniques described in Sect. 7.5.

Nucleic Acids

General stains include Feulgen-silver methamine for DNA and uranyl acetate with EDTA differentiation for RNA. However, the development of in situ hybridization techniques for electron microscopy will enable localization of specific RNA and DNA species (Wilkinson 1992).

Carbohydrates

One should distinguish between cell surface- and general carbohydrate staining.

Cell surface carbohydrates can be visualized by the following **Reagents** reagents:

1. Ruthenium red (1 mg/ml) at pH 7–7.4. Ruthenium red is selective for acidic mucosubstances of the glycocalyx. Standard fixation procedures are followed by ruthenium red included in the primary aldehyde and secondary osmium fixatives.

2. Incorporation of 1% (w/v) tannic acid into glutaraldehyde (less than 6 h of use) can give improved definition of the glycocalyx.

3. Specific surface sugar residues can be labelled by lectin-colloidal gold preparations; many are now commercially available. Cells are incubated with the lectin-gold prior to fixation.

A protocol for general carbohydrate staining [Periodic acid (PA)-silver technique] is given below. In this technique the PA oxidizes aldehyde exposing groups which interact via intermediates with silver ions leading to metal deposit. The PA-thiosemicarbazide (TSC)-silver protein method produces a fine distribution of carbohydrate.

Carbohydrate staining

1. Follow the standard fixation protocols although omission of osmium fixation will produce clearer results.

2. Collect sections on gold grids (Cu and Ni grids will interact with the PA)

3. Incubate grids with PA at room temperature for 20–25 min. Wash two or three times with H_2O dist.

4. Incubate grids with 0.2% TSC in 10% acetic acid for 12–72 h at room temperature.

5. Wash in 15% acetic acid 2×1 min

6. 5% acetic acid 1×20 min

7. 2.5% acetic acid 1×20 min

8. 1% acetic acid 1×20 min

9. H_2O dist. 2×5 min

10. Stain with 1% aqueous silver proteinate in dark at room temp for 30 min (maximum). Prepare 30 min before use in darkroom (red safe light Kodak Wratten No. 1). Filter before use.

11. Wash with H_2O dist. 2×5 min.

Lipids

Enhanced staining of lipid structures is best achieved through modification of fixation procedures, either through secondary fixation in OsO_4 in 0.2 M imidazole buffer (pH 7.5), or the $OsO_4 - KMnO_4$ method (see Sect. 7.2.2, standard fixation protocols).

7.3.4
Enzyme Stains

Since enzyme histochemistry was first introduced by Gomori in 1939, techniques for the localization of a variety of enzymes have been described. Recent volumes by Van Noorden and Frederiks (1992) and Lewis and Knight (1992) list techniques for many enzymes, but until recently, few had relevance to pathophysiological processes. Methods to localize enzymes must be precise (reaction product precipitated at the correct subcellular site), specific (reaction product only generated by the enzyme under study), reproducible, valid (formation of the product proportional to enzyme activity) and able to generate electron-dense reaction product.

In studies on parasites, a series of metal capture methods have been used to localize phosphatases, sulphatases and nucleotidases in studies on lysosomal and phagosomal vacuoles (Antoine et al. 1987). Recent studies on cell invasion have been concerned with the role of proteases, but these enzymes have been localized by the immunogold technique (Sect. 7.5.3). The development of cerium methods to detect oxidases (sites of hydrogen peroxide production) and phosphatases has added a sensitive technique for research on host-parasite interactions and localisation of peroxisomes and lysosomes (Van Noorden and Frederiks 1993).

7.4
Preparation for Scanning Electron Microscopy (SEM)

To visualize an object by SEM, an electron beam is rastered across the surface of the specimen. In contrast to TEM, electrons do not penetrate the specimen, but are reflected as either primary backscattered electrons or secondary electrons. Primary backscattered electrons originate from the electron source itself, and are directly reflected after hitting the specimen surface. They are detected using a special backscattering detector. Secondary electrons originate from the specimen itself and are emitted after being pushed out from the electron shell by the electron beam. The signal produced is collected with an appropriate detector, amplified and displayed on a screen.

In principle, preparation of biological specimens for SEM involves fixation and dehydration, similar to that for TEM, followed by subsequent coating of the sample.

Fixation and Dehydration

Standard fixatives used for TEM can also be used for SEM.

Dehydration is usually performed by sequential washing in increasing concentrations of organic solvents such as acetone or methanol. The specimen is then dried either by critical point drying, or by other procedures. Evaporation by air is normally not recommended, except for tissue with a hard outer surface (Kennedy et al. 1989). Critical point drying is the most commonly used method, but requires a special critical point drying apparatus. As an alternative, we recommend sublimation using Peldri II (Ted Pella Inc. USA), a commercially available fluorocarbon, which is solid at room temperature, but becomes liquid at temperatures higher than 24–25 °C. It is non-flammable, but can cause skin irritation and its fumes are toxic. It should be handled in a fume hood. However, the procedure is very easy to perform and this advantage outweighs the disadvantages For a recent application of Peldri II on parasitic protozoa see Hemphill and Ross (1995).

Special cryofixation techniques have been developed, such as the propane-jet technique where samples are fixed without any chemical pretreatment (Mueller and Moor 1984) The sample is rapidly frozen by immersing it in liquid nitrogen, propane or freon 22 and subsequently dried at –80 °C for 30–60 min at 10^{-5} Pa. However, this procedure requires special equipment such as a vacuum freeze dryer (e.g. a Balzers BAF 300, Balzers, Liechtenstein).

Coating of Specimens

For routine SEM, coating the specimen is a standard procedure. After drying, the sample must be mounted on the type of holder appropriate for the SEM available. It is usually attached to the holder by double-sided cellotape, or silver or carbon paint. Although the cellotape is easier to use, the paint will provide a much better contact. The electron beam striking a specimen surface requires a conducting path to earth in order to remove any electron charge that results.

Specimen damage during coating is one of the most common causes of SEM artefacts. Some coating units will heat up the specimen extensively, causing irreversible damage. Metals used for coating of biological specimen are gold, platinum or tungsten. Evaporation is most commonly achieved using the sputter-coating technique, where the metal target is bombarded with heavy gas atoms (usually argon). Metal atoms ejected from the target are then deposited on the specimen surface. Carbon or carbon/platinum is also often used for coating specimens. Carbon fibres heated to 2000 °C in a vacuum will emit atoms, and carbon films are formed on adjacent areas. Although carbon is not a good emitter of electrons, it does provide a continuous conducting path over the specimen, no matter how thin the layer (Chapman 1986).

More recent developments have taken the direction of viewing specimens without coating, provided they are prepared correctly and the electron microscope is operated at low accelerating voltage. However, the required magnification and resolution may not be achieved without coating the specimen.

7.5
Immunogold-Labelling Techniques

7.5.1
Basic Information

Historical Background

Before the development of colloidal gold as a marker for immunoelectron microscopy, a variety of electron-dense tags such as iron dextran and hamocyanin have been coupled to antibodies to determine the distribution of the corresponding antigens within cells or tissues (Horrisberger 1984). Ferritin-antibody conjugates were first introduced in 1959 by Singer (Singer 1959). By 1971 a major breakthrough was achieved with the development of colloidal gold as a particulate marker for the localization of microorganisms, cell surface associated and intracellular antigens (Faulk and Taylor 1971). Today, colloidal gold conjugates have become the most widely used tools for immunolocalization in electron microscopy.

Properties of Gold Conjugates

Colloidal gold has distinct advantages which makes it uniquely suited for EM immunolocalization. It is particularly electron-dense, and can easily be prepared in various, relatively monodisperse, sizes (1–100 nm). The ability of gold probes to produce strong secondary as well as backscattered signals makes them suitable not only for TEM, but also for immunolocalization in SEM (Hyatt 1989, 1990). For applications of immunogold-SEM in parasitology refer to Tetley et al. (1987) and Erlandsen et al. (1990). Silver enhancement procedures allow the use of smaller-sized gold conjugates (1–10 nm) for lower resolution SEM (see also Lang et al. 1991; Shakibaei and Frevert 1992). Also double- or even triple-labelling experiments can be performed on the same specimen.

Colloidal gold can be easily coupled to a variety of ligands such as lectins, many enzymes, Protein-A, streptavidin and immunoglobulins. For the preparation of lectin- and glyco-protein-gold probes see Egea (1993). Bendayan has pioneered the use of enzyme-gold complexes and procedures for their preparation and use are given in Bendayan (1985). An excellent review on preparation of Protein A-, streptavidin- and immunoglobulin-gold complexes is provided by Lucocq (1993). In this chapter we will focus on the application of antibodies conjugated to colloidal gold.

The size of the gold particle conjugated to an antibody molecule will influence the activity of this molecule. Smaller particles coupled to an antibody impose less steric hindrance, which will increase the immunolabelling efficiency of the conjugate (Hyatt 1990). Also, the stoichiometry of colloidal gold particles to IgG molecules varies from 0.2 to 10, depending on the size of the particle (Lucocq and Baschong 1986). Colloidal gold is non-covalently linked to its carrier proteins, and repeated freezing and thawing will dissociate the conjugates. Therefore, freshly prepared or purchased conjugates should be kept in 50% glycerol or sucrose, aliquoted, rapidly frozen and stored in liquid nitrogen or at –70 °C. Once thawed, conjugates are stored at 4 °C where they will retain their activity for several months.

A new generation of gold antibody-probes, nanogold particles, have recently come on the market (Hainfeld and Furuya 1992). The 1.4-nm nanogold probes are not colloidal gold particles, but solid gold compounds covalently linked to immunoglobulins or Fab fragments at a ratio of 1 gold particle to 1 antibody. They are used in immunoelectron microscopy in conjunction with the silver enhancement procedure.

Fixation and Processing of Samples

The classical dilemma in immunoelectron microscopy is the need to preserve the fine structure of cellular components by fixation, dehydration and resin embedding, and at the same time to retain epitopes still accessible for immunoglobulins. Initial fixation is a very critical step, and generally, high GA concentrations (>1%) and OsO_4 fixation should be avoided. However, there exists no protocol which is suitable for all applications, or all parasites, since all parameters are highly dependent on the nature of both epitopes and antibodies. It is therefore important to test as many parameters as possible by light microscopy using either fluorescent probes or gold-conjugated probes followed by silver enhancement (Barta and Corbin 1990) prior to processing cells or tissue for immunoelectron microscopy.

Immunoreagents

We strongly recommend that affinity-purified antibodies are used for immunolabelling, since contaminating components might produce unnecessary background staining. In addition, it is important that the optimal dilution of antibodies is determined. A general rule is to select the largest dilution which maintains as low as possible background level. Ideal antibody concentrations will probably vary between 2–20 μg/ml. For 15 nm Protein A gold an OD_{525} of 0.44, and for 6–10 nm Protein A gold an OD_{525} of 0.06 is recommended (Hyatt 1990).

There are three types of immunolabelling techniques which are commonly used:

a) the direct method which involves antibodies directly conjugated to colloidal gold;

b) the indirect method, carried out by detection of bound primary antibody with a gold-conjugated secondary antibody or Fab fragment, which reacts specifically with the first one; and

c) the detection of the first antibody with Protein A- or Protein G-gold conjugates.

The direct method (a) has the advantage that only one antibody incubation step is needed, resulting in reduced background staining. However, relatively large amounts of antibodies are necessary for efficient conjugation to gold. Indirect labelling (b) is most commonly used (see Fig. 1b–c, 2b). The fact that several gold-conjugated antibodies can bind to one first antibody will enhance the staining compared to the direct labelling technique. Protein A- and G-gold conjugates (c) are also widely used, and it has been postulated that they cause less background staining than secondary antibody conjugates (Griffiths 1993). However, only one Protein A or G-molecule binds to one primary antibody molecule, which will result in a lower staining intensity as compared to technique (b). Care must be taken to ensure that Protein A or -G-complexed gold is compatible with the species of which the first antibody has been derived.

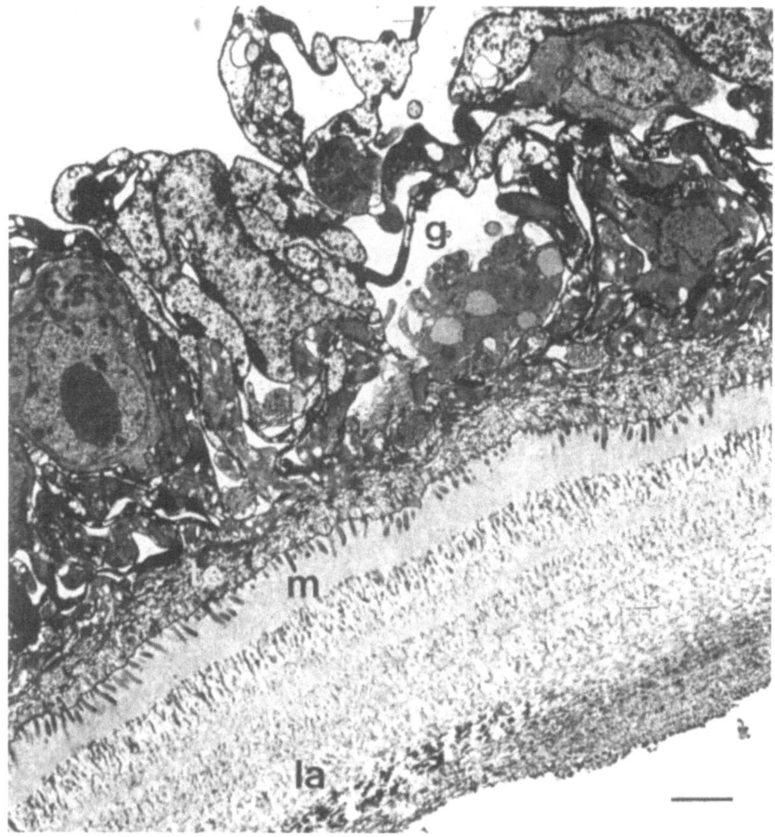

Fig. 2a

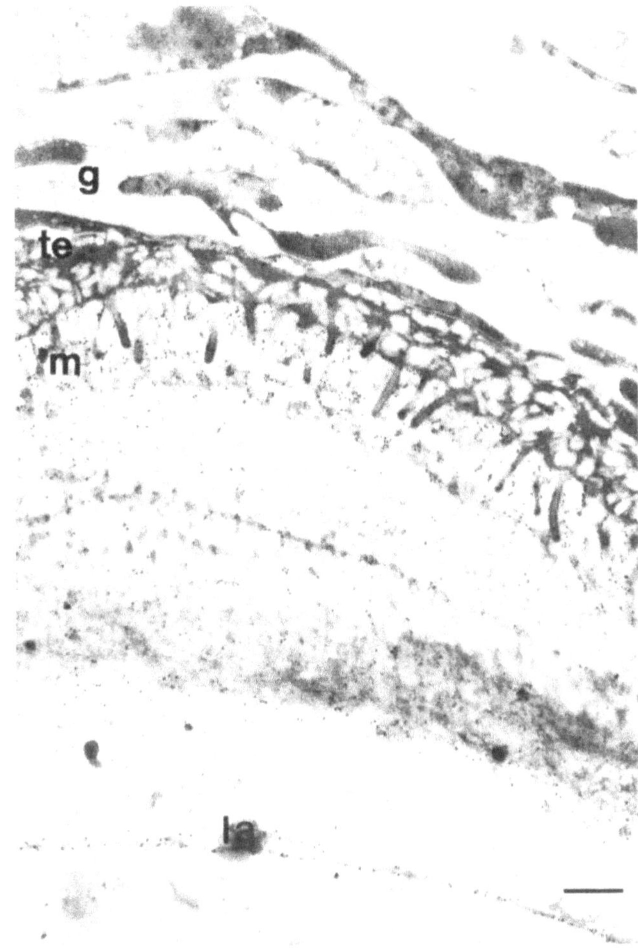

Fig. 2a,b. In vitro-generated *Echinococcus multilocularis* metacestode (larval stage) tissue. *g* Germinal layer; *te* tegument; *m* microtriches; *la* laminated layer. **a** Fixed in 3% glutaraldehyde containing 0.5% tannic acid, followed by osmium tetroxide fixation, epon 812 resin. Uranyle acetate and lead citrate-stained. *Bar* 1.8 μm. **b** Fixed in 3% paraformaldehyde/0.1% glutaraldehyde, embedded in LR-White and immunogold-labelled with a monoclonal antibody directed against the laminated layer-specific Em2 antigen and goat anti-mouse conugated to 10-nm gold particles. Stained with uranyle acetate and lead citrate. *Bar* 1 μm. (Hemphill and Gottstein 1995)

Non-Specific Interactions

In every immunolabelling experiment, non-specific interactions, caused by aldehyde fixation, protein denaturation and electrostatic interactions due to the incomplete protein coating of colloidal gold, will inadvertently result in background staining. In order to prevent this, a number of treatments have been developed.

Free aldehyde groups, especially present after fixation in GA, must be quenched. After fixation, specimens should be incubated in PBS containing either 50 mM NH_4Cl, 50 mM glycine, lysine or other amino acids for 1 h. Since pFA fixation is reversible, especially at concentrations below 4% (Roos and Morgan 1990), it is advisable to keep an amino acid (e.g. 50 mM lysine) present at all steps during immunolabelling in order to block free aldehyde groups.

Hydrophobic interactions can be inhibited by including a detergent such as 0.1% Triton X-100 or Tween 20. Care must be taken with the use of detergents during preembedding labelling and for immunolabelling of cryosections, since they will have a disruptive effect on membranes. Other components which are commonly used are ovalbumin (1%) bovine serum albumin (BSA, 1%), gelatin (1–5%), and normal serum from the species of which the second antibody has been derived, e.g. normal goat serum (1–5%).

Electrostatic interactions are overcome by increasing the ionic strength in the buffer by adding up to 500 mM NaCl.

Silver Enhancement

As mentioned above, it is an advantage to use smaller gold conjugates which will label antigens much more efficiently. However, gold particles of a diameter below 5 nm are often difficult to detect, especially on an intensely stained specimen. By using silver enhancement technique, where electron-dense silver grains are deposited around small gold particles, this problem can be easily overcome. Silver enhancement kits are commercially available, and it is possible to use this procedure in conjunction with very small gold conjugates, e.g. 1 nm (Burry et al. 1992). For further information on silver enhancement see Lucocq (1993) and Barta and Corbin (1990). A recent application of the silver enhancement procedure is described in Baker et al. 1994. Examples are shown in Fig. 1b,d.

7.5.2
Approaches for Immunogold-Labelling

Three approaches used for immunostaining in TEM will be described in this chapter:

- On-grid labelling of whole mount preparations, where thin objects such as cells, organelles, bacteria, viruses or single molecules are adsorbed onto a grid, processed and treated with an antibody (see Sect. 7.5.2.1)

- Thin-section preembedding labelling, where cells or tissue are incubated with the immunoreagents prior to conventional resin embedding (see Sect. 7.5.2.2)

- Thin-section postembedding labelling of either resin embedded quick-frozen cells/tissues or cryosections (see Sect. 7.5.2.3)

All three methodologies have been succesfully applied to parasites (see Aikawa and Atkinson 1990; Bannister and Kent 1993). Which procedure to use for a particular problem strongly depends on the biological material to be investigated, the nature of the antigen, its distribution, and on the questions which are being asked in an experiment.

Resin-based postembedding labelling is clearly the most popular approach. It is used when thin sections of intact cells and tissues are needed in order to examine the intracellular distribution of antigens.

Ultrathin-cryosection cytochemistry (Tokuyuasu 1973, 1978, 1984) involves a partial and temporary dehydration in that water is partly replaced by sucrose, the specimen is rapidly frozen in liquid nitrogen and further processed at low temperature. Cryosectioning is regarded as the most efficient technique in terms of labelling intensity, and is the method of choice when antigens exceptionally vulnerable to dehydration and embedding are investigated (see Fig. 1b). However, this technique requires special equipment and skills, and is best learned in a laboratory where the method has been established. For more information on cryosectioning see Griffiths (1993).

7.5.2.1
On-Grid Labelling of Whole-Mount Preparations

This approach is usually restricted to single cells, bacteria, organelle preparations and viruses. It requires only a small amount of material as opposed to preparing and processing a pellet for thin sectioning. Bacteria, viruses and many organelles may be viewed unfixed. The procedure itself can be carried out in less than one working day. Any specific antibody can be used, and labelling is usually followed by negative staining of the specimen (see Sect. 7.3).

A suitable processing protocol can be worked out by immunofluorescence with objects previously adhered to a glass cover slip (see also Sherwin and Read 1993). However, when GA is used for fixation at a conc. higher than 0.1%, it will cause extensive autofluorescence. In this case, fixation should be followed by a brief (10–15 min) incubation in PBS containing 0.1% $NaBH_4$, which quenches free aldehyde groups and eliminates autofluorescence. This solution should be freshly prepared since the half-life of $NaBH_4$ at neutral pH is very short.

Adsorption of bacteria, organelles and viruses onto EM grids

1. Prepare a concentrated suspension of the sample in a suitable buffer or H_2O dist.

2. Place a droplet of 15 μl on a sheet of parafilm, float a coated grid on the droplet for about 1 min.

3. Remove the grid with a pair of tweezers, and dry it with the edge of a piece of filter paper.

4. Perform negative staining by placing the grid for 30 s on a droplet of heavy metal stain. For bacteria 1.5% ammonium molybdate pH 6.8 is recommended, while viruses and cell organelles are ideally stained in 3% potassium phosphotungstate pH 6.8 (Monaghan et al. 1993).

5. Remove the grid with a pair of tweezers, dry with the edge of filter paper, and examine in the electron microscope.

6. Adjust the sample to a suitable concentration. If the stain is too weak or to intense, adjust the concentration of the stain accordingly.

1. Wash, and prepare a concentrated suspension of parasites in PBS.

2. Place some double-sided tape onto a parafilm, and place the grids onto the tape with only their edges touching the tape. Alternatively, grids can be held at their edges with a pair of fixed tweezers.

3. Place a droplet (10–15 μl) of cell suspension onto the grid, and leave for 1–10 min. For prolonged incubation times, use a moist chamber to prevent drying of the sample.

4. Place the grids for 5–10 min on a drop of a cytoskeleton stabilizing buffer (e.g. 60 mM PIPES, 25 mM Hepes, 5 mM EGTA, 2 mM $MgCl_2$, pH 6.9) containing 0.2–0.5% Triton X-100. Leave for 5–10 min. The detergent decreases the surface tension and this will result in flattened drops.

5. Wash by placing grids onto several drops of buffer, 10–20 s each. Note that after two or three washes the surface tension increases again.

6. Place the grids on a drop of buffer containing 1% GA for 10 min, followed by several washes with H_2O dist.

7. Perform negative staining of specimens in 1% uranyl acetate for 30 s, dry the grids with filter paper, and examine in the electron microscope.

8. Adjust the concentration of cells and the intensity of the stain as desired.

Note. Although it is also possible to load whole cells onto a coated grid, the cells themselves will be mostly far too thick to be viewed by whole-mount TEM. Thus adsorption is mostly followed by a permeabilization step using a non-ionic detergent. The adsorption of cells is increased by placing the droplet of cell suspension *on top* of the grid. It is also possible to detergent-extract and fix the parasites in suspension. However, this will require multiple centrifugation steps and might be deleterious to the ultrastructure of the samples.

Samples are adsorbed onto the coated grids at a density ideal for immunolabelling. Cytoskeletons should be fixed according to a protocol previously determined by immunofluorescence. When GA is used as a fixative, quenching of

Adsorption and Permeabilization of Protozoa

Immunolabelling of Whole-mount Preparations

free aldehydes is performed on droplets of PBS containing 0.1% NaBH$_4$ for 10–15 min. Alternatively, grids can also be incubated in 50 mM NH$_4$Cl or 50 mM glycine in PBS for 30–60 min prior to immunolabelling. It is not advisable to let the grids dry completely prior to labelling, since this could destroy epitopes. A typical labelling protocol is shown below.

1. Float grids, specimen side down, on a drop of PBS/BSA/lysine, 15–30 min.

2. Place grid, specimen side down, on a 15–20 μl droplet of antibody solution suitably diluted in PBS/BSA/lysine. Incubate for the time required.

3. Wash grids 3 × 5 min on drops of PBS/BSA/lysine.

4. Incubate grids, specimen side down, on droplets of colloidal gold probe, diluted in the same buffer.

5. Wash grids six times 5 min in PBS, then 4 × 1 min in H$_2$O dist.

6. Perform negative staining.

Note. Immunolabelling is performed on droplets of reagents placed on a sheet of parafilm. All reagents are diluted in PBS/ 0.5% BSA/50 mM lysine.

7.5.2.2
The Preembedding Labelling Approach

Cells or tissues are immunostained before they are postfixed dehydrated and embedded in a suitable supporting resin. The advantage of this approach is, that the loss of immuno-reactivity by harsh fixation and embedding conditions can be avoided. Preembedding labelling is the ideal method to use for staining of surface antigens in tissue culture or single cell isolates. To study intracellular antigens by preembedding labelling, cells or tissues are either chemically or physically permeabilized, with the disadvantage of destroying the ultrastructure and the risk of epitope dislocation caused by the permeabilization procedure. It is generally not advisable to use this method for immunolabelling of intracellular components, except when relatively "resistant", e.g. cytoskeletal components are studied. A major setback of the preembedding labelling approach is that a relatively high amount of antibody is needed to produce a workable specimen.

For further information see Monaghan et al. (1993), Bannister and Kent (1993), and Hemphill et al. (1996).

Outer membrane antigens of prokaryotic and eukaryotic cells can be conveniently labeled in suspension with immunoreagents prior to fixation, dehydration and embedding in epoxy resins. If necessary, cells can be immunolabelled with or without prior prefixation, and the whole surface of a cell is available for the antibody-antigen interaction to take place. If no fixative is used prior to immunostaining, gold labelling must be carried out at 4 °C in order to prevent capping of surface antigens (dePetris 1978). Also, bivalent antibodies can cause clumping of the cells, which in turn could prevent penetration of gold probes. Centrifugation steps should be carried out at minimal speed in order to avoid ultrastructural distortion. A typical protocol for cell surface immunolabelling is shown below. **Surface labelling**

1. Fix cells (optional).

2. Wash in PBS containing 50 mM glycine, 15–60 min (optional).

3. Resuspend in a suitable buffer containing 1% BSA/50 mM lysine, incubate for 15–60 min.

4. Wash and incubate in at least three times the cell volume of primary antibody, diluted in PBS/BSA/lysine, 15–60 min.

5. Wash in PBS, 3 × 5 min.

6 Incubate in second antibody, diluted in PBS/BSA/lysine, 15–60 min.

7. Wash in PBS, 3 × 10 min.

8. Fix in 3% GA in 100 mM phosphate buffer pH 7.3.

9. Osmicate and process as described for epoxy resin embedding (see Sect. 7.2).

Several techniques for rendering cells or tissues permeable to antibodies and small gold probes have been described. Physical methods include infiltrating the sample with a cryo- **Labelling of intracellular antigens** protectant (sucrose, glycerol, dimethylesulphoxide) followed by repeated freezing and thawing in liquid nitrogen, and passing the sample through an ascending and descending series of ethanol dilutions in a suitable buffer. Chemical methods involve brief exposures of samples to low concentrations of detergents such as Triton X-100, CHAPS or saponin (Aikawa and Atkinson 1990). These methods render cell membranes permeable, but they will also result in extensive extraction of cytoplasmic material. An excellent application of the preembedding approach is given in Crossley et al. (1986), where cytoskeletal elements in *Giardia* are investigated using this technique. For further information on preembedding immunolabelling refer to Langanger and deMey (1989), Griffiths (1993) and Oliver (1994a).

7.5.2.3
The Postembedding Labelling Approach

The resin-based postembedding labelling approach is applied when intact cells and tissues are needed to examine the intracellular distribution of antigens. Samples are embedded and sectioned prior to immunolabelling. The main advantages of this technique are that

(1) cells are not extensively damaged prior to labelling,

(2) a relatively small amount of antibody is needed for an experiment,

(3) a single sample can be probed with several antibodies, and

(4) double- or triple-labelling techniques using different particulate markers can be performed on a single section.

Further information on postembedding labelling has recently been published (Oliver 1994b), also with respect to malaria parasites (Aikawa and Atkinson 1990), (Bannister and Kent 1993).

Embedding at ambient temperature using LR-White LR-White is an acrylic resin which has been developed for immunolabelling of sections in both electron- and light microscopy (Barta and Corbin 1990). The resin has a low viscosity and can infiltrate partially hydrated specimens, so that samples can be transferred to resin from 70% alcohol dehydrating agent. This may be an advantage for antigens which are vulnerable to exposure to organic solvents. Acetone will interfere in the polymerization, and should not be used for dehydration. LR-White is available as three grades (soft, medium, hard) of which the hard grade is usually used for biological tissues and cells. (see Fig. 2b). The resin is a single component, which is cross-linked under anaerobic conditions by heat (55 °C) or UV for 24 h each. Chemical cross-linking involves the addition of an accelerator. This reaction is exothermic and the temperature will exceed 60 °C for a short time. Chemical cross-linking, although not recommended for immunolabelling by the manufacturers, should therefore be carried out in ice-cooled capsules. A typical processing schedule for LR-White is given below.

1. Fix tissues or cells with an appropriate fixative (e.g. 4% pFA/0.1% GA) for the time required (15 to 60 min).

2. Rinse in PBS, and incubate in 50 mM NH_4Cl in PBS, 30 min.

3. Dehydration   40% ethanol   10 min
                 60% ethanol   10 min
                 70% ethanol 2×10 min

4. Embedding   LR-White/70% ethanol 2/1   30 min
               LR-White   60 min
               LR-White   Overnight

Dehydration and embedding are best carried out at 4 °C.

5. Cross-linking

- chemical: embed in BEEM capsules on ice in LR-White + fresh accelerator (10 ml+15 μl, stirred vigorously before use). Polymerization is complete within 30 min.
- UV: embed in LR-White+0.5% benzoyl methyl ether, expose to UV light for 24 h.
- heat: embed in fresh LR-White, and keep at 55 °C for 24 h.

Note. Incubation times are given for ideally sized tissue blocks (1 mm^3). Upon polymerization, LR-White exhibits up to 10% shrinkage. Polymerized blocks can be stored indefinitely. Preservation of ultrastructure with LR-White is acceptable, but better results can be achieved by low-temperature embedding methods.

Low temperature embedding in LR-gold resin

LR-gold was initially developed for the use with unfixed material, but at least a brief aldehyde fixation is recommended. Polymerization of this resin is achieved by the addition of 0.5% benzoyl methyl ether and exposure to UV (see Fig. 1c,d).

A typical schedule for LR-Gold embedding is given below.

1. Fixation of samples in a suitable fixative.

2. Dehydration

30% methanol	10 min	4 °C
50% methanol	30 min	–25 °C
70% methanol	30 min	–25 °C
90% methanol	30 min	–25 °C
100% methanol	30 min	–25 °C

3. Embedding

LR-Gold/methanol 1/1	45 min	–25 °C
LR-Gold/methanol 7/3	45 min	–25 °C
LR-Gold	60 min	–25 °C
LR-Gold+0.5% benzoyl methyl ether	24–48 h	–25 °C

4. Cross-linking. Add fresh LR-Gold+0.1% benzyoyl methyl ether, and polymerize anaerobically at –25 °C by exposure to UV, or to a 12-V 100-W quartz-halogen projector bulb for 48–72 h.

Note. Ultrastructural preservation of cellular components is enhanced compared to LR-White due to low-temperature processing. This procedure should not be used for darkly pigmented tissue or eythrocytes, as UV light will not penetrate, and polymerization will be incomplete.

Low temperature embedding in Lowicryl resin

The Lowicryl resins are highly cross-linked acrylate- and methacrylate-based media with low viscosity, usable at very low temperatures. There are four types of Lowicryl resins:

Name	Characteristics	Usable temp.
K4M	Hydrophilic	–35 °C
HM20	Hydrophobic	–40 °C
K11M	Hydrophilic	–60 °C
HM23	Hydrophobic	–80 °C

K4M and K11M are capable of polymerization with up to 5% water content, while HM20 and HM23 require complete dehydration. It has been shown that the sensitivity of immunolabelling is practically identical for all four resins. However, the ultrastructural preservation seems to be improved at lower temperatures (Monaghan et al. 1993). Processing and embedding are carried out in a similar manner for all four resins, with the exception of the minimum temperature.

A typical protocol for embedding in K4M is given below.

1. Fixation of samples in a suitable fixative.

2. Dehydration

30% ethanol	30 min	0 °C
50% ethanol	60 min	–20 °C
70% ethanol	60 min	–35 °C
95% ethanol	60 min	–35 °C
100% ethanol	2 × 60 min	–35 °C

3. Embedding

K4M/ethanol 1/1	60 min	–35 °C
K4M/ethanol 2/1	60 min	–35 °C
K4M	Overnight	–35 °C

4. Cross-linking UV-light for 24 h at –35 °C, often followed by another 2 days at room temperature

Note. Care must be taken to embed the samples in gelatin capsules only, since the plasticizer tends to be leached out from BEEM capsules.

Freeze substitution was initially developed for morphological studies of dynamic cellular processes which are not preserved by chemical fixation. This technique differs from other resin embedding techniques in that it precludes the use of aldehydes. Instead, the sample is rapidly frozen, which will stabilize cellular components within microseconds, and preserves the cell as close to the in vivo situation as possible. This method is advantageous when antigens are studied which are not stabilized, or would be completely destroyed, by aldehyde fixation (Monaghan and Robertson 1990) After rapid freezing, the ice in the sample is replaced by methanol or acetone, at –80 °C, and the sample is then embedded in one of the Lowicryl resins.

The freeze-substitution technique

Freezing of tissue samples should be undertaken as soon as they are obtained in order to prevent postmortem changes, and mechanical damage induced by trimming of blocks should be minimized. For freezing of cells it is advantagous to prepare a concentrated suspension of cells in a 4% gelatin solution in PBS, and handle them as for solid tissues. In any case it is important to know that the well frozen region of a specimen will extend maximally 10–15 μm into the sample. Care must therefore be taken not to damage the surface of the frozen sample during handling.

Samples can be rapidly frozen in several ways. The easiest is to plunge the sample into a cryogen cooled with liquid nitrogen, but the most popular procedure is to slam the sample onto a liquid nitrogen-cooled copper block (Roos and Morgan 1990). The equipment necessary is commercially available, but expensive (Leica, U.K., RMC, USA), and it is not normally available in a routine EM laboratory. In addition, a high-pressure freezing apparatus is commercially available from Balzers (Liechtenstein), where the sample is rapid frozen with liquid nitrogen at a pressure of 2100 bar. This increases the depth of well frozen regions to 400 μm.

For morphological studies, substitution is usually done with methanol or acetone containing fixatives such as aldehydes and osmium. For immunolabelling, fixatives are best omitted (Monaghan and Robertson 1990). Substitution is best done at –80 to –90 °C in a chamber capable of maintaining these temperatures. Suitable equipment is commercially available (Balzers, Liechtenstein). Methanol will substitute faster (within 30 h) than acetone (2–8 days). After substitution is completed, the sample is warmed up to the temperature suitable for infiltration with Lowicryl resin. The protocol follows the one described above. For excellent reviews on freeze substitution and immunolabelling see Humbel and Schwarz (1989) and Schwarz and Humbel (1989).

Sectioning and immuno-labelling The acrylic resin blocks (LR White, LR Gold, and Lowicryl resins) are more brittle than Epon resin blocks, and good sectioning requires that the block is carefully trimmed, with a clean block face. Sections can be cut with glass or diamond knives, and should be picked up onto nickel or gold grids. Copper grids are not suitable since they will react with reagents during subsequent immunolabelling, affecting pH and ionic strength, and rendering conditions uncontrollable. It is of advantage to use plastic coated grids since

- sections are easily damaged during immunolabelling and

- most acrylic resins used for immunolabelling are not as stable under the electron beam as conventional resins (Smith and Croft 1990)

The immunolabelling procedure for resin embedded tissues or cells are very similar for LR-White, LR-Gold and Lowicryl resins.

These resins have been especially developed for immunocytochemical studies; however, one should keep in mind that about 90% of potential epitopes present on a section are probably no longer accessible due to loss of reactivity from fixation, dehydration and steric hindrance (Griffiths 1993). The ideal concentration of immunoreagents has to be tested by trial and error. Incubation times will also depend largely on antibody concentration and the amount of antigen present.

Theoretically, the antibody-antigen reaction occurs when the first 30–60 min of incubation at room temperature. However, for the detection of low-level antigens, it is often advantageous to perform incubation with antibody over night at 4 °C.

Thus, protocols for immunolabelling have to be worked out separately for each antibody. This can be very time-consuming and tedious. However, it is worthwhile to invest the time and effort to work out the ideal parameters.

A typical immunolabelling protocol for the acrylic resin sections is given below:

1. Incubate grids in blocking buffer (e.g. PBS, 1% BSA, 50 mM lysine or another suitable mixture) 1–2 h at room temp.

2. Apply the first antibody, diluted suitably in blocking buffer. Incubate 1–2 h at room temperature or at 4 °C overnight in a moist chamber.

3. Wash grids in PBS, 3 × 5 min.

4. Apply gold-conjugated second layer at a suitable dilution in blocking buffer. Incubate 1–2 h at room temp.

5.   Wash grids in PBS, 6 × 5 min, at room temperature.

6   Dry the grids at room temp. and stain with uranyl acetate and lead citrate (see Sect. 7.3).

Some antigens have been shown to be still recognized by antibodies even after GA- and OsO_4 treatment and embedding in epoxy resin (Bendayan and Zollinger 1983; Vickerman et al. 1988). However, these sections exhibit a very smooth surface, thus the actual surface area exposed to the antibody during labelling of epoxy sections is significantly diminished when compared to acrylic resins. Consequently, epoxy sections are etched with hydrogen peroxide or sodium metaperiodate prior to successful immunolabelling:

1.   Cut conventional sections and mount them onto nickel or gold grids.

2.   Treat sections with 10% H_2O_2 (10–20 min) or saturated $NaIO_4$ (30–60 min).

3.   Rinse in PBS (6 × 3 min).

4.   Proceed with immunolabelling as described above.

7.5.3
Applications of Immunogold EM Techniques

The earlier review by Aikawa and Atkinson (1990) illustrated some of the diversity of uses of the immunogold technique in parasitology. Since then, the technique has become part of the routine of immunological and molecular studies. Studies describing antigen distribution, for example in *Plasmodium* (Baker et al. 1994) and structures, for example myosin in *Acanthamoeba* (Baines and Korn 1990), are now outnumbered by those looking at functional molecules. A selective list illustrates the range of topics including: the localization of cysteine proteases in *T. cruzi* (Souto-Padron et al. 1990), purine salvage enzyme in *P. falciparum* (Shahabuddin et al. 1992), a calcium-binding protein in *S. mansoni* (Ram et al. 1994), replication sites of kinetoplast DNA in *T. brucei* (Robinson and Gull 1994), endosomal and lysosomal labelling in *T. brucei* (Brickman and Balber 1993) and macrophages infected with *Leishmania* (Russell et al. 1992) and estimation of lysosomal pH (Antoine et al. 1988). One study to localize surface nucleotidases on *Leishmania* used both the immunogold and the cerium cytochemical techniques (Corte-Real et al. 1993). In other studies, both TEM and SEM immunogold techniques were used to examine the distribution of the surface molecule LPG during different stages of the *Leishmania* life cycle (Lang et al. 1991). The immunogold technique has even reached chemotherapy studies, in particular those on the antiprotozoal drug, chloroquine, in relation to distribution of the drug (Moreau et al. 1986) and the P-glycoprotein involved in resistance (Cowman et al. 1991).

The immunogold technique has also been applied in studies of helminths. Just a few examples are listed here: monoclonal antibodies were used to localize phosphorylcholine-associated antigens in *Trichinella spiralis* (Takahashi et al. 1993), and other immunodominant antigens were localized using sera of rats infected with *Trichinella spiralis* (Takahashi et al. 1994). More recent studies include the identification and characterization of myophilin, a muscle-specific antigen of *Echinococcus granulosus* (Martin et al. 1995) and the immunolocalization of the Em2 antigen, the immunodominant antigen situated on the laminated layer of *Echinococcus multilocularis* (Hemphill and Gottstein 1995).

References

Aikawa M, Atkinson T (1990) Immunoelectronmicroscopy of parasites. In: Baker JR, Muller R (eds) Advances in parasitology 29. Academic Press, London, pp 151–210

Antoine JC, Jouanne C, Ryter A, Zilberfarb V (1987) *Leishmania mexicana*: a cytochemical and quantitative study of lysosomal enzymes in infected rat bone marrow derived macrophages. Exp Parasitol 64:485–498

Antoine JC, Jouanne C, Ryter A, Benichou JC (1988) *Leishmania amazonensis*: acidic organelles in amastigotes. Exp Parasitol 67:287–300

Baines IC, Korn ED (1990) Localization of myosin IC and myosin II in *Acanthamoeba castellanii* by indirect immunofluorescence and immunogold electron microscopy. J Cell Biol 111:1895–1904

Baker DA, Daramola O, McCrossan MV, Harmer J, Targett GAT (1994) Subcellular locaization of Pfs16, a *Plasmodium falciparum* gametocyte antigen. Parasitology 108:129–137

Bannister LH, Kent AP (1993) Immunoelectron microscopic localization of antigens in malaria parasites. In: Hyde JE (ed) Methods in molecular biology: protocols in molecular parasitology. Humana Press, Totowa, pp 415–429

Barta JR, Corbin J (1990) Use of immunogold-silver staining to visualize antibody-antigen complexes on LR-White embedded tissues prior to electron microscopy. J Electron Microsc Tech 16:83–84

Bendayan M (1985) The enzyme-gold technique: a new cytochemical approach for the ultrastructural localization of macromolecules. In: Bullock GR, Petrusz P (eds) Techniques in immunocytochemistry 3. Academic Press, London, pp 180–201

Bendayan M, Zollinger M (1983) Ultrastructural localization of antigenic sites on osmium tetroxide-fixed tissues applying the protein A-gold technique. J Histochem Cytochem 31:101–109

Brickmann MJ, Balber AE (1993) *Trypanosoma brucei rhodesiense*: membrane glycoproteins localized primarily in endosomes and lysosomes of bloodstream forms. Exp Parasitol 76:329–344

Burry RW, Vandre DD, Hayes DM (1992) Silver enhancement of gold antibody probes in pre-embedding electron microscopic immunocytochemistry. J Histochem Cytochem 40:1849–1856

Chapman SK (1986) Working with a scanning electron microscope. Lodgemark Press, London

Corte Real S, Porrozzi R, De Nazareth M, De Meirelles L (1993) Immunogold labeling and cerium cytochemistry of the enzyme ecto-5'-nucleosidase in promastigote forms of *Leishmania* species. Mem Inst Oswaldo Cruz 88:407–412

Cowman AF, Karcz S, Galatis D, Culvenor JG (1991) A P-glycoprotein homologue of *Plasmodium falciparum* is localized on the digestive vacuole. J Cell Biol 113:1033–1042

Crossley R, Marshall J, Clark JT, Holberton D (1986) Immunocytochemical differentiation of microtubules in the cytoskeleton of *Giardia lamblia* using monoclonal antibodies to alpha tubulin and polyclonal antibodies to associated low molecular weight proteins. J Cell Sci 80:233–252

DePetris S (1978) Immunoelectron microscopy and immunofluorescence in membrane biology. In: Korn ED (ed) Methods in membrane biology. Plenum Press, New York, pp 1–201

De Priester W (1991) Techniques for the visualization of cytoskeletal components in *Dictyostelium discoideum*. Electron Microsc Rev 4:343–376

Egea G (1993) Lectin cytochemistry using colloidal gold methodology. In: Gabius HJ, Gabius S (eds) Lectins and glycobiology. Springer, Berlin Heidelberg New York

Elbers PF (1966) Ion permeability of the egg of *Lymnala stagnalis* L. on fixation for electron microscopy. Biochem Biophys Acta 112:318–329

Erlandsen SL, Bemrick WJ, Schupp DE, Shields JM, Jarroll EL, Sauch JF, Pawley JB (1990) High resolution immunogold localization of *Giardia* cyst wall antigens using field emission SEM with secondary and backscatter electron imaging. J Histochem Cytochem 38:625–632

Faulk W, Taylor GM (1971) An immunocolloidal method for the electron microscope. Immunocytochemistry 8:1081–1083

Fujikawa S (1990) Freeze-fracture techniques. In: Harris JR (ed) Electron microscopy in biology. A practical approach. IRL-Press, Oxford, pp 173–201

Griffiths G (1993) Fine structure immunocytochemistry. Springer, Berlin Heidelberg New York

Hainfield J, Furuya F (1992) A 1.4-nm gold cluster covalently attached to antibodies improves immunolabelling. J Histochem Cytochem 40:177–184

Harris R, Horne R (1990) Negative staining. In: Harris JR (ed) Electron microscopy in biology. A practical approach. IRL-Press, Oxford, pp 203–228

Hayat A (1981) Principles and techniques of electron microscopy: biological applications, 2nd edn, vol 1. University Park Press, Baltimore

Hemphill A, Gottstein B (1995) Immunology and morphology studies on the proliferation of in vitro-cultivated *Echinococcus multilocularis* metacestodes. Parasitol Res 81:605–614

Hemphill A, Ross CA (1995) Flagellum-mediated adhesion of *Trypanosoma congolense* to bovine endothelial cells. Parasitol Res 81:412–420

Hemphill A, Lawson D, Seebeck T (1991a) The cytoskeletal architecture of *Trypanosoma brucei*. J Parasitol 77:603–612

Hemphill A, Seebeck T, Lawson D (1991b) The *Trypanosoma brucei* cytoskeleton: ultrastructure and localization of microtubule associated and spectrin-like proteins using quick-freeze, deep-etch, immunogold electron microscopy. J Struct Biol 107:211–220

Hemphill A, Frame I, Ross CA (1994) The interaction of *Trypanosoma congolense* with endothelial cells. Parasitology 109:631–641

Hemphill A, Stager S, Sottstein B, Müller N (1996) Electron microscopical investigation of surface altrations on *giandia lambhia* trophozoites after exposure to a cytotoxic monoclonal antibody. Parasitol Res 82:206–210

Hopwood D, Milne G (1990) Fixation. In: Harris JR (ed) Electron microscopy in biology. A practical approach. IRL-Press, Oxford, pp 1–15

Horrisberger M (1984) Electron-opaque markers: a review. In: Polak JM, Varndel IM (eds) Immunolabelling for electron microscopy. Elsevier, Amsterdam, pp 17–26

Humbel B, Schwarz H (1989) Freeze-substitution for immunocytochemistry. In: Verkleji AJ, Leunissen JLM (eds) Immuno-gold labelling in cell biology. CRC Press, Boca Raton, pp 115–134

Hyatt AD (1989) Immunogold-labelling techniques. In: Hayat MA (ed) Colloidal gold: principles, methods, and applications vol 2. Academic Press, London, pp 19–31

Hyatt AD (1990) Immunogold-labelling techniques. In: Harris JR (ed) Electron microscopy in biology. A practical approach. IRL Press, Oxford, pp 59–81

Karnovsky MJ (1965) A formaldehyde-glutaraldehyde fixative of high osmolarity for use in electron microscopy. J Cell Biol 27:137A

Kellenberger E, Villiger W, Carlemalm E (1986) The influence of surface relief of thin sections of embedded unstrained material on image quality. J Micron Microsc Acta 17:331–348

Kennedy JR, Williams RW, Gray JP (1989) Use of Peldri II (a fluorocarbon solid at room temperature) as an alternative to critical point drying for biological tissues. J Electron Microsc Tech 11:117–125

Kok LP, Visser PE, Boon ME (1987) Histoprocessing with the microwave oven: an update. Histochem J 20:323–328

Lang T, Warburg A, Sacks DL, Croft SL, Lane RP, Blackwell JM (1991) Transmission and scanning EM-immunogold labelling of Leishmania major lipophosphoglycan in the sandfly Phlepotomus papatasi. Eur J Cell Biol 55:362–372

Langanger G, DeMey J (1989) Detection of cytoskeletal proteins in cultured cells at the ultrastructural level. In: Bullock GR, Petrusz P (eds) Techniques in immunocytochemistry, vol 4. Academic Press, Harcourt Brace Jovanovich Publ, New York, pp 47–66

Lewis PR, Knight DP (eds) (1992) Cytochemical staining methods for electron microscopy. Elsevier, Amsterdam

Login GR, Dvorak AM (1988) Microwave fixation provides excellent preservation of tissues, cells and antigens for light and electron microscopy. Histochem J 20:373–387

Lucocq J (1993) Particulate markers for immunoelectron microscopy. In: Griffiths G (ed) Fine structure immuno cytochemistry. Springer, Berlin Heidelberg New York, pp 279–306

Lucocq JM, Baschong W (1986) Preparation of protein colloidal gold complexes in the presence of commonly used buffers. Eur J Cell Biol 42:332–337

Martin RM, Gasser RB, Jones MK, Lightowlers MW (1995) Identification and characterization of myophilin, a muscle specific antigen of *Echinococcus granulosus*. Mol Biochem Parasitol 70:139–148

Minassian H, Huang S-N (1979) Effect of sodium azide on the ultrastructural preservation of tissues. J Microsc 117:243–253

Monaghan P, Robertson D (1990) Freeze-substitution without aldehydes or osmium fixatives: ultrastructure and implications for immunocytochemistry. J Microsc 158:355–363

Monaghan P, Robertson D, Beesley JE (1993) Immunolabelling-techniques for electron microscopy. In: Beesley JE (ed) Immunocytochemistry. A practical approach. IRL Press Oxford, pp 43–76

Moreau S, Presnier G, Maalla J, Fortier B (1986) Identifcation of distinct accumulation sites of 4-aminoquinoline in chloroquine sensitive and resistant *Plasmodium berghei* strains. Eur J Cell Biol 42:207–210

Mueller M, Moor H (1984) High pressure freezing. In: Bailey GW (ed) Proc 42nd Annu Meet Electron Microscopy Society of America, San Francisco, pp 6–9

Oliver C (1994a) Pre embedding Labelling Methods. In: Javois LC (ed) Methods in molecular biology 34: Immunocytochemical methods and protocols. Humana Press, Totowa, pp 315–319

Oliver C (1994b) Postembedding labelling methods. In: Javois LC (ed) Methods in molecular biology 34: Immunocytochemical methods and protocols. Humana Press, Totowa, pp 321–328

Pinto da Silva P (1984) Freeze-fracture cytochemistry. In: Polak J, Varndell I (eds) Immunolabelling for electron microscopy. Elsevier, Amsterdam, pp 179–188

Ram D, Romano B, Schechter I (1994) Immunochemical studies on the cercarial-specific calcium binding protein of *Schistosoma mansoni*. Parasitology 108:289–300

Robinson DR, Gull K (1994) The configuration of DNA replication sites within the *Trypanosoma brucei* kinetoplast. J Cell Biol 126:641–648

Roos N, Morgan AJ (1990) Cryopreparation of thin biological specimen for electron microscopy: methods and applications. Royal Microscopical Society Microscopy Handbooks, Oxford Science Publications, Oxford

Russell DG, Xu S, Chakraborty P (1992) Intracellular trafficking and the parasitophorous vacuole of *Leishmania mexician*-infected macrophages. J Cell Sci 103:1193–1210

Schwarz H, Humbel BM (1989) Influence of fixatives and embedding media on immunolabelling of freeze-substituted cells. Scanning Microsc Suppl 3:57–64

Shahabuddin M, Gunther K, Lingelbach K, Aikawa M, Schreiber M, Ridley RG, Scaife JG (1992) Localisation of hypoxanthine phosphoribosyl transferase in the malaria parasite *Plasmodium falciparum*. Exp Parasitol 74:11–19

Shakibaei M, Frevert U (1992) Cell surface interactions between *Trypanosoma congolense* and macrophages during phagocytosis in vitro. J Protozool 39:224–235

Sherwin T, Gull K (1989) Visualization of detyrosination along single microtubules reveals novel mechanisms of assembly during cytoskeletal duplication in trypanosomes. Cell 57:211–221

Sherwin T, Read M (1993) Immunofluorescence of parasites. In: Hyde JE (ed) Methods in molecular biology 21: Protocols in molecular parasitology. Humana Press, Totowa, pp 407–414

Simionescu N, Simionescu M (1976) Galloylglucoses of low molecular weight as mordant in electron microscopy. J Cell Biol 70:608–621

Singer SJ (1959) Preparation of an electron dense antibody conjugate. Nature 183:1523–1524

Slayter H (1990) High resolution shadowing. In: Harris JR (ed) Electron microscopy in biology. A practical approach. IRL-Press, Oxford, pp 151–172

Smith M, Croft S (1990) Embedding and thin-section preparations. In: Harris JR (ed) Electron microscopy in biology. A practical approach. IRL-Press, Oxford, pp 17–37

Souto-Padron T, De Souza W, Heuser JE (1984) Quick-freeze, deep-etch rotary replication of *Trypanosoma cruzi* and *Herpetomonas megasiliae*. J Cell Sci 69:167–178

Souto-Padron T, Campetella OE, Cazzulo JJ, De Souza W (1990) Cysteine proteinase in *Trypanosoma cruzi*: immunocytochemical localization and involvement in parasite-host cell interaction. J Cell Sci 96:485–490

Takahashi Y, Homan W, Lim LP (1993) Ultrastructural localization of the phosphorylcholine-associated antigen in *Trichinella spiralis*. J Parasitol 79:604–609

Takahashi Y, Mizuno N, Araki T, Okuda H (1994) Immunocytochemical localization of antigens in adult worms of *Trichinella spiralis* recognized by Fischer rats. Parasitol Res 80:291–296

Tetley L, Vickerman K (1985) Differentiation in *Trypanosoma brucei*: host-parasite cell junctions and their persistence during acquisition of the variable antigen coat. J Cell Sci 74:1–19

Tetley L, Turner CMR, Barry JD, Crowe JS, Vickerman K (1987) Onset of expression of the variant surface glycoproteins of *Trypanosoma brucei* in the tsetse fly studied using immunoelectron microscopy. J Cell Sci 87:363–372

Tokuyasu KT (1973) A technique for ultracryotomy of cell suspensions and tissues. J Cell Biol 57:551–565

Tokuyasu KT (1978) A study of positive staining of ultrathin frozen sections. J Ultrastruct Res 63:287–307

Tokuyasu KT (1984) Immuno-cryoultramicrotomy. In: Polak JH, Varndell IM (eds) Immunolabelling for electron microscopy. Elsevier, Amsterdam, pp 71–82

Van Noorden CJF, Frederiks WM (1992) Enzyme histochemistry: a laboratory manual of current methods. Oxford University Press, Oxford

Van Noorden CFJ, Frederiks WM (1993) Cerium methods for light and electron microscopical histochemistry. J Microscopy 171:3–16

Vickerman K, Tetley L, Hendry KAK, Turner MCR (1988) Biology of African trypanosomes in the tsetse fly. Biol Cell 64:109–119

Wilkinson DG (1992) In situ hybridization. A practical approach. IRL-Press, Oxford

Studies of the Surface Properties, Lipophilic Proteins and Metabolism of Parasites by the Use of Fluorescent and "Caged" Compounds

J. Modha, C.A. Redman, S. Lima, M.W. Kennedy and J. Kusel

8.1
Introduction

Fundamental problems in parasitology, such as the metabolic changes occurring during transition between intermediate and final hosts or the effects of drugs on surface membranes, can be studied by sensitive techniques involving the use of fluorescent compounds coupled with fluorescence microscopy or spectrofluorimetry. Fluorescence is the absorption of light energy which raises a molecule to a higher energy state, and the subsequent release of energy in the form of light, but always at a lower energy than that of the exciting light. In practical terms, this means that the wavelength of the excitatory light is always less than the emitted light. Thus, excitation in the ultraviolet leads to fluorescence in the visible range (usually in the blue or green) and excitation in the blue or green will usually lead to emission in the red. The advantages of using fluorescence is that it is possible to quantify the uptake of a probe into an organelle or a surface membrane, or to examine its distribution in the living parasite, without damaging the organism. The observation of the parasite population by fluorescence microscopy allows an assessment of the variability of the process to be measured, and regional differences on or within the parasite can be studied. The surface of many parasites may be very impermeable to some compounds. These compounds can be rendered membrane permeant by the formation of an ester, which is hydrolysed when inside the parasite. The impermeable compound can also be "caged" and thus made hydrophobic and released within the parasite by brief exposure to light at a wavelength of less than 360 mμ. Parasite surfaces have a variety of functions during the interaction of the parasite with host tissues. Receptors, and transport molecules may be located in membrane or membrane-like structures. Surface antigens, often of unknown function, interact with the immune system, and certain drugs act within the surface (Kusel and Gordon 1989). All parasite surfaces contain lipid, protein and carbohydrate, although the nature of the interaction between surface macromolecules may vary greatly according to the parasite. Nematode surface lipids do not appear to exist as a conventional bilayered membrane (Wright and Hong 1988).

To investigate the interaction of surface membrane lipids and proteins, fluorescent compounds can be incorporated into surfaces. In this chapter, we suggest protocols for the use of a variety of fluorescent and caged compounds, which have been used in our laboratory, and give some evidence of their success, sometimes of unexpected pleasure.

8.2
Incorporation of Fluorescent Compounds into Membranes

Compounds Fluorescent compounds for microscopy are given in Table 1; caged compounds in Table 2. Most of the reagents can be purchased from Molecular Probes, Eugene, Oregon, USA. Sigma market some phospholipids and the PKH-2 and PKH-26 membrane probes. Calbiochem sell caged Ca^{2+} (NITR-5) and caged IP_3. It is possible to synthesise some fluorescent compounds (Haughland and Larison 1992).

Equipment A fluorescent microscope fitted with a camera is a minimum requirement. Quantitative fluorescence can be carried out with a camera with automatic exposure, or a photomultiplier attachment. Commercially available microscopes can be purchased from Leica, Zeiss, Nikon.

8.2.1
Surface Membrane Incorporation
of Fluorescent Compounds (Table 1, Nos. 1–13)

Incorporation Procedure Anionic and cationic membrane probes can be incubated with a variety of parasites, and studies with trypan blue quenching (below) show some of these probes to be re-

Table 1. Fluorescent compounds for microscopy

No.	Full name of compound	Catalogue no. in Molecular Probes Inc.[a]	Abbreviations	Excitation/emission wavelength $m\mu$	Use in biological research
1	5-(N-octadecanoyl) amino fluorescein	0–322	$C_{18}Fl$	EX 495, EM 521 495/521	Inserts into lipid membranes. There is a range of compounds of shorter acyl chain lengths similar in structure to 1,2
2	5-(N-hexadecanoyl) amino fluorescein	H-110	$C_{16}Fl$	495/521	Inserts into lipid membranes. There is a range of compounds of shorter acyl chain lengths similar in structure to 1,2
3	11-Dioctadecyl 333'3' tetramethyl indocarbocyanine percholate	D-282	$DiIC_{18}$	547/571	Inserts into lipid membranes. There is a range of compounds of a variety of acyl chain lengths similar in structure to 3
4	Octadecyl rhodamine B chloride	0–246	Rh18(3)	556/577	
5	25-(NBD-methyl amino)-27-norcholesterol	N-237	NBD-Chol	484/543	Cholesterol distribution and exchange
6	NBD-hexanoic ceramide	N-1154	NBD-Ceramide	464/532	Stains Golgi apparatus
7	BoDipy-3-pentanoylceramide	D-3521	BoDipy-ceramide	505/5111	

Table 1 (*Contd.*)

No.	Full name of compound	Catalogue no. in Molecular Probes Inc.[a]	Abbreviations	Excitation/emission wavelength $m\mu$	Use in biological research
8	NBD-hexanoylsphingosyl phosphocholine	N-3524	C6 NBD-Sphingo-myelin	466/539	Labels membrane to study properties of sphingomyelin
9	BoDipy-pentanoylsphingosyl phosphocholine	D–3522	BoDipy sphingomyelin	505/512	Labels membrane to study properties of sphingomyelin
10	NBD dodecanoyl 1-hexade-canoyl-snglycero-3 phospho-choline	N–3787	NBD-phosphatidyl choline	466/531	Studies of distribution of phospholipids in membranes
11	NBD-1,2 dihexadecanoyl sn-glycero-3 phospho ethanola-mine, triethyl ammonium salt	N-360	NBD phosphatidyl ethanolamine	460/534	In (10) the acyl chain is la-belled, in (11) the head group
12	NBD-dodecanoic acid	N-678	NBD-dodecanoic acid	472/533	Uptake and metabolism of fatty acids
13	BoDipy hexadecanoic acid	D-3821	BoDipy hexadecanoic acid	503/512	
14	Hoechst 33258	H-1398	Hoechst 33258	346/460	Viability of cells
15	Hoechst 33342	H-1399	Hoechst 33342	346/460	Nuclear staining of living cells (hydrophobic form of 14)
16	Ethidium bromide	E-1305		526/605	Binds to DNA of damaged cells
17	Dihydroethidium	D-1168			Binds to DNA after oxidation by living cells

18	Fluorescein diacetate	F-1303	FDA		These compounds are hydrolysed by esterases and are retained by viable cells
19	(5 and 6) carboxy fluorescein diacetate	C-195	CFA		"
20	Fluorescein phalloidin	F-432	F-Phall	495/520	Bind to F-actin
21	Rhodamine phalloidin	R-415	Rh-phall	550/580	Bind to F-actin
22	BoDipy phalloidin	B-607	BD-phall	503/512	Bind to F-actin
23	Merocyanin 540	M-299	MC-540	555/581	Binds to fluid lipid
24	Nile red	N-1142	Nile red	551/636	Labels lipid rich vesicles and lipid globules
25	5-Iodoacetamido fluorescein	I-3	51AF	491/515	Reacts with sulphydryl groups
26	Fluorescein-5-maleimide	F-150	FM	490/515	Reacts with sulphydryl groups
27	Fluorescein – dextran. Many different derivatives of dextran can be purchased over a wide range of molecular weights	D-1822 (molecular weight 70,000)		495/520	Used to follow endocytosis
28	Rhodamine 123	R-302	R123	505/534	Vital stain for mitochondria
29	Carboxy SNARF-1	C-1270	SNARF	488/560,620	pH indicator
30	Carboxy SNARF-1 acetomethoxyacetate	C-1271	SNARF-AM		pH indicator, membrane permeant

[a]Haughland and Lavison (1992).

Table 2. Caged compounds

Caged compound	Catalogue no.	Abbreviation	Biological activity after light flash < 360 mμ
ATP,3-0-1(4,5 dimethoxy-2-nitrophenyl)ethyl ester, Na$_2$	A-1049	DMNPE-caged ATP	Release of ATP
γ-amino-butyric acid, 4,5 dimethoxy-2-nitrobenzyl ester, HC1	A-2506	Caged GABA	Release of neurotransmitter GABA
N-((α-2-carboxy)-2-nitrobenzyl) carbamoylcholine, tri fluoroacetate	C-1791	Caged carbamoylcholine	Release of acetylcholine analogue, carbachol
cAMP (4,5 dimethoxy-2-nitrobenzyl)	D-1037	Caged cAMP	Release of intracellular cAMP
5-(N-dodecanoyl) amino fluorescein bis 4,5, dimethoxy-2-nitrobenzyl ether	D-2512	Caged C$_{12}$Fl	Release of fluorescent lipid probe
Diazo-2, AM	D-3035	Caged Ca^{2+} scavenger (membrane permeant)	Release of a compound which chelates free Ca^{2+}
NITR-5	Calbiochem 482477	Caged Ca^{2+}	Release of intracellular Ca^{2+}

stricted to the outer monolayer of adult schistosomes (Foley et al. 1986).

Different concentrations, labelling times and conditions (presence and absence of serum) should be tried. The following protocol has been used for schistosomes.

1. Adult worms; schistosomula. Wash with GMEM with 10% FCS 37 °C three times, and resuspend five adult worms or 1000 schistosomula in 1 ml of medium in wells of a 24-well plate.

2. Add 1–5 μl of an ethanol solution (1 mg ml^{-1}) of the fluorescent probe to the medium. The following have been found to be surface-restricted: C_{14} Fl, C_{16} Fl, C_{18} Fl; DiIC$_{14}$(3); DiIC$_{16}$ (3); DiICI$_{18}$(3); DiIC$_{22}$(3); RHC$_{18}$.

3. Incubate for 2–10 min, 37 °C, and then wash thoroughly (6×) with medium at 37 °C. Adult worms are washed by sedimentation within the well, but schistosomula must be transferred to a plastic or glass centrifuge tube and centrifuged gently (1000 g, 10 s).

4. Resuspend parasites in medium containing 50 μg carbamoyl choline (carbachol) to cause paralysis. Carry out all procedures at 37 °C.

5. Mount on a microscope slide in about 80 μl medium (without serum) plus carbachol contained in a silicone grease square, transferring adult worms carefully with a fine paint brush to avoid damage. Seal by addition of a cover slip. Observe using the appropriate filter block (FITC for C_{14}Fl–C_{18}Fl; Rhodamine for DiIC$_{14}$–DiIC$_{22}$ and RHC$_{18}$).

• The fluorescent probes in the surface will exchange with serum components, or serum albumin within 10 min. Thus the parasites should be placed under the cover slip in medium lacking serum if long term observation is required. Cercariae (*Schistosoma mansoni*) can be labelled with C_{14}Fl–C_{18}Fl as above, but incubation is at room temperature in water.   **Notes**

• Cercariae should be mounted in water containing 50 μg ml^{-1} carbachol. Nematodes can be labelled with C_{18}Fl; other probes do not enter the epicuticle (Kennedy et al. 1987). *Hymenolepis diminuta, H. nana; H. microstoma, Leishmania* sp, *Giardia lamblia*, have also been labelled with $C_{18}Fl$.

Results The morphology of adult worms (*S. mansoni*) can be readily observed before and after cooling or praziquantel treatment (Fig. 1,2). Dissolution of surface membrane morphology with formation of fragments during detergent treatment can be seen (Fig. 1)

Vesicle formation after high salt treatment of schistosomula is readily studied (Fig. 3). Regional labelling at the head end of schistosomula can be observed during transformation.

8.2.2
Examination of Location of Fluorescent Probe in Membrane

Trypan blue quenching Fluorescent membrane probes (e.g. $C_{14}Fl$, $C_{16}Fl$, $C_{18}Fl$) insert into the surface membrane of a number of parasites. Trypan blue (0.25% w/v final concentration) will quench surface fluorescence if the fluorophore is exposed at the parasite surface to the external environment, since the energy of fluorescence emission from the fluorophore is transferred to the trypan blue molecules in contact with the fluorophore (by a process of fluorescence resonance energy transfer; Blatt and Sawyer 1984).

1. Trypan blue (Sigma 0.4% in saline) is diluted in GMEM medium containing carbachol (50 μg ml^{-1}) to a final concentration of 0.25%.

2. The trypan blue solution (0.25%) is added to the fluorescently labelled parasites, and mounted on a silicone grease square on a microscope slide.

3. Observation of the parasites under the fluorescent microscope will reveal that they will not fluoresce if the fluorophore is restricted to the outer surface monolayer.

Fig. 1. a $C_{18}Fl$ incorporated into the surface membrane of normal adult *Schistosoma mansoni* and adult worms cooled for 1 h at 0 °C. Similar folds can be seen after praziquantel treatment. **b–f** Progressive dissolution of adult *S. mansoni*. Surface revealed by staining with DiIC$_{18}$. Dissolution was induced by incubating labelled worms in 1% Tween 20 + 1 mg ml^{-1} retinol. Adult male worms were labelled with DiIC$_{18}$ and the surface folds appear as in **b**. The surface becomes unfolded **c, d**, and appears to break down into small particles **e, f**. *Bar* 100 μm

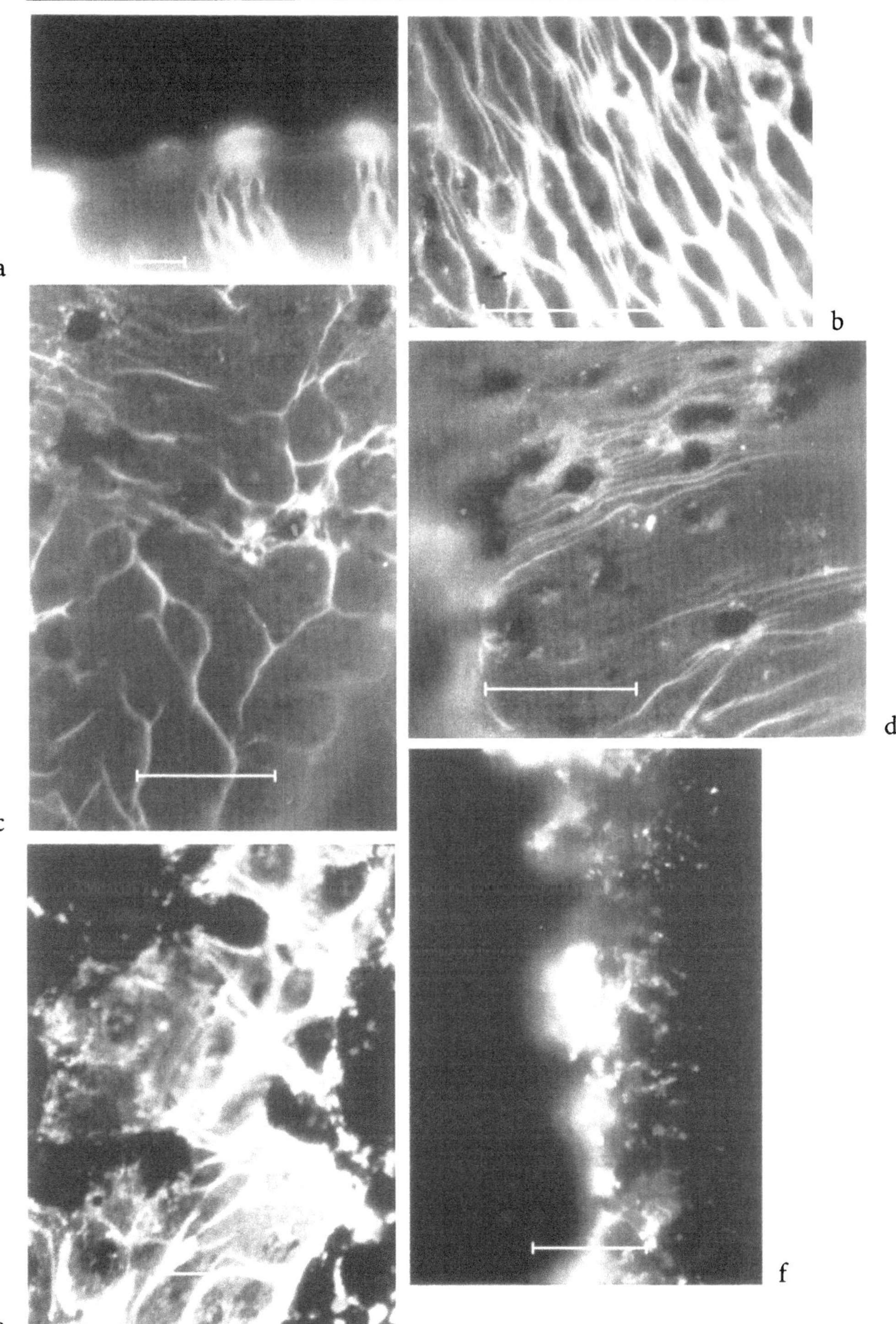

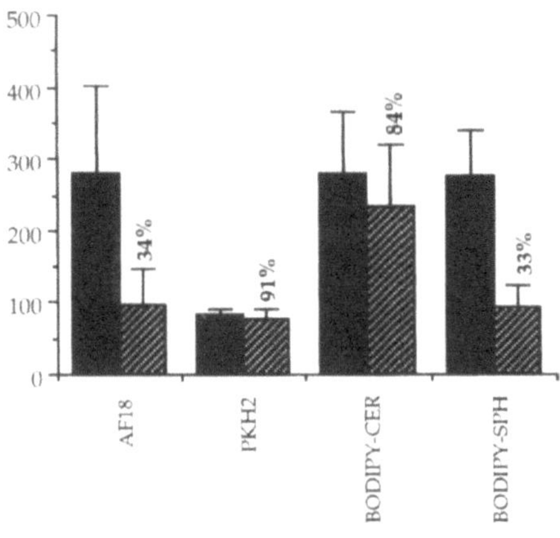

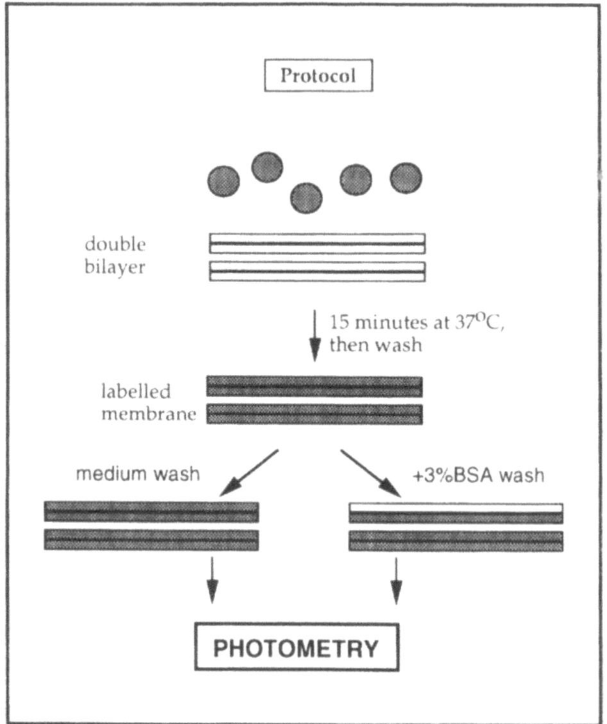

Fig. 2. Uptake of fluorescent probes into adult *S. mansoni* (■), and percentage insertion into structures underneath the outer monolayer(■). Worms were incubated for 15 min with dfBSA/probe at 37 °C, washed and incubated for a further 15 min with medium +3% dfBSA at 37 °C. Photometry was performed immediately after immobilisation with carbachol

Fig. 3. a,b Surface membrane vesicle formation from schistosomula as revealed by staining with DiIC$_{18}$(3). **c** Concentration of NBD-ceramide in oesophageal gland of adult *S. mansoni* after labelling and incubating for 4 h GMEM + 10% FCS (37 °C). **d** *Moniliformis moniliformis* stained with NBD-dodecanoic acid; stain concentrates in probosis and menisci. *Bar* 50 μm

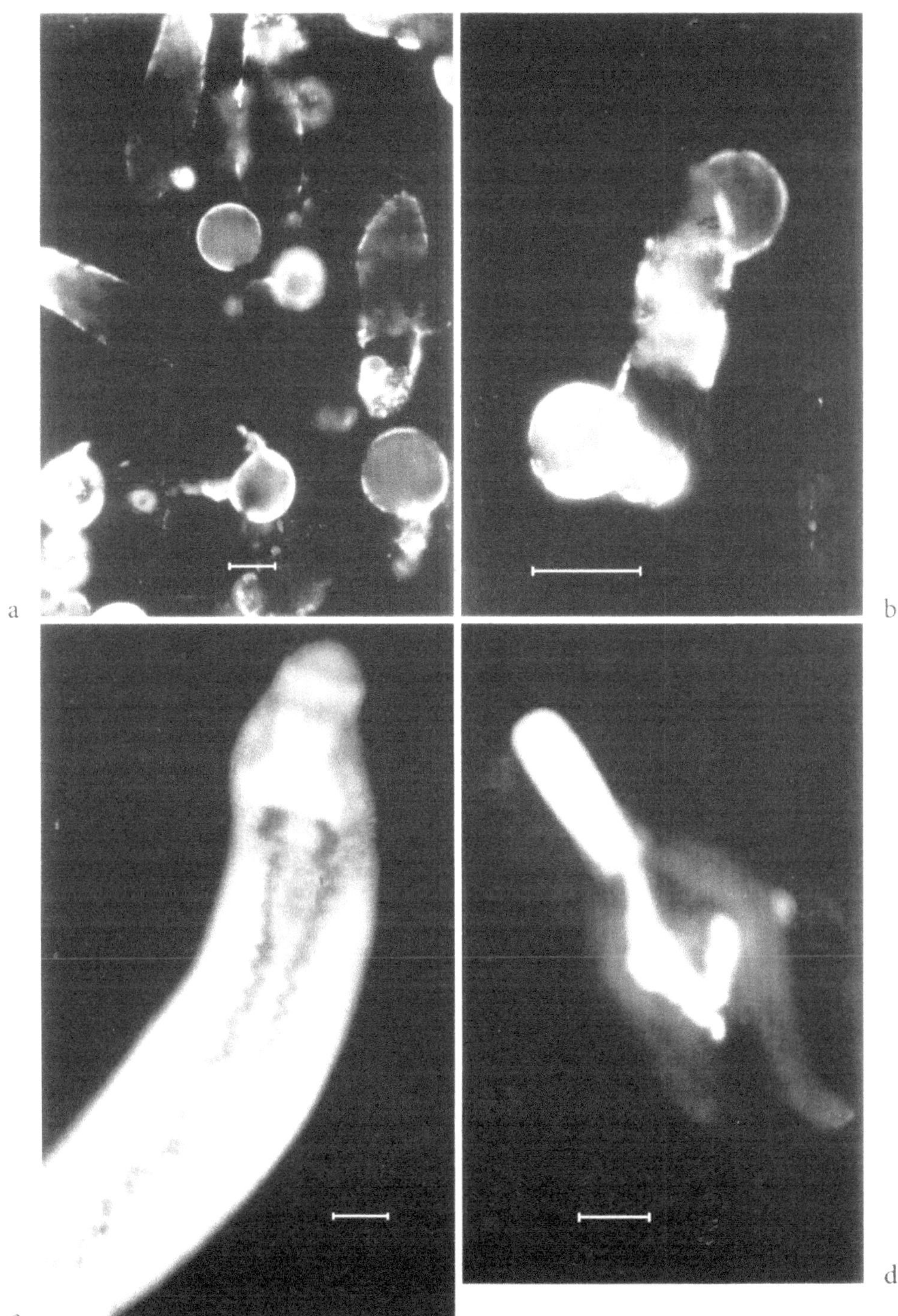

Results The fluorescence of schistosome adults labelled with $C_{14}Fl$, $C_{16}Fl$ and $C_{18}Fl$ is quenched. Quantitative fluorescence measurements show this quenching to be almost complete. The fluorescence of schistosome adult labelled with $DiIC_{18}(3)$ is not quenched, indicating a distribution of the dye in the inner monolayers. Fluorescent fatty acids, cholesterol and ceramides can be shown to be located in both the outer surface monolayer and the internal monolayer, since quenching is between 44 and 68% (Moffat and Kusel 1992). Differences in the distribution of $C_{18}Fl$ (surface-restricted) and in fluorescent phospholipids have been observed in nematodes (Proudfoot, L. PhD Thesis 1991); Proudfoot et al. 1993).

Removal of the Fluorescent Probe from the Outermost Membrane Monolayer by Defatted Bovine Serum Albumin (dfBSA)

Removal of the Probe In order to quantify the percentage of the fluorescent probe which is localised in the outer monolayer, the organisms can be washed, after labelling, with medium containing dfBSA, which extracts the probe off the surface (Fig. 2). This method has proved particularly successful in monitoring trafficking of fluorescent ceramide derivatives from the Golgi to the plasma membrane and to other bodies as sphingolipids (Lipsky and Pagano 1985; Pagano and Martin 1988; Pagano et al. 1991 van Meer 1993; Hackstadt et al. 1995). In order to monitor organelle-mediated transport events, organisms are labelled at 4 °C with fluorescent ceramide, washed with dfBSA in medium and then warmed to 37 °C. The transport and sorting of the probe is monitored with time, with respect to the amount of fluorescence that is washed off with further washes of dfBSA. Derivatives of the fluorescent ceramide can also be analysed using thin layer chromatography (Hackstadt et al. 1995). This method can also be used to monitor flip–flop of certain fatty acids from the outer to the inner membrane.

Preparing the lipid for labelling The fluorescent label is complexed to dfBSA prior to labelling. The following method can be used (Pagano and Martin 1988):

1. Dry 50 nmol of stock lipid and redissolve in 200 μl ethanol.

2. Inject this into 10 ml medium containing 0.34 mg ml^{-1} dfBSA while vortexing.

3. Dialyse overnight against 500 ml medium at 4 °C.

4. Freeze at –20 °C as 1-ml aliquots which contain lipid and dfBDA at 5 μM.

This can be done at a variety of temperatures depending on whether one is interested in distribution or trafficking. The organisms can be labelled for 30–60 min at 4 °C with, for instance BoDipy-C6-ceramide (dfBSA complex), and then warmed to the temperature suited for the organism. Where basic data are desired on the sites to which lipids are directed, labelling can be carried out for 15 min at 37 °C using 1-ml aliquots of the above preparation.

Labelling of organisms

Extraction of the probe is accomplished by incubating the organisms in 1–2 ml of 2% dfBSA for 15 min at 37 °C. The organisms are washed with medium prior to further analysis, e.g. photometry. For EM studies of probe localisation, washing can be done at 4 °C using four washes of 0.34% dfBSA (Pagano et al. 1991).

Washing of the label

Data are presented in Fig. 2 for localisation of four fluorescent probes in the surfaces of adult male schistosomes. A clear difference can be observed in the localisation of AF18 and BoDipy-sphingomyelin, which are in the main surface-restricted, and PKH2 and BoDipy-ceramide. The latter probes cross over the outer membrane, entering structures in the tegument.

Results

8.2.3
Uptake of Fluorescent Fatty Acids and Cholesterol

Fluorescent derivatives of fatty acids and cholesterol can be used to label parasites. The label is not surface-restricted and rapidly disperses within the body, and can concentrate in certain organelles and organs (Moffat and Kusel 1992).

1. Stock solutions of 1 mg ml^{-1} of fluorescent compounds in ethanol are stored under nitrogen.

Uptake Procedure

2. The condition of labelling can be varied according to the parasite. 1 to 5 μl of stock solution is added to 5 μl of GMEM plus 10% FCS in wells of a 6-well plate. Incubation can be varied from 2 to 60 min at 37 °C.

3. Parasites are washed 6× with GMEM plus 10% FCS, mounted on a microscope slide in a silicone grease square and observed.

4. Pulse chase experiments can be carried out; a short pulse of labelled compound followed by a period of culture in the absence of the label. The concentration of the probe in particular regions of the parasite can then be observed.

Results In adult schistosomes, NBD-dodecanoic acid, and BoDipy hexadecanoic acid are taken up through the surface and transported to the gut within hours of labelling. Diffusion through the mouth into the pharynx, which becomes strongly labelled, and the gut can also occur (Fig. 3). Labelling of the oesophageal gland is particularly strong with fluorescent ceramides. Thin layer chromatography of lipid extracts has shown the presence of metabolites. (BoDipy ceramide is converted to BoDipy glucosyl ceramide – P. Gissen, unpubl.).

The acanthocephalan parasite *Moniliformis moniliformis* concentrate NBD-dodecanoic acid in the proboscis and menisci (Fig. 3).

8.2.4
Labelling Parasite Surfaces with a "Stable" Cell-Labelling Reagent (Sigma, PKH2 and PKH26)

Sigma (UK) are marketing reagents, PKH2 and PKH26, which are prepared for them by Zynaxis Cell Science Inc. 371 Phoenixville Pike, Malvern, PA 19355, USA. These reagents insert into cell membranes of mammalian cells and, unlike the lipid probes described in Table 1, remain bound to the membrane for some time in vivo, and can be used to study the fate of labelled cells in vivo. We have been able to label *Trichinella*, *Globodera rostochiensis* and schistosome adult worms, and schistosomula, and have observed surface shedding in the case of *Trichinella* (Fig. 4).

The method of labelling parasites is very similar to that described in the Sigma kit for mammalian cells.

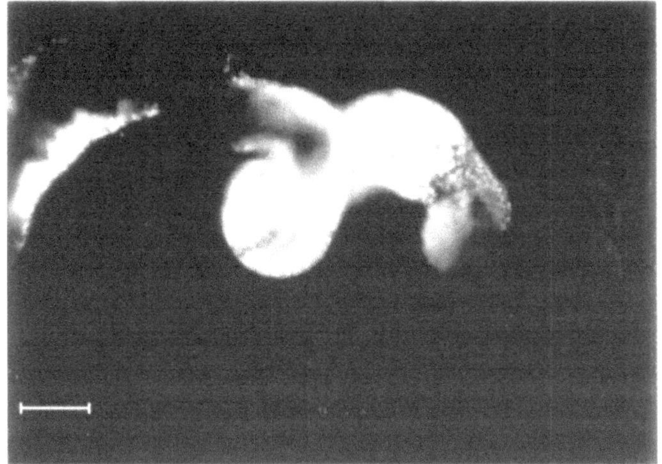

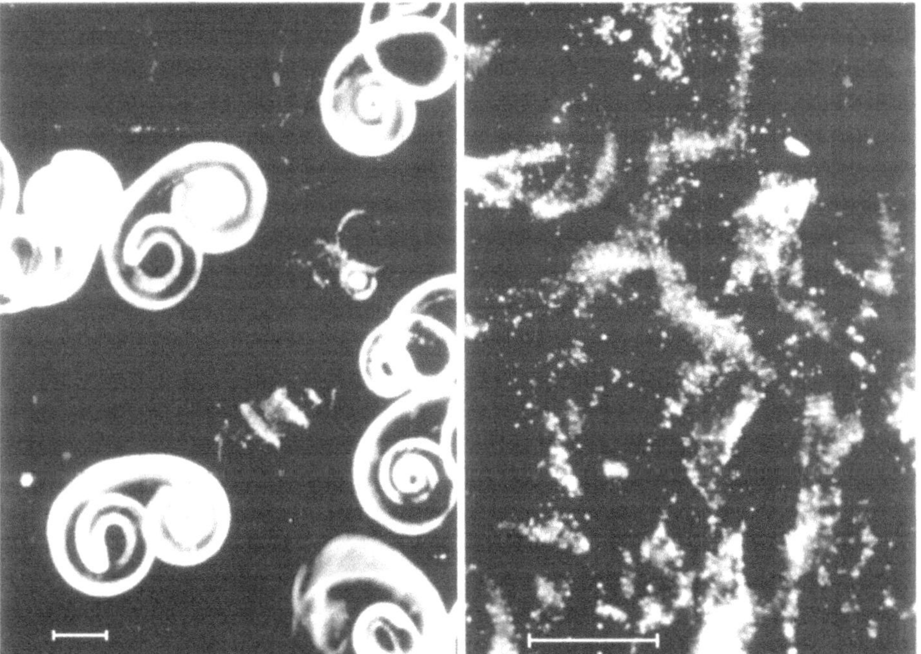

Fig. 4. a Damage to surface of adult *S. mansoni* revealed by Hoechst 33258. *Bar* 1000 μm. **b, c** *Trichinella spiralis*, prepared from infected mouse muscle, were incubated with the Sigma cell labelling compound, PKH-26. The red fluorescent surface is rapidly shed. *Bar* 50 μm

Labelling 1. Add 10 μl of PKH2 or PKH26 to 1.0 ml of diluent just before you are ready to carry out the labelling procedure.

2. Wash parasites 2× (with 500-μl portions of the diluent provided. The diluent is non-toxic and isotonic. We have found that thorough washing of the parasite is essential.

3. Add 100 μl of the solution produced in **1** to the parasites contained in a minimal volume (50 μl) of diluent.

4. Incubate for 30 min, room temperature, or 37 °C.

5. Wash parasites 3× in diluent and observe under the fluorescent microscope. PKH2 (fluorescein filter) PKH26 (rhodamine filter).

8.3
Using Fluorescent Techniques to Measure Membrane Fluidity

Fluorescence recovery after photobleaching (FRAP), a microscope-based technique, can be applied to measure the diffusion of a wide range of molecules presented in biological or artificial membranes, as well as to study the motion in the cytoplasm.

Equipment For FRAP, an attenuated laser beam is focused to a small spot on the sample (Fig. 5). The initial fluorescence level inside the spot is measured and the laser is momentarily deattenuated to bleach irreversibly the molecules within the spot. The recovery of fluorescence inside the spot is a consequence of the diffusion of unbleached molecules from outside the spot. In biological membranes only a fraction of molecules is free to move; therefore just a partial recovery of fluorescence is observed. FRAP thus provides two measures of diffusibility; the fraction (f) of molecules that are able to move and the diffusion coefficient (D_L) of that fraction.

In FRAP the molecule whose diffusion is to be studied must be fluorescently labelled directly in a non-cross-linking manner. Fluorescein isothiocyanate (FITC) and tetramethylrhodamine isothiocianate (TRITC) conjugated with hydrophobic molecules as well as to antibody fragments and lectins remain probably the most widely used reactive fluorescent dyes.

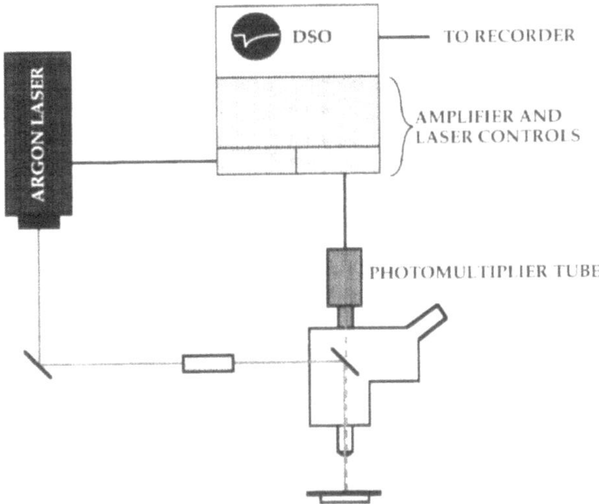

Fig. 5. This figure shows a simplified scheme of the FRAP apparatus used in this laboratory. The laser beam is provided by a 1.0-W Argon laser with a acoustooptic modulator, and is passed through a series of adjustable mirrors and spatial filters to the microscope, where it is reflected onto the sample. The beam is reflected once more through the objective to the photomultiplier in order to amplify the signal. The signal is monitored on a digital storage oscilloscope (DSO), and can be recorded on a chart recorder or personal computer. The laser power can be altered to suit the level of parasite fluorescence in order to give a stable baseline. The power is deattenuated by the laser controller GUED 729 (Glasgow University Electronic Dept.) in order to bleach the sample

Procedure

In FRAP experiments some controls are necessary: (1) to calculate the diameter and the position of the laser beam in the field of view of the microscope and (2) to show the surface localisation and stability of the probe. The first control is done by performing the FRAP technique in a fluorescent solution of known viscosity like fluoresceinated bovine serum albumin in thin films of glycerol (Foley et al 1986).

The second, confirming the superficial localisation of the probe and that it is not being internalised by the specimen, can be done by measuring the ability of a non-permeant molecular like trypan blue (0.25% w/v; see below) to quench probe fluorescence by resonance energy transfer (Foley et al. 1986) after different times of probing or by the use of de-fatted serum albumin.

If possible, it is often more convenient to strip label from the surface using either buffers or low pH or proteases.

Trypsin is the most useful way for stripping labelled antibodies and lectins from cells, while acidic buffers are useful for stripping smaller molecules, peptide hormones, and *beta*$_2$-microglobulin.

Short times of labelling reduce the possibilities for internalisation and degradation of the label. These unwanted events are also reduced by labelling, when possible, at low temperature (0–4 °C).

It was shown that damaged areas present different D_L and f (Lima et al. 1994). Then another control has to be done in order to assure that the surface of the specimen is not damaged. The bisbenzimide dye Hoechst 33258 is relatively non-toxic and water-soluble. Hoechst dyes can be excited with UV lines of the argon laser and most conventional fluorescence excitation sources. The Hoechst dye can be used as a control for membrane integrity because, due to its hydrophilic nature and because it fluoresces upon binding to DNA, it binds preferentially to contiguous AT base pairs (Lima et al. 1994, see below).

Results Membrane fluidity is important in a number of processes, including cell signalling, antigen binding and complement-mediated damage. Therefore the investigation of membrane fluidity of parasites may yield important information on signalling pathways caused by different environmental cues, as well as aspects of membrane formation and regeneration.

FRAP has revealed that the physical natures of the inner and outer bilayers of the adult stage of *S. mansoni* are substantially different (Foley et al. 1986). The outer bilayer shows that some areas are restricted in mobility, while the inner layer is completely restricted in mobility, suggesting the presence of gel-phase lipid domains in both, but especially the inner, layers. The mobility of the lipid probe, $C_{18}Fl$, the localisation of which is restricted to the outer monolayer in the adult *S. mansoni*, is decreased by the antihelminthic drug praziquantel (Lima et al. 1994a). However, the mobility of another probe, PKH2, which locates in outer and inner membranes, is increased by praziquantel. This suggests that praziquantel affects different membranes or membrane domains in different ways.

While $C_{18}Fl$ inserted into fluid domains in *S. mansoni* (4.9 $\times 10^{-9}$ cm^2 s^{-1}, Lima et al. 1994a) and the nematode *A. Viteae*

$(7.4 \times 10^{-8}$ cm^2 s^{-1}, Proudfoot et al. 1993), for nematode *Trichinella spiralis* no fluidity was detected at all using C_{18}Fl, in both resting and trypsin/bile-activated stages (Modha et al. 1996). Interestingly, activation of resting stages using caged cAMP resulted in increased fluidity of this probe in the surface (8.3×10^{-11} cm^2 s^{-1}). However, this value is 100–1000-fold less than for values found in the other organisms, suggesting that the probe may have inserted into a hydrophobic protein, as opposed to a lipid environment. This protein may be linked to the cytoskeleton.

As well as studying lipids, it is also possible to measure the fluidity of proteins and carbohydrates using fluorescently labelled antibodies and lectins. In the case of *S. mansoni*, praziquantel damaged areas of the adult surface show increased binding of fluorescently labelled lectins due to unmasking of carbohydrate residues, possibly in the lower bilayer (Lima et al. 1994b). The lateral mobility of these residues was found to be 3.53×10^{-8} cm^2 s^{-1} (Lima et al. 1994b). This is remarkably fast for an integral membrane protein (normally in the range of 10^{-10} to 10^{-11}, Jacobson et al. 1987). More likely is the possibility that these residues are associated with the GPI-linked glycoprotein antigens, exposed by praziquantel treatment (Redman et al. 1996). One other GPI-linked protein (Thy-1) has been found to have a mobility of 2–5×10^{-9} cm^2 s^{-1} in fibroblasts and lymphoid cells (Jacobson et al. 1987).

Of course, one has to be careful, with FRAP studies, to ensure that any lipid probes used do not alter the intrinsic properties of the membrane, resulting in artifacts (Zachowski 1993). Therefore it is important to check D_L and f, with respect to probe concentration used and amount taken up into the surface.

8.4
Distribution of Actin

Fibrous actin binds to phalloidin, but cell membranes cannot be penetrated by phalloidin. Fluorescent phalloidin may stain actin very close to the surface, when associated with special membranes (as in the schistosome spines) but, in general, staining can only be achieved if there is an extensive increase in permeability of the surface membrane. This can be accomplished by detergent or lysophosphatidyl choline treatment.

Staining after Permceabilisation

1. Fix the parasite in 0.5% formaldehyde in saline for 30 min at 4 °C or 37 °C, according to the experiment.

2. Wash the fixed parasites 6× with GMEM (without FCS).

3. Treat fixed parasites for 30 min 37 °C with 1% (w/v) Triton X-100 made up in GMEM. This disrupts the lipid phase of the surface membrane and exposes tegumental and muscle actin.

4. Wash parasites 6× with GMEM (without FCS). Resuspend in 1.0 ml GMEM in the well of a 24-well plate.

5. Add 10 μl of Rhodamine or Fluorescein phalloidin (stock solution 600 units ml^{-1} in ethanol). Incubate for 30 min and, without washing, observe the parasites under the fluorescent microscope. Washing is not necessary, as unbound probe does not fluoresce. Leave parasite under the sealed coverslip and observe over the next few days. As the phalloidin diffuses into the body muscle systems, more and more detail will be revealed (Fig. 6).

Results

Actin has been demonstrated in schistosomula membrane vesicles and surface preparations. Schistosomula muscle systems have been demonstrated (Fig. 6a). Adult muscle blocks before and after cooling the parasite at 4 °C can be readily observed; actin associated with the spines of the tubercles have been observed (Fig. 6b).

8.5
Assessment of Parasite Membrane Damage

Stock solutions

– 1 mg ml^{-1} in water. Hoechst 33258
– ethidium bromide, 1 mg ml^{-1} in acetone
– fluorescein and carboxyfluorescein diacetate

Damaged parasites can be readily detected by either the uptake or release of membrane impermeable compounds and through the damaged surface. DNA-binding compounds (Hoechst 33258, ethidium bromide) and fluorescein diacetate and carboxyfluorescein diacetate (hydrolysed within the cell

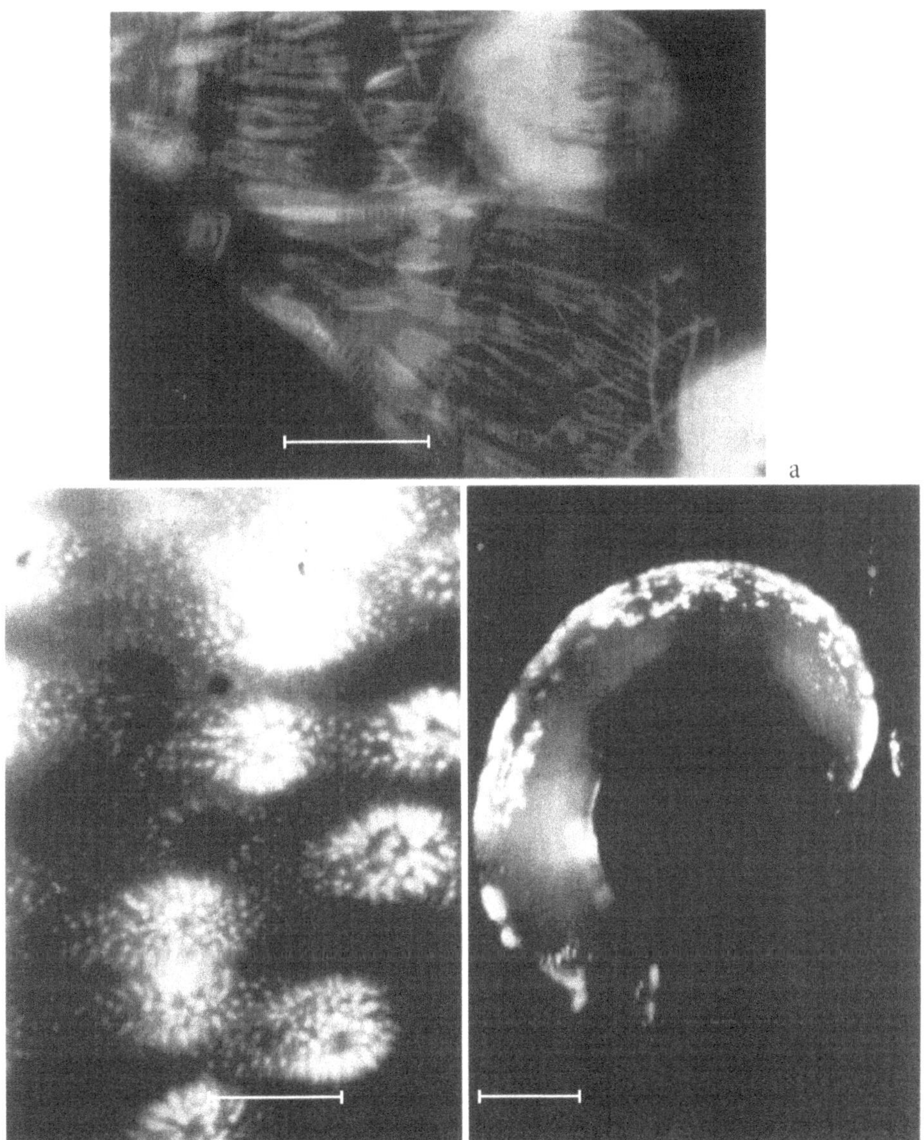

Fig. 6. a Actin found in muscle blocks of schistosomula treated with saponin (1%), and stained with rhodamine phalloidin. *Bar* 25 μm. **b** Actin in spines and **c** muscles of adult worms, stained by rhodamine phalloidin. *Bar* 25 μm **b,** 1000 μm **c**

to yield intracellular fluorescein or carboxyfluorescein) have been used effectively.

Damage revealed by nuclear fluorescence Parasites are incubated in Hoechst 33258 (1 μg ml^{-1}) or ethidium bromide (10 μg ml^{-1}) for 15 min and the parasites mounted for fluorescent microscopy without washing since the dyes do not fluoresce unless they bind to DNA. The damaged areas of the parasite is observed to give a blue fluorescence with Hoechst 33258 (Fig. 4) and a red fluorescence with ethidium bromide.

Damage revealed by lack of intracellular fluorescence Parasites are incubated in GMEM 10% FCS with 10 μg ml^{-1} fluorescein diacetate or carboxylfluorescein diacetate for 30 min at 37 °C. The parasites are washed six times with warm medium and examined by fluorescence microscopy. Parasites whose membranes are damaged appear non-fluorescent or partially fluorescent, while undamaged parasites appear very brightly fluorescent due to cytoplasmic fluorescein or carboxyfluorescein released by esterase activity from the membrane permeant diacetate form.

Damage revealed by uptake of fluorescent phalloidin Fluorescent phalloidin is membrane impermeable when damage to the surface membrane occurs, fluorescent phalloidin diffuses through the membrane and localised heavy staining of actin in cytoplasm or underlying muscle can be observed. Incubate parasites in GMEM plus FCS with 10 μl of rhodamine or fluorescein phalloidin (stock solution 600 units ml^{-1}) for 30 min 37 °C. Mount and observe fluorescence. Damaged areas will show actin staining (Fig. 6).

8.6
Localisation of Probes in Organelles

In Golgi apparatus NBD and BoDipy ceramides are concentrated in the Golgi apparatus. The BoDipy fluorophore alters its fluorescence emission maximum when concentrated. Red fluorescence detectable in the rhodamine filter set can be observed when concentrated, e.g. in the Golgi apparatus.

Parasites are incubated in 5 μg ml (stock solution 1 mg ml^{-1} in ethanol) for 30 min, 37 °C in GMEM plus 10% FCS. The parasites are washed six times in medium and cultured in

GMEM plus 10% FCS for 24 h. The ceramides are found to be concentrated in the Golgi apparatus of subtegumental cells and of the oesophageal glands of adult schistosomes (Moffat and Kusel 1992).

Rhodamine 123 is ethanol-soluble and stains mitochondria, which appear green/yellow through the fluorescein filter but red through the rhodamine filter. Stock solution 1 mg ml^{-1} in ethanol. Parasites are incubated in 10 μg ml^{-1} in water (cercariae) or GMEM plus 10% FCS for 30 min at room temperature (cercariae) or at 37 °C. **In mitochondria**

Wash the labelled parasites six times with the appropriate medium; mount in medium plus 50 μg ml^{-1} carbachol and observe.

Endocytosis can be studied by incubating cells or parasites with fluorescent beads or fluorescent dextrans. Dextrans coupled to pH-sensitive dyes have been used to study vesicle pH in *Plasmodium falciparum*. Incubation of the parasite with 10 μg ml^{-1} fluorescein dextran for 30 min in GMEM without serum followed by washing six times with medium with serum will reveal intracellular endocytosis vesicles. Using this method, endocytosis was not detected in schistosomes in vitro (Cushley and Kusel 1987). **In endocytotic vesicles**

8.7
Measurement of pH at the Surface of Parasites and Intracellular pH

A wide range of pH-sensitive dyes are available. Fluorescein itself alters fluorescence intensity with pH, but other dyes allow ratios of intensity to be made at two different wavelengths and these ratios change with pH. For example, carboxy SNARF-1, excited at 488 nm, shows high emission at 520 nm but low at 640 nm at pH 6 but low emission at 520 nm and high at 640 nm at pH 9. The appropriate filters are required for the microscope (Glen Spectra, Ltd., Wigton Gardens, Stanmore, Middlesex), which would need to be set up for quantitative fluorescence.

For surface pH measurements, parasites are incubated in 100 μg ml^{-1} dye (Carboxy-SNARF) in medium and mea- **pH measurements**

surements taken at 520 and 640 nm. pH measurements with buffer solutions of known pH allow a standard curve to be constructed and the extracellular or surface pH calculated from this curve.

Intracellular pH can be measured by using the membrane permeant carboxy-SNARF-AM (the ester of carboxy-SNARF), which is hydrolysed by intracellular esterases to the free salt.

Results In our experience, the SNARF dyes bind in their acid form to the schistosome surface, and gave misleading results on surface pH. This problem may also occur inside the parasite if dye binding occurs to certain structures. Dextran coupled to SNARF may not suffer the same drawbacks.

8.8
The Use of Caged Compounds

Many biological processes and systems involve the second messengers inositol trisphosphate (IP_3), Ca^{2+} ions and cyclic AMP (or cGMP), which directly, or more commonly indirectly (via binding proteins) exert an effect. We have been studying the changes in surface of larvae of the parasitic nematode *Trichinella spiralis* which, when activated by incubation in simulated mammalian culture conditions, are preferentially able to insert the fluorescent lipid probe octadecanoyl-aminofluorescein ($C_{18}Fl$) into their surface (Proudfoot et al. 1993). Thus, $C_{18}Fl$ can be used to monitor changes in surface lipophilicity following different treatments of the parasite, and since electron microscopy demonstrates that an increase in surface lipophilicity correlates with a loss of layers (the surface coat and the epicuticle) at the larval surface (Modha et al. 1994), $C_{18}Fl$ insertion may be used as a marker for a loss of these layers. *T. spiralis* larvae are also able to shed their surface coat and recently we have been able to observe this phenomenon using another fluorescent lipid probe called PKH-26 (Fig. 8b, c).

A dissection of the mechanism behind the phenomenon of the loss of surface layers following activation of larvae has necessitated the use of caged compounds (McCray and Trentham 1989), since the cuticle of the larvae is generally impermeable to regular compounds (Wright 1987). These are effector molecules that are rendered inert and more hydrophobic by linkage to a photosensitive group which is released (once the molecule is inside the cell) by a brief flash of ultraviolet light (λ = 300–370 nm) to release the active compound and produce an effect in milliseconds (Fig. 7 for some structures).

We have investigated the mechanism of $C_{18}Fl$ insertion into the surface of *T. spiralis* larvae using Diazo-2AM, a caged Ca^{2+} chelator (Adams et al. 1989) and caged cAMP (Nerbonne et al. 1984). Thus, larvae were incubated in a microtitre plate with 20 μM Diazo-2 AM and/or caged 10 μM cAMP (in medium RPMI 1640) or medium alone for 1 h at room temperature, to allow the caged molecules to permeate through the nematode cuticle, before flashing for 10 s from a Leitz UV light source (power: 150 $\mu W/cm^2$ at 310 nm and 13 mW/cm^2 at 365 nm, measured using a Macam UV 103 digital radiometer) 20 cm away from the parasites. This protocol for photoactivating the caged compounds has been worked out empirically. Using cells the preincubation time is of the order of minutes or the caged compound is microinjected into cells, with the subsequent photoactivation requiring a brief (ms) flashing at $\lambda<360$ nm. However, since the *T. spiralis* cuticle is much thicker and more robust than the cell membrane, the preincubation and flashing times are longer than usual to provide sufficient time for the compounds to permeate and sufficient energy to become photoactivated, bearing in mind the thickness of the cuticle, which would absorb some of the energy from the UV source. After photoactivation of the compounds, the larvae were immediately washed three times in PBS (containing 1 mM EGTA, to prevent exogenous Ca^{2+} entering the parasites and interfering with the experiment) and stimulated by incubation in simulated mammalian culture conditions for 30 min. The larvae were again washed three times and incubated in 10 $\mu g/ml$ $C_{18}Fl$ (containing 1 mM EGTA). Following extensive final washing, fluorescence on the surface of the larvae was quantified. Briefly depletion of Ca^{2+} with Diazo-2AM prior to stimulation of the larvae abolished $C_{18}Fl$ insertion. However, addition of caged cAMP to Ca^{2+}-depleted larvae restored probe insertion. This not only indicated that Ca^{2+} have a role in $C_{18}Fl$ insertion into the larval surface following

Procedure and results

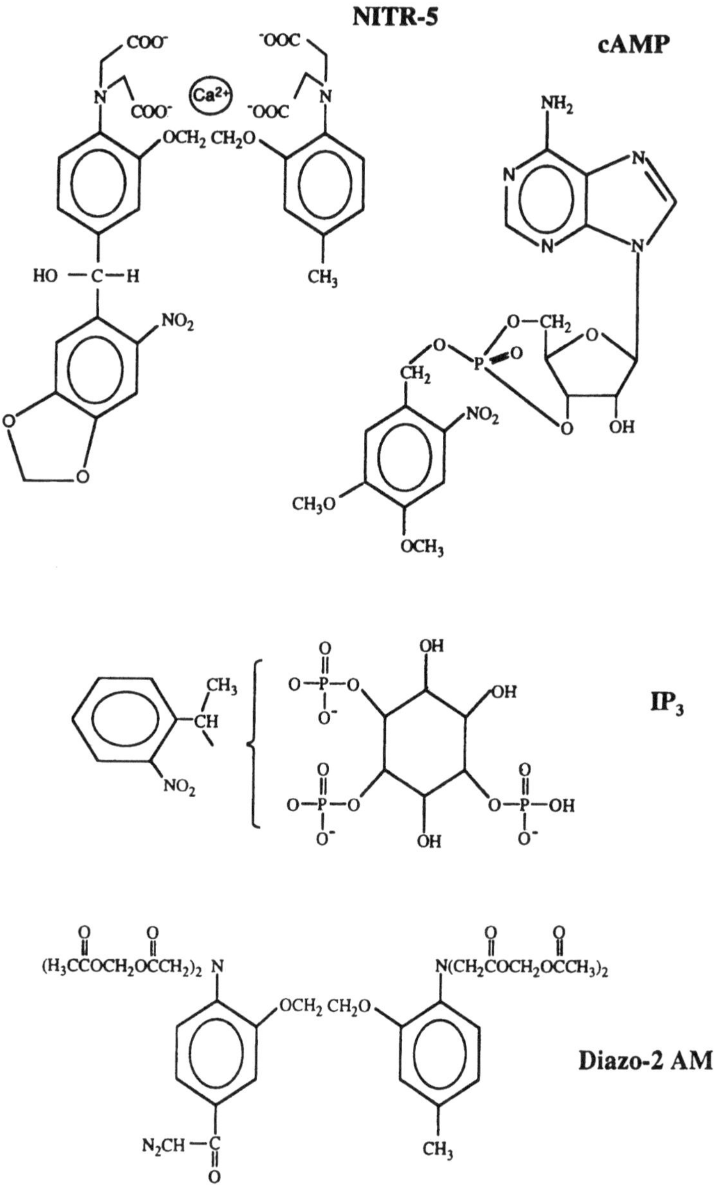

Fig. 7. Structure of caged compounds

stimulation of the larvae in culture, but that cAMP was also involved, and most probably downstream of Ca^{2+}, since it enabled the larvae to overcome the inhibition of probe uptake due to Ca^{2+} depletion.

The involvement of IP_3, Ca^{2+} and cAMP in insertion of $C_{18}Fl$ into the larval surface was subsequently confirmed by incubation of inactivated larvae (i.e. larvae in buffer alone) with 10 μM caged cAMP, 10 μM inositol trisphosphate (IP_3) and 0.5 mM NITR-5 (a caged Ca^{2+} releasant). Following photoactivation of the molecules, washing and labelling with $C_{18}Fl$ as described above, significant increases in probe insertion were observed, often with levels comparable to those of trypsin- and bile-activated larvae. It seems that stimulation of the IP_3-Ca^{2+}-cAMP pathway may lead to changes in surface lipophilicity and $C_{18}Fl$ insertions during activation of *T. spiralis* larvae (Modha et al. 1995).

The phenomenon of surface shedding in *T. spiralis* larvae (Fig. 4b, c) was also investigated using caged cAMP. Larvae incubated with 10 μM caged cAMP in PBS (containing 1 mM IBMX, to inhibit unwanted phosphodiesterase activity) and photoactivated as described above, when labelled with PKH-26, demonstrate an increased rate of surface shedding, indicating that cAMP plays a role in altering the surface properties of larvae leading to loss of the surface coat.

8.9
Use of Fluorescent Probes in the Analysis of Parasite Proteins

One aspect of the use of fluorescent probes which has yet to find widespread application in parasitology is the analysis of parasite proteins and enzymes. In this section we will deal with the use of fluorescence in examining binding proteins and will concentrate on the example of those which bind hydrophobic molecules.

Fluorescence, in its simplest application, can be used to examine binding of a fluorescent probe to a parasite component, but it can also provide a great deal more information. The principle is that the fluorescence of some probe molecules can change upon binding in two ways; the intensity can increase markedly (either up or down) or the wavelength (and therefore colour) of the emitted light can change. The latter property can be particularly useful in that it can provide information on the environment of the probe within a binding site. Probes which behave in this way are usually termed polarity- or environment-sensitive.

The human eye of an investigator with normal colour vision is an extremely good spectrophotometer in that it can obtain information on both comparative light intensity and wavelength. Much of what follows can therefore be achieved by merely holding test tubes containing mixtures of a fluorescent probe and the protein of interest over a transilluminator (take care to wear appropriate eye protection) or in the beam of a fluorescent microscope. This can provide useful guidelines, but in order to obtain quantitative and qualitative information, a fluorimiter is required. The commonest of these can be set for a given wavelength of excitation light (λ_{Exc}) and to measure the intensity of the emitted light (λ_{Em}). The settings needed can be obtained from the literature, from the manufacturer of a given fluorescent probe, or from biological data books in the case of fluorescent natural compounds. In cases where an environment-sensitive probe is in use, then a spectrofluorimeter is better, which will display the full spectrum of the emitted light and any unexpected changes to it. Such machines can also be used to measure the excitation spectrum of a probe in order to establish which λ_{Exc} is the best to use, but this information is usually obtainable from the literature on the probe to be used.

Equipment
- Fluorimeter (for single wavelength measurements) or spectrofluorimeter (for analysis of spectrum of emission and changes in it).
- Cuvettes, usually about 2–3-ml capacity and 1 cm^2 in cross section. These need to be specially purchased for fluorescence. The best cuvettes are quartz glass, but plastic disposable can be used, particularly if dealing with a dangerous probes or proteins. Manufacturers of cuvettes can usually provide an absorption spectrum of their cuvettes and this can be consulted to ensure that neither the excitatory nor the emitted light will be significantly absorbed. Polystyrene cuvettes are only useful down to approximately 340 nm, but polymethacrylate (acrylic) will operate to about 290 nm, in contrast to quartz, which will extend to 220 nm. The choice essentially depends on the probes used. Cuvettes should always be placed in the sample holder in the same orientation with respect to incident light and measuring beams.

The widest selection of probes is available from Molecular **Fluorescent**
Probes Inc, Eugene, Oregon, USA, who can provide a hand- **probes**
book listing their products and the uses to which they can be
put. Having decided on a particular probe, it is worthwhile
checking out a few different manufacturers for the most
economically favourable supplier of these expensive items,
although high quality and purity is mandatory.

Probes are available to examine a wide range of properties,
but we will focus here on binding proteins. Fluorescence has
been used to examine, for example, the environment of
haem-binding sites (MacGregor and Weber 1986) and this
might be valuable for the study of nematode haemoglobins
and myoglobins ("nemoglobins" Blaxter 1993). Here we will
concentrate on a type of binding protein which is currently
receiving increasing attention in parasitology, and provides a
good example of the power and adaptability of fluorescent
techniques. These are fatty acid-binding proteins, many of
which will also bind retinol (vitamin A). Binding can be
detected by fluorescence using three different principles depending on the probe used; enhancement of fluorescence,
quenching of fluorescence, or enhancement plus a change in
the wavelength of emission. A change in the wavelength is the
property of polarity-sensitive probes, which are usually used
because the binding sites for hydrophobic ligands are
themselves highly apolar, and the change in the environment
of the probe upon binding will radically alter its fluorescence.
The types of probe used are usually fatty acids to which are
covalently attached fluorescent reporter groups, such as
dansyl (Wilkinson and Wilton 1986) or anthracene (Wootan
et al. 1990). The fluorescence of both is increased and shifted
to the blue end of the spectrum upon binding.

The following is an example of the methods used for examining fatty acid binding proteins, using serum albumin as
a control protein (Fig. 8). The probe is 11-[(5-dimethylami-
nonaphthalene-1-sulphonyl)amino]undecanoic acid, usually
abbreviated to DAUDA. This is the most commonly used
method for fatty acid-binding proteins, but your protein may
be more specific and require a different probe. The principles
are, however, the same.

1. The protein needs to be purified from the parasite using **Protein**
   methods which do not lead to denaturation. This, **preparation**

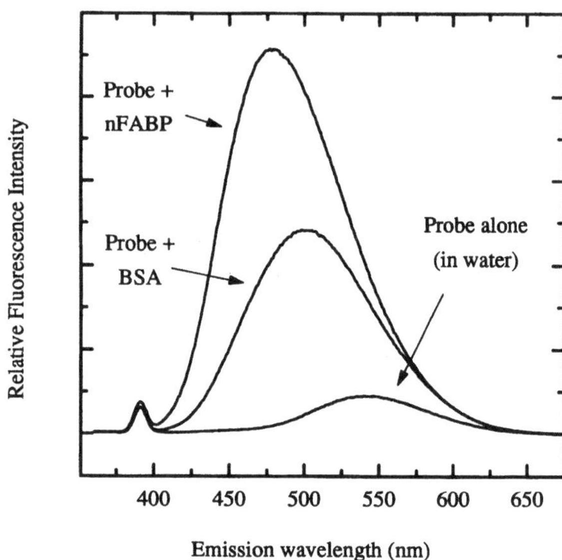

Fig. 8 Examining fatty acid binding proteins

therefore, usually excludes excision and electroelution from SDS-PAGE gels, unless recovery of fully functional protein can be achieved, and all detergent can be removed. The purification of recombinant protein produced in, for example, *E. coli*, often uses detergents, and detergents may also remain in binding sites after purification.

2. Detergent can be removed by passage of the protein sample down small disposable columns such as the Extracti-Gel columns supplied by Pierce, Rockford, Illinois, following the manufacturer's instructions. The higher the protein concentration, the more efficient the removal of detergent on a molar basis.

3. For the fluorescence assays, it is desirable to have stock protein concentrations is excess of 1 mg/ml. If the protein is less than this, it can be concentrated by centrifugal ultrafiltration using Centricon devices supplied by Amicon, or similar products from other manufacturers.

4. Knowing the precise concentration of the protein is not critical unless affinity measurements are to be made. The concentration of the protein can be determined by

conventional techniques, but we have found that most methods, such as dye-binding assays or even amino acid analysis, are error-prone. The best method we have found is to produce the extinction coefficient for the protein and to use this to calculate the concentration by absorbance at 280 nm. This requires a knowledge of the amino acid content of the protein, which is routinely available for recombinant material. This method requires complete removal of detergent, because many detergents absorb at wavelengths close to 280 nm. The formula for extinction coefficients is, as per Gill and Hippel (1989):

Molar extinction = (number of tyrosines $\times$ 1280)
+ (number of tryptophans $\times$ 5690).

So, if the protein has five tyrosines and one tryptophan, then the molar extinction coefficient (ε_{280}) will be 10810 $M^{-1}cm^{-1}$. For absorbance of a 1 mg ml^{-1} solution at 280 nm (A_{280}), divide by the molecular weight (MW) of the protein.

So, if the A_{280} is 1.28, then its molarity will be 0.118 mM. If the protein's MW is 15257, then its concentration will be 1.8 mg ml^{-1}.

Dilution of the protein will probably be necessary in order to bring the absorbance within the linear measurement range of the spectrophotometer being used. Most laboratory spectrophotometers give best results if the absorbance reading is within the range 0.3 to 0.6, but, for better machines, 0.2 to 1.5 may suffice.

1. Set the fluorimeter to the appropriate λ_{Exc} and λ_{Em}. For DAUDA, these would be 345 and 500 nm, respectively. For scanning spectrofluorimeters, set the machine to scan a range of λ_{Em}, which encompasses both the wavelength of probe in buffer and the greatest shift it is known to undergo upon binding, with a margin on either side for the whole spectrum to be recorded. For DAUDA, this would be 350 to 650 nm.

Fluorescence measurements

2. Scan the buffer to be used to check for any intrinsic fluorescence. For the same reason, now add protein (about 10 or 20 μl of a 10 mg ml^{-1} solution, or equivalent) and remeasure. Ideally, neither should be fluorescent, but if this cannot be avoided, then the data obtained can be

used for subtraction from data obtained with fluorescent probe.

3. Prepare the fluorescent probe in a stock solution of 1 to 5 mg ml^{-1} and store at −20 °C. Dilute freshly for use to 1 μM in the buffer solution used for the test protein. Load 2 ml into a fresh cuvette and measure the fluorescence or carry out an emission scan.

4. Add 5 μl of test or control protein (bovine serum albumin is a readily available positive control protein) solution, mix by gentlepipetting with a 200-μl automatic pipette or stirring bar if stirrer fitted to base of sample holder, avoiding making bubbles (which will cause light scatter and can stick to the inside of cuvettes) and remeasure. In the case of no, or little, effect on fluorescence, add 10, then 20, then 50 μl protein solution. Remember to correct for dilution if absolute fluorescence values are required.

5. If changes in fluorescence are observed, then test for specificity of binding by adding some of the suspected ligand, e.g. oleic acid. The stock solution of this needs to be dissolved in ethanol, and freshly diluted in buffer to about 0.1 or 1 mM. Add 10 μl of this and remeasure. Competition will result in a reduction in fluorescence. Add progressively more to observe stepwise loss of fluorescence (Fig. 6).

Affinity measurements These will not be described in detail here, but measurement of the affinity of probe for the protein, or the suspected ligand (such as oleic acid), can be obtained by progressive additions of protein to probe solution, or addition of competitor to a probe:protein mixture. This can also provide information on the number of binding sites per protein molecule. For such experiments, the concentration of protein, probe and competitors needs to be known as accurately as possible. Usually, concentrations of protein and probe can be determined directly by absorbance as long as the extinction coefficients are known.

Problems • Fluorescence experiments can often be confounded by the intrinsic fluorescence of the proteins concerned. Tryptophan and tyrosine will both fluoresce, but this can be avoided by performing experiments with λ_{Exc} other than close to the absorbance maximum of these amino acids. It is usually, therefore, a good idea to measure the fluorescence of the protein in buffer alone at the excitation wavelength to be used, which will also be valuable to detect other sources of fluorescence such as ligands already bound, or contaminants.

- Water will itself fluorescence, at a wavelength which is related to the excitation wavelength. This is termed Raman fluorescence, and the peak can be readily identified by measuring the fluorescence of water alone. It is usually a problem only at very low fluorescence levels.

- Scatter of the excitatory light will be observed at that wavelength, or multiples thereof. This is particularly a problem in turbid solutions of protein or with lipid micelles.

- Avoid using high concentrations of probe, because the inner filter effect can result in aberrant readings. We have found that 1 μM is ideal for most purposes.

References

Adams SR, Kao JPY, Tsien RY (1989) Biologically useful chelators that take up Ca^{++} upon illumination. J Am Chem Soc 111:7957–7968

Blatt E, Sawyer H (1984) Depth-dependent quenching in micelles and membranes. Biochim Biophys Acta 822:43–62

Blaxter M (1993) Nemoglobins; Divergent nematode globins. Parasitol Today 9:353–360

Cushley W, Kusel JR (1987) Interactions of the plant toxin ricin with different life-cycle stages of *Schistosoma mansoni*. Mol Biochem Parasitol 24:67–71

Foley M, MacGregor A, Kusel JR, Garland PB, Downie T, More I (1986) The lateral diffusion of lipid probes in the surface membrane of *Schistosoma mansoni*. J Cell Biol 103:807–818

Gill SC, von Hippel PH (1989) Calculation of protein extinction coefficients from amino acid sequence data. Anal Biochem 182:319–326

Hackstadt T, Sadmore MA, Rockey DD (1995) Identification of location of the probe in the outer membrane: (1995) Lipid metabolism in *Chlamydia trachomatis*-infected cells directed trafficking of Golgi-derived sphingolipids to the chlamydial inclusion. PNAS USA 92:4877–4881

Harris DA, Bashford CL (1987) Spectrophotometry and spectrofluorimetry: a practical approach. IRL Press, Oxford

Haughland RP, Larison KD (1992) Handbook of fluorescent probes and research chemicals 5th edn (1992-1994). Molecular probes Inc Eugene, Oregan

Jacobson K, Ishihara A, Inman R (1987) Lateral diffusion of proteins in membranes. Annual Reviews in Physiology 49:163-175

Kennedy M, Foley M, Kuo Y-M, Kusel JR, Garland PB (1987) Biophysical properties of the surface lipid of parasitic nematodes. Mol Biochem Parasitol 22:233-240

Kusel JR, Gordon JF (1989) The properties of surfaces of parasites under drug and immune attack. Parasite Immunol 11:431-451

Lima SF, Vieira LQ, Harder A, Kusel JR (1994a) Effects of culture and praziquantel on membrane fluidity parameter of adult *Schistosoma mansoni*. Parasitology 109:57-64

Lima SF, Vieira LQ, Harder A, Kusel JR (1994b) Altered behaviour of carbohydratebound molecules and lipids in areas of the tegument of adult *Schistosoma mansoni* worms damaged by praziquantel. Parasitology 109:469-477

Lipsky NG, Pagano RE (1985) A vital stain for the Golgi apparatus. Science 228:745-747

McCray JA, Trentham DR (1989) Properties and uses of photo reactive caged compounds. Annu Rev Biophys Chem 18:239-270

MacGregor RB, Weber G (1986) Estimation of the polarity of the protein interior by optical spectroscopy. Nature (Lond) 319:70-73

Modha J, Kusel JR, Kennedy MW (1995) A rôle for second messengers in the control of activation-associated modification of the surface of *T. spiralis* infective larvae. Mol Biochem Parasitol 72:141-148

Modha J, Robertson WM, Kennedy MW, Kusel JR (1994) The biology of activation of *Trichinella spiralis* muscle-stage larvae. In: Proceedings of the Eighth International Conference on Trichinellosis, WC Campbell, E Pozio, F Bruschi (eds) Istituto Superiore di Sanita Press, Rome, p 161-166

Modha J, Roberts MC, Kennedy MW, Kusel JR (1996) Induction of surface fluidity in *Trichinella spiralis* larvae during penetration of the host intestine: simulation by cyclic AMP in vitro. Parasitology 113 (in press)

Moffat D, Kusel JR (1992) Fluorescent lipid uptake and transport in adult *Schistosoma mansoni*. Parasitology 105:81-89

Nerbonne JM, Richard S, Nargoet J, Lester HA (1984) New photoactivable cyclic nucleotides produce intracellular jumps in cyclic AMP and cyclic GMP concentrations. Nature (Lond) 310:74-76

Pagano RE, Martin OC (1988) A series of fluorescent N-acylsphingosines: synthesis; physical properties and studies in cultured cells. Biochemistry 27:4439-4445

Pagano RE, Martin OC, Kang HC, Haughland RP (1991) A novel fluorescent ceramide analogue for studying membrane traffic in animal cells. J Cell Biol 113:1267-1279

Proudfoot L (1991) Dynamic aspects of the nematode surface. Ph.D. Thesis, University of Glasgow

Proudfoot L, Kusel JR, Smith HV, Harnett W, Worms MJ, Kennedy MW (1993) Rapid changes in the surface of parasitic nematodes during transition from pre-to post-parasitic forms. Parasitology 107:107-117

Redman CA, Robertson A, Fallon PG, Modha J, Kusel JR, Doenhoff MJ, Martin RJ (1996) Praziquantel: An urgent and exciting challenge. Parasitology Today 12:14–20

Van Meer G (1993) Transport and sorting of membrane lipids. Curr Opinion Cell Biol 5:661–673

Wilkinson CI, Wilton DC (1986) Studies on fatty acid-binding proteins. The detection and quantification of the protein from rat liver by using a fluorescent fatty acid analogue. Biochem J 238:419–424

Wootan MG, Bass NM, Bernlohr DA, Storch J (1990) Fatty acid binding sites of rodent adipocyte and heart fatty acid binding proteins. Characterization using fluorescent fatty acids. Biochemistry 29:9305–9311

Wright KA (1987) The nematodes cuticle – its surface and the epidermis: function, homology, and anology. J Parasitol 73:1077–1083

Wright KA, Hong H (1988) Characterization of the accessory layer of the cuticle of muscle larvae of *Trichinella spiralis*. J Parasitol 74:440–451

Zachowski A (1993) Phospholipids in animal eukaryotic membranes: transverse assymetry and movement. Biochem J 294:1–14

Chapter 9

Detection and Identification
of Parasite Surface Carbohydrates by Lectins

George A. Ingram

9.1
Introduction

Lectins (often termed agglutinins) are proteins, usually glycoproteins, derived from both animal and plant material, that recognise and bind to specific sugar moieties (or certain glycosidic linkages) of polysaccharides, glycoproteins and glycolipids present on cell membrane surfaces or soluble in biological fluids (Lis and Sharon 1986; Sharon and Lis 1989). Lectin specificity is usually defined in terms of the monosaccharides or simple oligosaccharides that inhibit lectin-mediated agglutination (cells), precipitation (soluble molecules) or binding (membranes).

In order to detect membrane-bound carbohydrates, commercial lectins are employed either unconjugated (agglutination/precipitation reactions) or conjugated with a marker (cell or tissue surface membrane). The labels used include fluorescein isothiocyanate, enzymes (e.g. peroxidase) and biotin. These lectin-based techniques encompass the use of light, ultraviolet and electron microscopy or fluorimetry.

Lectins have been widely used in parasitology (Jacobson and Schnur, 1996) to detect and ascertain the types of saccharide residues on the surface of trypanosomatid flagellates, e.g. *Crithidia* (Esteves et al. 1987) *Leishmania* (Guegnot et al. 1984) and *Trypanosoma* species (Mutharia and Pearson 1987), to demonstrate differences between parasite growth and stationary phases (Grogl et al. 1987; Jacobson and Schnur 1990) and to distinguish between the different stages of the trypanosomatid life cycle (Andrade et al. 1991). In addition, lectins have been employed to differentiate between parasite species (Maraghi et al. 1989) and to identify various strains

(Schnur and Jacobson 1989) and stocks (Schottelius 1989) of these protozoans. Additionally, lectins have been used, albeit less often, to elucidate the nature of glycoconjugates on the external surface of nematodes (Furman and Ash 1983), trematodes (Coles et al. 1988) and cestodes (Ham et al. 1988; Leducq et al. 1990).

The following methods are as applied to kinetoplastid flagellate parasite species and insect vector tissues but can be adapted to study different parasitic organisms, e.g. helminths and other protozoans. Examples of lectins routinely used in investigations on parasites are given in Table 1, together with their sugar-binding specificities.

9.2
Lectin-Mediated Trypanosomatid Parasite Agglutination

Parasites, e.g. trypanosomes, *Leishmania* and *Crithidia*, would usually have been isolated from host/vector tissue and placed into an appropriate nutrient medium for cultivation purposes (Taylor and Baker 1978; Schnur and Jacobson 1987). Alternatively, blood stream forms can be directly obtained from infected animals possessing a high level of parasitaemia by passage of blood through a DEAE cellulose anion exchange chromatography column according to the method of Lanham and Godfrey (1970).

Various workers use different culture media for certain parasite species and in-depth detail regarding types of parasite growth media is not within the scope of this chapter. However, the following direct agglutination technique is as applied to *Trypanosoma brucei brucei*, *Leishmania hertigi hertigi* and *Crithidia fasciculata* cultured procyclic, promastigote and choanomastigote forms, respectively.

Cunningham's medium containing *T.b. brucei* or Locke's **Agglutination** solution containing *Leishmania* or *Crithidia* is centrifuged at 1000 *g* for 10 min in a conical base tube. The parasite pellet obtained is resuspended by agitation in phosphate buffered saline-PBS (containing Ca^{2+} and Mg^{2+} ions) pH 7.3 (see Note 5), washed in the same buffer and the suspension recentrifuged.

Table 1. Common lectins (agglutinins) routinely used in parasitology together with their sugar-binding specificities and carbohydrates used as inhibition controls

Lectin	Designation	Specificity	Inhibitors
Canavalia ensiformis (Jack bean)	Con A	α-D-mannose and α-D-glucose	α-Ch$_3$-mannopyranoside, α-CH$_3$-D-glucopyranoside, D-mannose and D-glucose
Ulex europaeus (Gorse)	UEA-I	α-L-fucose	α-L-fucose
Limulus polyphemus (Horseshoe crab)	LPA	NeuNAc[a]	NeuNAc, Fetuin
Triticum vulgaris (Wheat germ)	WGA	(D-GlcNAc)$_2$	N, N^1,-diacetylchitobiose, GlcNAc
Glycine max (Soybean)	SBA	D-GalNAc	GalNAc, D-galactose
Sophora japonica (Japanese pagoda tree)	SJA	β-D-GalNAc	α-Lactose, GalNAc
Lens culinaris (Lentil)	LCA	α-D-glucose and α-D-mannose	As per Con A
Arachis hypogaea (Peanut)	PNA	β-D-gal(1–3) Gal-NAc	GalNAc, D-galactose
Solanum tuberosum (Potato)	STA	(D-GlcNAc)$_3$	N, N^1, N^{11}-triacetyl-chitotriose, N,N^1-diacetylchitobiose, GlcNAc
Tetragonolobus purpureas (Lotus)	TPA	α-L-fucose	α-L-fucose
Dolichos biflorus (Horse gram)	DBA	a-D-GalNAc	GalNAc
Bandeiraea simplicifolia (Bandeiraea)	BS-I	α-D-galactose and α-D-GalNAc	GalNAc, D-galactose

[a]NeuNAc = N-acetylneuraminic acid.
GlcNAc = N-acetyl-D-glucosamine.
GaNAc = N-acetyl-D-galactosamine.
gal = galactose.

The process is repeated a further three times with washing followed by centrifugation.

The cells are enumerated using an improved Neubauer haemocytometer and the parasite suspension finally adjusted with PBS to approximately $4\text{--}5 \times 10^6$ *T.b. brucei* or 1×10^7 *Leishmania* or *Crithidia* cells/ml (see Note 2).

To each well of a flat-bottomed 60-well microtitre (Terasaki) plate are added 5 μl PBS pH 7.3 (containing 3 mM $CaCl_2.2H_2O$ and either 17 mM $MgCl_2.6H_2O$ and/or $MnCl_2.4H_2O$ depending upon lectin divalent cation requirement; see Note 1).

Two-fold serial dilutions of each lectin, initial concentration 1000 μg/ml in PBS (with appropriate metal ion added) ranging from 500 to 0.49 μg/ml are prepared with thorough mixing. To all wells are dispensed 5 μl adjusted parasite suspension.

Control wells comprise parasites plus buffer alone to check for the presence of non-specific parasite agglutination, occasionally associated with cultured forms (see Note 3).

The plates are kept internally humid (to avoid drying out of the well mixtures) by lining the internal margin of each plate with four filter paper strips moistened with distilled water. The plates are covered and incubated at 27 °C for 1 h.

The degree of parasite agglutination with positive lectins is visually assessed using an inverted microscope at ×200 magnification. The minimum concentration of each lectin (i.e. highest dilution) that just causes agglutination of the trypanosomatid flagellates is recorded and regarded as the endpoint titre value. The degree of agglutination can be scored using a relative scale of 4+ (100% agglutination), 3+ (75%), 2+ (50%), 1+ (25%), tr (trace) and O (negative; see Note 4).

9.3
Fluorescein Isothiocyanate (FITC)-Conjugated Lectins

Culture or blood stream forms of parasites are washed four times each by centrifugation at 1000 g for 10 min in PBS pH 7.4 containing 0.5% bovine serum albumin (BSA) and 1 mM glycerol and adjusted to a final concentration of 1×10^6 cells/ml.   **Isolated parasites**

Equal volumes (usually 75 μl) of parasites and FITC-conjugated lectin (100 μg/ml), prepared in PBS/1 mM glycerol pH 7.4 buffer are added together in a test tube and mixed thoroughly. The mixtures are incubated in the dark at 5 °C for 1 h (or on ice) with occasional mixing.

The cells are washed three times by centrifugation using 10 mM Tris-HCl buffer pH 8 and finally resuspended in a test volume of 75 μl in Tris-HCl/glycerol (1:1 v/v).

A drop of the treated cell suspension is placed on a microscope slide and examined initially by light microscopy for parasite presence and then using a UV fluorescence microscope under incident fluorescence. The fluorescent intensity of FITC-lectin binding to the parasite cell surface is designated as either 3+ (intense), 2+ (intermediate), 1+ (weak to trace) and 0 (no fluorescence).

As an alternative, the relative intensity of yellow-green cell surface associated fluorescence can be performed using flow cytofluorimetry or fluorescence-activated cell sorting according to manufacturer's instructions. Should these methods be employed, the treated and washed parasites are fixed in 1–2% formaldehyde solution for 5–10 min and rewashed twice in PBS/glycerol/BSA buffer prior to examination. These techniques would give an objective evaluation and comparative indication of the amounts of fluorescence caused by the binding of lectins to specific sugars, if present, on the parasite membrane surface (see Note 10).

Blood-parasite smears Thin blood smears from animals infected with trypanosomatid flagellates are prepared; a high level of parasitaemia is normally required. The smears are fixed in 95% methanol for 10 min and the microscope slides washed in PBS pH 7.3 (containing Ca^{2+}, Mg^{2+} and/or Mn^{2+} ions) three times, each for 3 min.

The slides are treated with FITC-lectin conjugate (75 μg/ml) or FITC alone (100 μg/ml) and incubated for 45 min at 5 °C or 30 min at 21 °C.

They are then thoroughly washed three times, each for 5 min, with PBS/glycerol (1:1 v/v) and finally washed with Tris-HCl/glycerol buffer pH 8.

The sample is mounted in Tris-HCl/glycerol buffer under a cover slip and examined for fluorescent activity using a Leitz Dialux E-21 fluorescence microscope.

Whole tissue The surfaces of intact vector tissues to which parasites may attach in order to undergo morphogenesis during transformation from one life-cycle stage to the next or attachment sites prior to transmission may also be studied. These investigations, in part, enable detection of carbohydrate moieties on a membrane surface that might act as binding sites or facilitate parasite adhesion (Ingram and Molyneux 1991). For example, tsetse fly (*Glossina* spp.) and mosquito (*Anopheles* spp. and *Aedes aegypti*) salivary glands and midguts have been used with FITC-conjugated lectins (Perrone et al. 1986; Okolo et al. 1988, 1990; Rudin and Hecker 1989; Mohamed et al. 1991; Mohamed and Ingram 1993).

Mosquito midguts and salivary glands are dissected according to the methods of Hayes (1953) and Goma (1966), respectively. With the aid of a binocular microscope, tsetse fly salivary glands are withdrawn from the thorax and abdomen by carefully pulling the head away from the thoracic region with fine forceps whilst placing a seeker needle, using light pressure, on the abdominal surface. The head is withdrawn away from the thorax very slowly to allow the paired salivary glands to be exposed and removed from the body into saline on a microscope slide.

To dissect tsetse midguts, preferably 24 h post-blood meal, into PBS pH 7.3 (containing divalent cations), two pairs of fine forceps are used. One pair holds the thoracic region whilst the other pair enables grasping of the posterior abdominal segments for pulling out the alimentary tract.

The hindgut (and Malpighian tubules) are removed from the posterior end of the midgut prior to examination.

The tissues are placed on a cavity slide and fixed in 250 mM glutaraldehyde in 100 mM sodium cacodylate buffer pH 7.3 for 45 min to 1 h at 21 °C.

The tissues are washed three times, each 5 min, with care using PBS pH 7.3 and a Pasteur pipette. The tissues can either be stored at 4 °C in sterile PBS (24–72 h) or used immediately.

The tissues are incubated with FITC-conjugated lectins at concentrations of 30 µg/ml (see Note 9) in cacodylate buffer for either 1 h at 4 °C or 45 min at room temperature. Following incubation, the tissues are washed three times in PBS, each 10 min, and finally in Tris-HCl buffer pH 8.

The samples are mounted in a drop of Tris-HCl/glycerol buffer (1:1 v/v) pH 8, covered with a cover slip and examined as before for fluorescence.

9.4
Peroxidase-Labelled Lectins

Detection using Peroxidase-Labelled lectins Insect vector salivary glands and gut tissue are fixed and washed exactly as per FITC-labelled lectins except that they are treated with 30 μg/ml peroxidase-labelled lectins (see Note 9) prepared in either sodium cacodylate buffer or PBS (containing divalent cations) pH 7.4.

The mixtures are incubated at 25 °C for 1 h, after which time the samples are washed in PBS three times, each 4 min.

The substrate 3-amino-9-ethylcarbazole is prepared fresh by dissolution of a tablet with stirring (Sigma) in 1 vol dimethylformamide mixed with 19 vol 50 mM acetate buffer pH 5 (see Note 5).

To this mixture add 30 μl fresh 30% H_2O_2 immediately before the tissues are incubated with the substrate solution for 1 h at 25 °C in the dark and then wash thoroughly as before.

The tissues are then mounted in PBS/glycerol buffer and visual assessment of the degree of colour intensity of the red-brown insoluble end product performed by light microscopy (see Notes 8 and 11).

9.5
Biotinylated Lectins

Detection using Biotinylated lectins Tissues (e.g. salivary glands, peritrophic membrane or mid- and foregut samples from insect vectors) or parasite cells are refixed and washed as for fluorescence microscopy prior to detection of lectin binding and examination by electron microscopy. The samples are incubated with biotinylated lectin at a concentration of 30 μg/ml (see Note 9) in either PBS (containing divalent cations) or sodium cacodylate buffer, both at pH 7.3, for 1 h at 25 °C.

Appropriate controls (see Sect. 9.8) are included and treated in the same way as the biotin-lectin conjugate.

The samples following incubation with biotin-lectin are washed three times, each 5 min, with PBS (in the case of parasite cells by centrifugation at 1000 g). The samples are incubated for 1 h at room temperature in a 150 μg/ml solution of avidin-ferritin conjugate prepared in buffer and then rewashed as before.

The samples are postfixed in 1% (w/v) osmium tetroxide in sodium cacodylate buffer or PBS at 10 °C for 2 h and dehydrated through a graded ethanol series and embedded in Agar 100 resin (Taab).

Ultrathin sections, cut on a LKB Ultratome III, are stained in saturated uranyl acetate and Reynolds' lead citrate for 5 and 10 min each respectively to intensify contrast.

Treated samples are examined on an AEI Corinth 600 transmission electron microscope at 60 kV. The visual assessment of labelling density is similar to that for fluorescence microscopy (see Note 11). Variations and use of electron microscopical techniques are reported elsewhere in the Chapter 7 by Hemphill and Croft.

9.6
Fixatives and Enzymatic Treatments

Lectin binding usually requires specific terminal sugar determinants of complex cell membrane glycoproteins and glycolipids (Sharon and Lis 1989). Certain enzymes, e.g. neuraminidase, other glycosidases and proteases are used to react upon cell surface glycoconjugates to degrade protein or carbohydrate moieties. The action of glycosidases in some cases would remove specific terminal sugar residues of oligosaccharides containing certain glycosidic linkages that would confer lectin-binding activity. The use of proteolytic enzymes or neuraminidase would split appropriate surface ligand molecules and expose cryptic membrane receptors (tentative sugar-binding sites) that specific lectin would bind with more easily, depending upon the stereochemical configuration and nature of the newly revealed binding sites (Uhlenbruck and Rothe 1974; Schauer 1985). The use of proteases, usually trypsin, is of particular importance when salivarian trypanosomes, that synthesise variant antigen-specific surface coat glycoproteins (VSGs), are being tested, e.g. *Trypanosoma brucei* complex. It is often necessary to primarily remove VSG molecules that block oligosaccharide-lectin interaction on untreated, intact cells in order to expose new lectin-binding sites.

The use of fixative agents would disrupt the surface VSG layer and increase lectin binding. Treatment of tissues or parasites with fixative also denatures membrane surface protein and allows in some cases specific sugar residues to be exposed, thus enabling better binding of lectins (see Note 7).

Fixative materials

Prior to examination with lectins, parasite cells can be fixed usually with 1.5% formaldehyde or 2.5% glutaraldehyde in PBS/glycerol pH 7.3 or sodium cacodylate buffer or other suitable fixative. The cells are incubated in the fixative solution for either 1 h at 5 °C or 30 min at 25 °C and washed three times with buffer as before.

The cells are then incubated with PBS containing 100 mM NH_4Cl (to prevent non-specific lectin binding) at either room temperature for 15 min or 5 °C for 45 min and rewashed three times before conjugated-lectin application.

Fixation of vector tissue (e.g. salivary gland or guts) can likewise be undertaken as above; 2.5% glutaraldehyde in either sodium cacodylate or PBS buffers pH 7.2 to 7.4 is routinely used with insect vectors (Rudin and Hecker 1989; Mohamed et al. 1991; Okolo et al. 1990; Mohamed and Ingram 1993).

Enzyme-treated samples

As further confirmation of binding specificity determined by saccharide inhibition, unfixed parasite cells or tissue can be exposed to selected glycoside hydrolases (e.g. glucosidase, galactosidase). Unfixed parasite cells or tissues or blood smears that have been washed with appropriate buffer pH 7.3 are incubated in PBS pH 7.8 containing 500 μg/ml trypsin for 25 min at 25 °C or 50 μg/ml from 45 min to 1 h at 15 °C.

The samples are then washed three times with buffer appropriate for the test and then applied to the desired lectin method.

In the cases of neuraminidase and glycoside hydrolases, these enzymes are used at concentrations of 500 and 200 μg/ml prepared in PBS at pH 6.5 and 7 respectively. Incubation of sample and enzyme is performed at 25 °C for 30 min followed by washing in buffer prior to assay (see Note 6).

9.7
Histological Sections

To demonstrate/detect carbohydrate moieties on cells of vector tissues, or parasites within tissues/cells, basic histological sectioning of tissue is performed. This process can vary according to different laboratories with regard to fixative type, fixation time, histological preparation and section thickness. Once processed, the tissue can be treated with lectin conjugates in a protocol similar to those cited in Sections 9.3, 9.4 and 9.5 with appropriate controls (Sect. 9.8) or optional enzyme treatment (Sect. 9.6).

9.8
Controls

Competitive carbohydrate inhibition is performed to confirm specificities of both lectin-mediated parasite agglutination and conjugated lectin binding to the parasite/tissue surface. Stock solutions of various sugars, each of equivalent molarities (0.3 M), are prepared in PBS (containing divalent cations) pH 7.3. Polysaccharides and selected glycoproteins with known terminal chain sugar types (e.g. fetuin) are used at 5 and 10 mg/ml respectively.

Examples of various lectin-specific sugars used are given in Table 1. However, different structural variations of the same sugar types can also be used to enable more accurate specific sugar configuration determination. Non-specific sugar choice is left to the discretion of the investigator(s).

Parasite Agglutination

Equal volumes (usually 2 μl) of lectin dilution and either non-specific or specific sugar(s) or glycoprotein(s) final concentrations 150 mM are mixed and incubated for 30 min at 25 °C before use in the agglutination test. **Controls for saccharide and glycoprotein inhibition**

Four microlitres of adjusted parasite suspension are then added to all wells and conditions of agglutination are as described above. Inhibition of agglutination is considered as the reduction in end-point titre, upon comparison to the controls, by at least 2 wells and is scored as 1+. A reduction in titre by three wells is designated 2+, four wells as 3+, five wells as 4+, six or more wells, 5+ and total inhibition as 6+. Inhibition of one well or no inhibition is designated ± and –, respectively.

Minimum inhibitor concentration

Only the inhibitors which cause parasite-mediated agglutination inhibition are used in this experiment. The agglutination titres for positive lectin against the parasite cells are evaluated and the lectin adjusted with PBS to a concentration that just gives 100% agglutination.

Twofold serial dilutions of appropriate sugar inhibitor(s) are prepared in PBS to give final concentrations ranging from 150 to 0.15 mM. To each dilution is added adjusted lectin and all plates are incubated as above.

After addition of the appropriate parasite suspension, the plates are incubated as before and the degree of agglutination assessed. The minimum inhibitor millimolarity is regarded as that which causes a reduction in agglutination from 100 to 10% or less as compared to the appropriate controls. The sugar(s) that cause the highest degree of agglutination inhibition is the predominant carbohydrate moiety on the parasite surface against which the lectin is the most specific/reactive.

Conjugated Lectins

Several controls are employed in the FITC-conjugated, biotinylated and peroxidase-labelled lectin assays and are treated under exactly the same conditions as the lectin-conjugate exposed tissue samples (e.g. salivary glands, midgut), parasites (e.g. trypanosomes) or infected blood smears. The controls which are selected according to each experiment include:

Controls for conjugated lectins

a) Unfixed, isolated tissues or parasite cells or blood smears that are washed and incubated only in appropriate buffer.

b) The above samples fixed (with glutaraldehyde or appropriate fixative) and buffer washed only.

c) Parasite cells incubated with specific or non-specific sugar alone.

d) Samples unfixed, washed, incubated with either FITC, biotin, avidin, substrate solution or peroxidase alone and then rewashed. (FITC alone is used at an incubation concentration of 75 μg/ml for parasite cells and 50 μg/ml for tissues. Peroxidase enzyme, biotin and avidin are used at 50 μg/ml for all samples.)

e) Unfixed samples washed and reacted with a previously incubated mixture (30 min at 25 °C in the dark) of either FITC-, biotin- or peroxidase-labelled lectin plus specific sugar (used at 150 mM).

f) As per e except that a non-specific sugar (at 150 mM) is used.

g) As per d, but using fixed or enzyme-treated samples.

h) Similar to e, but with fixed or enzyme-treated samples.

i) Other appropriate controls can be employed depending upon the nature of the assay.

- Note 1. Many plant lectins require the presence of divalent **Notes** metal ions for both activity and structural stabilization of the molecule (Sharon and Lis 1989). Consequently, inclusion of divalent cations in the incubation buffer should normally be considered.

- Note 2. The concentration of parasites in the suspension can range from 10^5 to 10^8 cells/ml although the more concentrated suspensions ($>10^8$ cells/ml) might lead to pseudo-positive results.

- Note 3. Should agglutination of parasites be observed following washing of the cells, inclusion of either 175 mM NaCl or 1% albumin in the buffer should alleviate this problem.

- Note 4. The agglutination assay can also be performed in either small test tubes (e.g. Kahn tubes) or tissue culture plates in which case larger volumes of reactants may be employed. Objective evaluation of endpoints and degrees of agglutination may be achieved, although relatively time-consuming, by dispensing about 10 μl of each reacted mixture onto a microscope slide, covering with a cover slip and examining under dark ground illumination.

- Note 5. The type of diluent/reaction medium used in the lectin tests for parasite washing and for assay purposes differs according to various workers. The pH is usually between 7.2 and 7.4 but the buffer constituents can vary. However, it is essential that culture medium used as diluent should not contain any carbohydrate component that might interfere (or even specifically bind) with lectin in the sugar inhibition experiments. Moreover, coloured compounds, e.g. phenol red, if used as a medium pH adjustment indicator should be omitted from a medium diluent because they may partially obscure light transmission through the microscope and result in erroneous interpretation of agglutination titre end points.

For example, diluent variations have been reported with the use of trypanosomes as agglutinogens. In the case of the *T. brucei* complex, saline (0.85% NaCl + 2% dialysed foetal calf serum – FCS) for culture procyclics; phosphate buffered saline pH 7.4 plus either 1 mM glycerol (for living parasites/bloodstream forms) or 0.1% bovine serum albumin for enzyme- or chemically treated bloodstream forms and RPMI medium without FCS for *T. cruzi* culture forms.

With *T. congolense* bloodstream trypomastigotes, Krebs-Ringer saline containing 25 mM Tris buffer + 20 mM glycerol pH 7.2 has been used, and for culture procylics, Cunningham's medium minus carbohydrates, FCS and phenol red but with 2% BSA and 7.2 mM glycerol, pH 7.2 has been employed.

The use of phosphate-buffered saline pH 7.3 is generally suitable for both *Crithidia* and *Leishmania* parasites (PBS: 7.53 mM $Na_2 HPO_4$; 1.5 mM $NaH_2 PO_4. 2H_2O$; 145 mM NaCl; 17.0 mM $MgCl_2. 6H_2O$ and 3 mM $CaCl_2. 2H_2O$).

Acetate buffer for use with peroxidase-conjugated lectins is prepared by dissolving 27.62 g $CH_3COONa.3H_2O$ to about 850 ml distilled water, adjusting the pH to 5.0 with glacial acetic acid and making the final volume to 1 l.

- Note 6. The choice of enzymes used to treat the parasites or tissues is variable. Trypsin is the most commonly used although other proteases, e.g. papain, chymotrypsin, pronase could equally be applied to modify cell surface at predetermined concentrations or units of activity.

- Note 7. Parasite cells and vector/host tissues should be fixed prior to electron microscope investigations and fixation normally would proceed enzyme treatment.

- Note 8. Alternative insoluble substrates for peroxidase-labelled lectins included 4-chloro-1-naphthol and 3,3'-diaminobenzidine (DAB) which produce blue-black and brown end products respectively.

- Note 9. Prior to use, optimum concentrations of enzyme-, biotin- and FITC-labelled lectins should be determined using an initial range of 5–200 μg/ml lectin conjugate in appropriate buffer diluent.

- Note 10. Tetramethylrhodamine isothiocyanate conjugated lectins may be used as an alternative to FITC-labelled types for examination by light microscopy.

- Note 11. For electron microscopy, lectins may be conjugated to horseradish peroxidase enzyme with DAB as substrate. Biotin-lectin conjugates may also be detected by avidin labelled with any enzyme, usually horseradish peroxidase, and examined using a light microscope. Biotinylated lectins can be detected by other electron-dense agents other than ferritin coupled to avidin for usage with the electron microscope. For example, colloidal gold-lectin complexes have been used to detect surface carbohydrates and descriptions of this technique for vector tissues are given by Peters et al. (1983) and Rudin and Hecker (1989).

References

Andrade De AFB, Esteves MJG, Angluster J, Gonzales-Perdomo M, Goldenberg S (1991) Changes in cell-surface carbohydrates of *Trypanosoma cruzi* during metacyclogenesis under chemically defined conditions. J Gen Microbiol 137:2845–2849

Coles GC, Jansson H-B, Zuckerman BM (1988) Lectin studies of surface carbohydrates and induction of gland secretion in the free-living stages of *Schistosoma mansoni*. J Chem Ecol 14:691–700

Esteves MJG, Andrade AFB, Alviano CS, Roitman I, De Souza W, Angluster J (1987) Cell surface carbohydrate differences in wild and mutant strains of *Crithidia fasciculata*. J Protozool 34:226–230

Furman A, Ash LR (1983) Analysis of *Brugia pahangi* microfilariae surface carbohydrates: comparison of the binding of a panel of fluoresceinated lectins to mature in vivo-derived and imature in utero-derived microfilariae. Acta Trop 40:45–51

Goma LKH (1966) The mosquito. Hutchinson tropical monographs, London, 122 pp

Grogl M, Franke ED, McGreevy PB, Kuhn RE (1987) *Leishmania braziliensis*: Protein, carbohydrate, and antigen diferences between log phase and stationary phase promastigotes in vitro. Exp Parasitol 63:352–359

Guegnot J, Guillot J, Damez M, Coulet M (1984) Identification and taxonomy of human and animal leishmaniasis by lectin-mediated agglutination. Acta Trop 41:135–143

Ham PJ, Smail AJ, Groeger BK (1988) Surface carbohydrate changes on *Onchocerca lienalis* larvae as they develop from microfilariae to the infective third-stage in *Simulium ornatum*. J Helminthol 62:195–205

Hayes RO (1953) Determination of a physiological saline solution for *Aedes aegypti*. J Econ Entomol 46:624–627

Ingram GA, Molyneux DH (1991) Insect lectins: role in parasite-vector interactions. Lectin Rev 1:103–127

Jacobson RL, Schnur LF (1990) Changing surface carbohydrate configurations during the growth of *Leishmania major*. J Parasitol 76:218–224

Jacobson RL, Schnur LF (1996) Lectin-parasite interactions. Parasitol Today 12:55–61

Lanham SM, Godfrey DG (1970) Isolation of salivarian trypanosomes from man and other mammals using DEAE-cellulose. Exp Parasitol 28:521–524

Leducq R, Berrada-Rkhami O, Gabrion J, Gabrion C (1990) Lectin analysis of glycoconjugate distribution during differentiation and strobilization of *Bothriocephalus gregarius* metacestode in a pathogenic host. Int J Parasitol 20:645–654

Lis H, Sharon N (1986) Lectins as molecules and tools. Ann Rev Biochem 55:35–67

Maraghi S, Molyneux DH, Wallbanks KR (1989) Differentiation of rodent trypanosomes of the subgenus *Herpetosoma* by lectins. Trop Med Parasitol 40:273–278

Mohamed HA, Ingram GA (1993) Salivary gland surface carbohydrate variation in three species of the *Anopheles gambiae* complex. Ann Soc Belge Med Trop 73:197–207

Mohamed HA, Ingram GA, Molyneux DH, Sawyer BV (1991) Use of fluorescein-labelled lectin binding of salivary glands to distinguish between *Anopheles stephensi* and *An. albimanus* species and strains. Insect Biochem 21:767–773

Muthuria LM, Pearson TW (1987) Surface carbohydrates of procyclic forms of African trypanosomes studied using fluorescence-activated cell sorter analysis and agglutination with lectins. Mol Biochem Parasitol 23:165–172

Okolo CJ, Molyneux DH, Wallbanks KR, Maudlin I (1988) Fluorescein-conjugated lectins identify different carbohydrate residues on *Glossina* peritrophic membranes. Trop Med Parasitol 39:208–210

Okolo CJ, Jenni L, Molyneux DH, Wallbanks KR (1990) Surface carbohydrate differences of *Glossina* salivary glands and infectivity of *Trypanosoma brucei gambiense* to *Glossina*. Ann Soc Belge Med Trop 70:39–47

Perrone JB, DeMaio J, Spielman A (1986) Regions of mosquito salivary glands distinguished by surface lectin-binding characteristics. Insect Biochem 16:313–318

Peters W, Kolb H, Kolb-Bachofen V (1983) Evidence for a sugar receptor (lectin) in the peritrophic membrane of the blowfly larva, *Calliphora erythrocephala* Mg. (Diptera). J Insect Physiol 29:275–280

Rudin W, Hecker H (1989) Lectin-bound sites in the midgut of the mosquitoes *Anopheles stephensi* Liston and *Aedes aegypti* L. (Diptera: Culicidae). Parasitol Res 75:268–279

Schauer R (1985) Sialic acids and their role as biological masks. Trends Biochem Sci 10:357–360

Schnur LF, Jacobson RL (1987) Appendix III–Parasitological techniques. In: Peters W, Killick-Kendrick R (eds) The leishmaniases in biology and medicine. volume-1: Biology and epidemiology. Academic Press, London, pp 499–541

Schnur LF, Jacobson RL (1989) Surface reaction of *Leishmania IV.* Variation in the surface membrane carbohydrates of different strains of *Leishmania major.* Ann Trop Med Parasitol 83:455–463

Schottelius J (1989) Contribution to the characterization of South American Trypanosomatidae: I. The importance of lectins, neuraminic acid and neuraminidase for the differentiation of trypanosomes and *Leishmania* from the New World. Zool Anz 223:67–81

Sharon N, Lis H (eds) (1989) The lectins. Chapman and Hall, London, 127 pp

Taylor AER, Baker JR (eds) (1978) Methods of cultivating parasites in vitro. Academic Press, London, 301 pp

Uhlenbruck G, Rothe A (1974) Biological alterations of membrane-bound glycoproteins by neuraminidase treatment. Behring Inst Mitt No 55:177–185

Immunological Analysis of Parasite Molecules

MICHAEL T. ROGAN

10.1
Introduction

The interaction between parasites and the host immune system is a complex series of events which may contribute to the death of existing parasites or those of a challenge infection; or produce immunopathological reactions which are detrimental to the host. Additionally, analysis of the acquired immune response to a parasitic infection may be of diagnostic importance, indicating that a person or animal has been exposed to a particular parasite.

Parasitologists therefore frequently ask questions as to what parasite molecules are recognised by the immune response at various stages in the parasite life cycle? What immune effector mechanisms are operating against these molecules? What is the consequence of recognition? Is there a difference in recognition patterns between immune and susceptible individuals? Can the immune response be manipulated in favour of the host (immunotherapy and vaccination)?

Work in the field of schistosomiasis, for example, has highlighted the potential importance of acquired immunity to this infection. Experiments on laboratory rodents have shown that an IgE response to schistosomular antigens can bring about an eosinophil-dependent parasite killing (Capron and Capron 1986). IgG4 or IgM responses to the same antigens, however, prevent eosinophil attachment and act as blocking antibodies, beneficial to parasite survival (Khalife et al. 1986; Hagan et al. 1991). Additionally, monoclonal antibodies to a 28 kDa antigen (glutathione-s-transferase) of *Schistosoma mansoni* can be cytotoxic to schistosomulae in vitro, providing further evidence that this molecule may be important as a candidate vaccine (Boulanger et al. 1991).

Regardless of the natural immune response to parasite molecules, it may be of benefit to develop antibody-based probes to selected parasite molecules in order to study changes in their expression; to screen cDNA libraries or quantify the presence of such molecules in host tissues or excreta. Such methods have been used in the development of diagnostic tests for circulating parasite antigen in blood and also coproantigens in faeces (Allan et al. 1992)

Identification of significant parasite molecules and their subsequent purification is often a prerequisite for further study and the application of immunological methods can be of use in achieving this. The present chapter concentrates on the analysis of parasite molecul es by enzyme-linked immunosorbent assays and immunoblotting, their isolation by affinity chromatography and the purification and development of antibody probes to selected parasite molecules. Although some of these techniques may be of diagnostic use, it is not the intention of this chapter to be a review of immunodiagnostic procedures. Such information can be obtained in other books (Gillespie and Hawkey 1994).

10.2
Parasite Extracts

In analyzing parasite molecules, the usual starting point is with the intact parasite itself. Parasites in this form are rarely useful for immunological analysis except in the case of detection of surface immunofluorescence. If parasite excretory-secretory substances (Chap. 4) are not being analyzed, then the organisms must be disrupted to release particular molecular components. Disruption can range from stripping surface components (McManus and Barnett 1985; Pritchard et al. 1988) to total disruption of the organism. Once disrupted, a crude extract may contain intact organelles, soluble cytosolic components and membrane-bound molecules. Some of these fractions may be removed by differential centrifugation techniques and further analyzed in isolation. In many cases, however, it is initially desirable to analyze the total cellular components of the parasite, and this will require the use of detergents to solubilise membrane components. In addition, the disruption of cellular components may release powerful parasite proteases which could degrade other parasite protein antigens. It is therefore usually recommended to include protease inhibitors in extraction protocols. Full details of parasite proteases and protease inhibitors are given in Chapter 5, but for general purposes, a cocktail of inhibitors, as indicated in Chapter 2, should be sufficient. It should be remembered that if enzymatic activity is to be studied, protease inhibitors should not be used.

Disruption of parasites The initial disruption of parasites to make a somatic extract will depend on the size and structure of the organism itself. Protozoa may be sufficiently disrupted by repeated freeze/thaw cycles in water whilst larger helminths may require homogenation in a ground glass homogenizer or sonication in PBS. In all cases, the extraction procedure should be done on ice, normally in the presence of protease inhibitors to reduce enzymatic activity. In all cases, the number or volume of parasites used should be recorded to maintain uniformity.

Detergents The decision to use detergents in an extraction procedure will depend on the nature of the extract being analyzed. A number of detergents are frequently used to extract membrane bound proteins and these are again discussed in Chapter 2.

Once a crude parasite extract has been made it can be further separated by differential centrifugation, molecular size chromatography or further (e.g. lipid) extraction. In addition, enzymatic digestion may be necessary to fractionate components of various structures such as the nematode cuticle (Pritchard et al. 1988).

Storage Generally parasite antigenic extracts should be stored in aliquots at $-20\ °C$ of if possible at $-70\ °C$. Prior to freezing, the protein concentration of the sample should be estimated by methods such as the BioRad method or the Lowry method (see Chap.1).

10.3
Production of Antibodies to Parasite Molecules

The production of antibodies to particular parasite molecules
provides us with a powerful tool with which to study many
aspects of parasite biology or host-parasite interactions. For
example an antibody probe to a parasite component of interest may be labelled with an isotope (Rogan et al. 1990),
enzyme (see Sect. 10.3.3) or a fluorochrome and used to lo-
calise the distribution of that particular molecule within the
parasite by microscopy (see Chap. 6) (Sanchez et al. 1991); it
may also be used to isolate that particular molecule by im-
muno-purification or to screen cDNA expression libraries; or
to study immunoglobulin classes and subclasses involved in
parasite killing (Butterworth 1984; Rogan and Craig 1992).

The major technical problems, however, arise with

- obtaining material with which to immunise

- production of an antibody which recognizes the molecule
  of interest and

- purifying the specific immunoglobulin from non-specific
  material.

The degree to which all of this will have to be carried out
depends on the purpose required, and may determine whether
polyclonal or monoclonal antibodies are to be produced.

Polyclonal antibodies raised against an immunogen are
usually directed against more than one epitope. This makes
them particularly useful for direct immunoprecipitation and
for binding to partially denatured proteins (e.g. after SDS-
PAGE or glutaraldehyde fixation) since, in such cases, all
epitopes are not usually destroyed. Polyclonal antibodies are
also more simple, quicker and less expensive to produce than
monoclonal antibodies.

Monoclonal antibodies, however, have a constant specificity and affinity and are theoretically available in unlimited
amounts. They will recognize only one epitope, which makes
them useful for mapping functional sites on proteins. This
may also make them more sensitive to modification of epi-

topes due to SDS denaturation, fixation or post-translational modifications, and this should be taken into account in immunoblotting or histological studies.

10.3.1
Production of Polyclonal Antibodies

Antigenic preparations for immunization Proteins of greater than 40 000 kDa or complex carbohydrates are most antigenic. Small peptides below a length of ten amino acids (1000 kDa) are generally not immunogenic, but can be made so by being coupled to carrier proteins (see Harlow and Lane 1988).

In order to avoid production of non-specific antibodies, immunogens should be of as high a purity as possible. Immunizing a rabbit with a crude parasite homogenate will therefore result in antibodies to many different parasite molecules. This may be of use, for example, in the detection of circulation parasite antigens (Craig 1986) but would be less useful for localisation or purification of a particular molecule. Purification of parasite components will depend of their molecular nature and may involve a range of biochemical methods including HPLC, FPLC or SDS-PAGE. Additionally, it may be possible to separate certain parasite enzymes (e.g. acetylcholine esterase) by substrate affinity chromatography (Pritchard et al. 1991).

Immunization procedure Rabbits are the most frequently used animal for polyclonal antibody production and typical immunization procedures involve injection of the antigen together with Freund's adjuvant to potentiate the immune response.

1. Take a 5-ml blood sample from the experimental rabbit prior to immunization to check for background antibody levels.

2. Take 0.5–1 ml of antigen containing 100–200 μg of protein in an aqueous solution (e.g. phosphate-buffered saline [PBS]) into a glass syringe (preferably with a Luer lock).

3. Take an equivalent volume of Freund's Complete adjuvant into another glass syringe and link the two syringes together with a three-way tap so that there is a channel between the two.

4. Emulsify the preparation by injecting it back- and for-
   wards between the syringes. After 5–10 min the mixture
   should be thick and white and give some resistance when
   pushed through the syringes.

5. Test the preparation by letting a small drop fall into a
   beaker of water. The drop should hold together rather
   than disperse in the water.

6. Immunize the rabbit by injecting 0.5 ml of the emulsified
   antigen subcutaneously into two or three sites on the side
   or the back of the animal.

7. Administer a booster injection of a similar amount of
   antigen in Freund's Incomplete adjuvant after 3 weeks.

8. Take a 10-ml blood sample 10 days later and assay for
   specific antibodies (ELISA or Immunofluorescence are
   probably the best methods for this).

9. Three weeks after the first booster injection give a second
   booster and take a blood sample 10 days later.

10. Repeat step 7, 3 and 6 weeks after the second booster
    until the antibody titres are maximum (usually giving a
    positive reaction in ELISA at dilutions of 1/400 or
    greater). At this point it may be best to kill the rabbit and
    exsanguinate it totally.

All blood samples collected should be allowed to clot for **Serum**
approximately 1 h at room temperature. The clot should then **preparation**
be separated from the wall of the tube by running a Pasteur
pipette around the edge. Leave the clot to retract overnight at
4 °C. Remove the bulk of the serum with a Pasteur pipette
without disturbing the clot and then centrifuge the tube at
10 000 g for 10 min before removing the remainder of the
serum.

The serum can be stored at –20 °C for several years.
Samples are, however, best stored aliquoted in smaller volumes, since repeated freezing and thawing may cause precipitation of the immunoglobulins. Thawed samples should
be kept at 4 °C for up to 1 week (0.1% sodium azide should
be added to prevent bacterial growth).

IgG separation For many applications where antibodies are to be used as probes it is necessary to label them with an enzyme, fluorochrome or isotope. Such labelling is easier to perform if non-reactive serum proteins, or culture supplements such as foetal calf serum, are removed to leave a semi-purified IgG fraction. In the case of polyclonal antibodies, however, it must be remembered that this fraction will contain a mixture of antibody molecules which are both non-specific and specific for the desired antigen.

IgG fractions can be isolated from serum, culture supernatants or ascites fluid by three basic techniques: ammonium sulphate precipitation; ion exchange chromatography; and protein A affinity chromatography. The yield and purity of the preparations vary for each technique with ammonium sulphate precipitation being the most productive but least pure and protein A affinity chromatography producing the purest preparation but in reduced quantities.

Ammonium sulphate precipitation **Reagents**

- Ammonium sulphate $(NH_4)_2SO_4$ saturated solution, pH 7; add 550 g to 950 ml of warm distilled water to dissolve and filter. Adjust pH to 7 and make up to 1 l. Store at 4 °C where crystals should form.
- Phosphate buffered saline (PBS); 8 g NaCl, 0.2 g KCl, 1.15 g Na_2HPO_4, 0.2 g KH_2PO_4 made up to 1 l with ddH_2O.
- Dialysis membrane with >100-kDa cut off

1. For each ml of serum or ascites add 1 ml of cold PBS.

2. Add 2 ml of cold saturated ammonium sulphate dropwise while stirring.

3. Stir for 30 min on ice while precipitate (γ-globulin) forms.

4. Centrifuge at 3000 g for 30 min (preferably in a cooled centrifuge)

5. Remove the supernatant and add 2 ml of PBS per ml of original serum to the precipitate. Repeat steps **2–5** twice more and take up the final precipitate in 1 ml PBS.

6. Dialyse against 3×1 litre of PBS at 4 °C overnight to remove all the ammonium sulphate. Centrifuge the final preparation at 200 g to remove any precipitate.

7. Store IgG fraction at 4 °C with 0.02% sodium azide for up to 1 week or frozen at −20 °C in aliquots.

Because antibodies have a more basic isoelectric point than the majority of other serum proteins, they can be purified by ion exchange using diethylaminoethyl (DEAE) cellulose (Whatman DE52). **Ion exchange separation (Batch method)**

1. Place 100 *g* DE52 in a 1-l flask and add 550 ml 0.01 M phosphate buffer, pH 8.

2. Titrate the mixture back to pH 8 with 1 M HCl.

3. Leave the slurry to settle for 30 min, then remove the supernatant.

4. Resuspend the cellulose in enough phosphate buffer to fill the flask.

5. Repeat this cycle twice.

6. Filter the slurry through two layers of Whatman No. 1 to leave a damp cake of cellulose.

7. Mix 10 ml of serum with 30 ml distilled water to lower its ionic strength.

8. Add 50 *g* of wet cellulose for every 10 ml of serum used and stir thoroughly every 10 min for 1 h at 4 °C.

9. Pour the slurry on to a Buchner funnel and suck through the supernatant containing the required IgG. Rinse the cellulose with 2×20 ml of 0.1 M phosphate buffer pH 8.

10. Store as above.

Protein A, extracted from the cell wall of the bacterium *Staphylococcus aureus*, has an affinity for IgG. However, not all species or subclasses of IgG bind with the same affinity. It is particularly useful for isolating rabbit or human IgG (with the exception of human IgG3) and for mouse IgG2a and IgG2b. Improved binding to other species can be obtained using a related molecule known as protein G. Further information on the specificities of these molecules is given by Richman et al. (1982) and Akerstrom and Bjorck (1986). **Protein A affinity chromatography**

Protein A can be commercially obtained coupled to Sepharose beads (Pharmacia, Protein A Sepharose CL-4B) and can be used to purify IgG by affinity chromatography, as described in Section 10.6. The gel matrix comes as a freeze-dried powder which must be swollen and washed with several aliquots of O.1 M phosphate buffer, pH 8. The gel is poured into a 1–5 ml chromatography column and run as in section (10.5.1). The Protein A content of the swollen gel is 2 mg/ml and the binding capacity for IgG is 10–20 mg/ml gel. Polyclonal serum has an average IgG content of 10 mg/ml; tissue culture supernatants with 10% FCS have a total antibody content of about 1 mg/ml, of which 5% is specific monoclonal antibody; ascites fluid will contain 1–10 mg/ml total antibody, of which 90% will be specific monoclonal (Harlow and Lane 1988). Protein A columns can be stored under 20% ethanol at 4 °C for up to 1 year.

Note   For large volumes of serum or ascites it is better to use the ammonium sulphate of DEAE cellulose methods. A combination of these methods and the Protein A method can give better results.

10.3.2
Production of Monoclonal Antibodies

A large number of laboratories have produced different protocols for monoclonal antibody production. The methods given here are among the most frequently adopted protocols. Production of monoclonal antibodies is a time-consuming procedure which can often lead to disappointing results. Productivity can be improved by having reasonably abundant antigen in a purified form. This is especially important during the screening of clones for antibody production.

Myeloma cell lines   The most frequently used myeloma cell line for fusion with murine B lymphocytes is the P3X63-Ag8.653 line, which has been derived from BALB/c mice. This line is selected using 8-azoguanine so that it is deficient in hypoxanthine guanine phosphoribosyl transferase (HGPRT) and does not secrete or synthesize imunoglobulin chains. Cells should be obtained from a supplier or taken out of cryopreservation and grown up in culture prior to the fusion.

Cells should be grown up in 4 × 30-ml vol of basic tissue culture medium (see below), initially at a concentration of

2×10^4/ml, 3 days before fusion. This should yield 5×10^6 total cells, of which 2×10^7 will be required for a single fusion.

A mammalian tissue culture medium such as RPMI 1640 (Gibco) is suitable for monoclonal antibody production. The basic medium is

Tissue culture media

- 400 ml RPMI 1640 with HEPES
- 5 ml 200 mM glutamine (1% final conc.) (Gibco)
- 5 ml pyruvate (1%) (Gibco)
- 5 ml penicillin-streptomycin (1%) (Gibco)
- 100 ml heat-inactivated foetal calf serum (FCS, 20%) (Myoclone, Gibco).

Optimem (Gibco) is a richer medium which only requires supplementation with antibiotics and 10% FCS.

The selection medium contains either aminopterin (A) or methotrexate (M) to block *de novo* purine and pyrimidine synthesis. These must be supplemented with hypoxanthine (H) and thymidine (T) to produce HAT or HMT. Alternatively, azerine, which only blocks purine synthesis may also be used if supplemented with hypoxanthine (AH). All can be obtained in solutions at ×50 strength and diluted accordingly in basic medium.

Immunization

BALB/c mice (about 6 weeks old) are used for production of most monoclonal antibodies. The dose of antigen and immunization regime may vary from lab to lab, but generally involves multiple subcutaneous or intraperitoneal injections. This regime will vary depending on the nature and amount of antigen available. Soluble proteins, for example, are best used at doses of 5–50 μg whilst complex carbohydrates should be used at 10–50 μg. the number of boosts (2–5) after the initial immunization should be sufficient to produce a high antibody titre, as checked by taking a tail bleed, but not so great as to induce tolerance. This is particularly relevant to carbohydrate antigens.

The route of immunization is also variable. Intraperitoneal and subcutaneous injections are frequently employed because relatively large volumes of soluble or particulate antigen can be used. Ideally, a combination of both of these routes (sometimes with the addition of a final intravenous boost) will provide suitable antigen detection. It must be remembered, however, that if intravenous injections are to be used, then they must not contain particulate antigens or adjuvant.

In most cases where the above immunizations are used, it is the spleen which is taken as a source of B cells. In some cases where antigen is limited, it may be preferential to immunize in the foot pad. Antigens administered here will be transported to the popliteal lymph nodes at the back of the leg. Surgical removal of these nodes can provide a source of B cells of which a high proportion will recognize the immunogen. Methods 1 and 2 below give standard protocols for the use of splenic and lymph node B cells, respectively.

Immunization, Method 1

1. Mark 3–5 female BALB/c mice individually using an ear punch and take a 50–100 μl sample bleed from the tail.

2. For each mouse, 250 μl of antigen should be emulsified with 250 μl of Freund's Complete adjuvant as in Section 10.3.1. Inject each mouse intraperitoneally (i.p.) with 0.2 ml of this preparation.

3. After 14 days repeat the injections using Freund's Incomplete adjuvant.

4. On day 24 repeat the immunizations as in **3** and take a tail bleed for antibody analysis.

5. On day 38 repeat the immunizations as in **4**.

6. On day 52 (3 days before fusion) inject the highest responding mouse with 100 μl of antigen (5–50 μg) in PBS intravenously via the tail vein and 100 μl intraperitoneally. The other mice can be maintained by injecting 100 μl of antigen in incomplete adjuvant, i.p. These can be used if the initial fusion is unsuccessful.

7. On day 55 sacrifice the mouse, bleed it and remove the spleen to serum-free culture medium under sterile conditions.

Immunization, Method 2

1. Carry out steps 1 and 2 as above.

2. Inject 0.1 ml of the prepared antigen, intradermally into each of the rear footpads. This can be best done using a 25G needle held parallel to the surface of the foot.

3. 14 days later repeat the immunizations using Freund's Incomplete adjuvant.

4. On day 28 repeat the immunizations using antigen in PBS.

5. On day 42 repeat step **4.**

6. On day 45 sacrifice three mice, bleed them, and remove the popliteal lymph nodes under sterile conditions, to serum-free culture medium. These should be about the size of a pea and are located at the back of the knee.

Early hybridoma cells may not grow well in culture especially **Feeder cells** at low densities. The addition of feeder cells can give improved growth. **Note.** This should be done the day before the fusion. Mouse or rat peritoneal cells are used for this purpose. Approximately three mice will be required for one fusion.

1. Kill the mouse by cervical dislocation and soak with 70% ethanol. Work under a laminar flow cabinet.

2. Make a long incision in the abdominal skin without puncturing the peritoneal wall.

3. Pull the skin back away from the peritoneal wall and again spray with ethanol.

4. Inject 5 ml of ice-cold serum free medium into the peritoneal cavity.

5. Withdraw the needle and massage the peritoneum for 2–3 min.

6. Tilt the mouse over to one side so that the fluid collects under the peritoneal wall.

7. Carefully put a 21G needle of a fresh syringe just under the peritoneum and remove as much of the fluid as possible. Be careful not to damage the internal organs or block the opening of the needle.

8. Pool all of the cells from each mouse and centrifuge at 400 g for 5 min.

9. Resuspend cells in serum-free medium and centrifuge again.

10. Resuspend the cells in 20 ml of selection medium and count, using a haemocytometer. 5×10^6 cells will be suf-

ficient for ten plates if suspended in 100 ml of selection medium and 100 μl is added to each well.

Fusion Protocol

The fusion and all cell manipulations must be carried out in a laminar flow cabinet using sterile equipment.

Splenocytes

1. Remove the spleen from the mouse and place in 10 ml of warm serum-free medium.

2. Pour the medium and spleen into a sterile 200-mesh sieve over a petri dish, and disrupt the spleen by pushing it through the sieve using the end of a sterile plastic syringe.

3. Transfer the spleen cells to a 50 ml tube, add another 20 ml of serum-free medium and allow to stand for 1 min. This will allow the larger clumps to settle.

4. Remove the supernatant to another 50 ml tube.

5. Centrifuge for 5 min at 400 g.

6. Remove the supernatant and resuspend the cells in another 20 ml of serum-free medium.

7. Repeat step **6** and leave cells until myeloma cells are ready.

Myeloma cells

1. Harvest the cells, which have been growing for 3 days, by gently scraping them from the bottom of the culture flask. Transfer them to a 50-ml tube.

2. Centrifuge at 500 g for 5 min, remove the supernatant and replace with fresh serum-free medium.

3. Resuspend pellet and repeat step **2**.

4. After the second wash, resuspend the cells in 10 ml of medium and count using a haemocytometer. (A viability check can also be done at this point by incubating a small sample of cells with 0.1% Trypan blue in PBS – dead cells stain blue within 30 s)

5. Remove 2×10^7 cells in medium to a new conical 50-ml tube.

6. Add the resuspended spleen cells and centrifuge the cells together at 500 g for 5 min.

7. Remove the supernatant to leave the pellet of cells just covered by medium.

Fusion is normally carried out using polyethelyne glycol 1500 (PEG). This can be obtained commercially ready to use (Boehringer) or can be made by adding 1 *g* of PEG 1500 to 1 ml of serum-free medium and autoclaving. Either way, the solution should be warmed to 37 °C before use.   **Fusion**

1. Take the tube with the mixed cells and gently loosen the pellet by tapping on the side of the tube.

2. Hold at 37 °C (either in a water bath or in the palm of the hand) for 5 min.

3. Add 1 ml of the PEG solution dropwise over 2 min. Mix continually but very gently.

4. Add 2 ml of serum free medium over 2 min as in step **3**.

5. Add a further 20 ml of serum-free medium gradually over 10 min.

6. Centrifuge the cells at 500 *g* for 5 min.

7. Remove the supernatant and add 20 ml of fresh serum-free medium.

8. Centrifuge at 500 *g* for 5 min and resuspend the cells in 200 ml of selection medium (HAT,HMT or AH)

9. Plate out cells in ten 96-well culture plates (200 μl per well) and incubate at 37 °C under 5% CO_2.

Cells should then be left without replenishing the medium for about a week. Any clones which have formed should start to become visible by 4–7 days. When clones have formed distinct circular colonies with several hundred cells, it is time for screening.

The first screening should be done around 7–10 days post-fusion, by which time unfused myeloma cells and lymphocytes should have died. Initial screening is best done using the ELISA and is carried out by aseptically removing 50 μl of medium from each culture well using a multipipette, and adding it to the corresponding well of a microtitre plate previously coated with the desired antigen, and blocked (as   **Screening** in Section 10.4). All wells in each plate should be screened simultaneously. It is important to include appropriate controls on the ELISA plate. Medium in which only myeloma cells have been grown and antigen-only wells should be incorporated as negative controls, and serum from the immunized mice should be included as a positive control.

The choice of enzyme-conjugated antibody probe used for screening will depend on the reasons for doing the fusion. If the class of monoclonal antibody produced is unimportant, then initial screening should be done using a polyvalent anti-mouse immunoglobulin recognizing IgG, IgA and IgM. Once cells have been cloned, a more accurate isotyping can be carried out using more specific reagents. Many workers, however, prefer to work only with IgG monoclonals, which are easier to purify, in which case, an anti-mouse IgG conjugate should be used.

An arbitrary, positive/negative, optical density cut-off value can be calculated from the ELISA by taking the mean plus two standard deviations of the OD values of at least 20 wells which, on microscopic examination, showed no signs of cell growth. Cell supernatants giving an OD value greater than this can be taken as positive for antibody.

Cell supernatants can also be screened by immunoblotting, where undiluted supernatants are incubated over nitrocellulose in a low-volume blotting chamber, as indicated in Section 10.6.2.

Cloning After a positive well has been identified, it is necessary to clone the antibody producing cell, as the original well may contain more than one population of hybridomas. It is best to try and clone from a well where there are only one or two colonies. For each cloning:

1. Prepare one 96-well culture plate with feeder cells as indicated above.

2. Remove 100 μl of cell suspension from the positive well and add to the top left-hand well of the plate (A1).

3. Do double dilutions of cells down the eight wells of column 1.

4. Using a multipipette, do doubling dilutions across the remainder of the plate.

5. Single clones will be screenable after 7–10 days by ELISA or immunoblotting.

6. The strongest growing and secreting colonies should then be subjected to a repeat cloning to ensure monoclonality.

Once antibody secreting clones have been produced, they can be expanded by culture. This should be done gradually by first taking actively growing cells from each well of a 96-well plate and transferring them to six wells of a 24-well plate containing 0.5 ml of basic medium.

When the bottom of the wells are almost confluent with cells, the contents of two wells can be added to 5 ml of basic medium in a 25-ml culture flask. This should then be incubated on its side until cells are almost confluent. The cells should then be carefully removed from the side of the flask using a sterile cell scraper. The contents of each flask can then be added to 50 ml of medium in a 250-ml flask. Cells can be maintained like this indefinitely by splitting the flasks about every 3 days by removing half of the cells and replacing with fresh basic medium. Supernatants from the removed cells should then be kept for antibody isolation. It is advisable to isotype each monoclonal antibody, as this will influence purification procedures.

Antibodies can be purified from supernatants by the methods given in Section 10.3.1, although the Protein A method is not suitable for purifying mouse IgG1.

Purified antibodies should be stored by freezing at –20 °C at relatively high concentrations (1–10 mg/ml). Concentration of antibodies may be necessary and can be done using ammonium sulphate precipitation or Amicon Ultrafiltration. If antibodies are not to be used for enzymatic labelling, they can be stored at lower concentrations with 1% BSA.

At each point in the production of monoclonal antibodies, it **Freezing cells** is advisable to freeze an aliquot of hybridoma cells, because of the risk of contamination or a loss of secretory activity. These can then be retrieved if problems arise with the final clones. Freezing in liquid nitrogen should be carried out slowly.

1. Freeze only cells which are healthy and actively growing.

2. Transfer the cells to a sterile , chilled centrifuge tube and centrifuge at 400 g for 5 min.

3. Remove the supernatant and gently resuspend the pellet in 8% DMSO/92% FCS at 4 °C. The final cell concentration should be 5×10^6–5×10^7/ml.

4. Transfer 0.5 ml of the cell solution to a chilled, sterile, freezing vial and store at −70 °C for 1–7 days in a polystyrene freezing rack. Label each tube.

5. Transfer vials to liquid nitrogen.

Recovering cells Cells should be removed from liquid nitrogen and thawed quickly.

1. Remove each vial to a 37 °C water bath, being careful to not submerge the top of the vial to reduce the risk of contamination.

2. When the cells are almost thawed, wipe the outside of the vial with 70% alcohol and remove the contents with a sterile Pasteur pipette and transfer to 10 ml of basic medium.

3. Centrifuge the cells at 200 g for 5 min.

4. Remove the supernatant and replace with a further 10 ml of medium.

5. Resuspend cells and plate out in a 24-well plate. Feeder cells can be used for poorly growing cells.

10.3.3
Labelling of Antibodies

Antibodies which are specifically labelled with an enzyme can be used as probes for a variety of assays. The nature of such antibodies may be varied. Commercially produced labelled antibodies are available for a range of animal immunoglobulins, but often parasitologists employ more obscure laboratory animals for experimental infections. Hamsters, for example, are frequently used as model hosts for hookworm infections, and some workers have found it necessary to develop their own antibody probes to molecules such as hamster IgA (Garside and Behnke 1989). Alternatively the use of capture ELISA systems for detection of parasite antigens in blood, urine or faeces (Allan et al. 1992) requires the use of labelled antibodies against the required antigen. The two most common enzymes to be linked to antibody molecules are horseradish peroxidase and alkaline phosphatase.

Conjugation of Antibodies to Horseradish Peroxidase
(Nakane and Kawaoi 1974)

1. Add 5 mg of horseradish peroxidase to 1.23 ml of distilled water. Add 0.3 ml of freshly prepared 0.1 M sodium periodate in 10 mM sodium phosphate (pH 7).   **Conjugation, Periodate Method**

2. Incubate for 20 min at room temperature.

3. Dialyse the solution against several changes of 1 mM sodium acetate (pH 4) overnight at 4 °C.

4. Make a 10 mg/ml solution of the antibody in 20 mM sodium carbonate (pH 9.5).

5. Remove the peroxidase from the tubing and add to 0.5 ml of the antibody solution.

6. Incubate at room temperature for 2 h.

7. Reduce any Schiff's bases that have formed by adding 100 µl of sodium borohydride (4 mg/ml in water). Incubate at 4 °C for 2 hours.

8. Dialyse against several changes of PBS overnight at 4 °C

Conjugation of Antibodies to Alkaline Phosphatase
(Avrameas 1969; Voller et al. 1976)

1. Centrifuge 5 mg of alkaline phosphatase (supplied in 65% saturated ammonium sulphate) at 4000 *g* for 30 min.   **Conjugation, glutaraldehyde method**

2. Remove the supernatant and resuspend the enzyme pellet in 1 ml of a 10-mg/ml solution of the antibody in PBS.

3. Dialyse the mixture against several changes of 0.1 M sodium phosphate buffer (pH 6.8) overnight at 4 °C.

4. Transfer the solution to a 5 ml bijou bottle containing a magnetic flea on a stirrer.

5. In a fume hood, slowly add 50 µl of a 1% solution of EM grade glutaraldehyde whilst stirring gently.

6. Stir for 5 min.

7. Leave for 3 h at room temperature and then add 1 ml of 1 M ethanolamine (pH 7)

8. Leave for a further 2 h and then dialyse overnight against several changes of PBS overnight at 4 °C.

9. Spin the mixture at 40 000 g for 20 min.

10. Store the supernatant at 4 °C in 50% glycerol containing 1 mM $ZnCl_2$, 1 mM $MgCl_2$ and 0.02% sodium azide.

Note This procedure can be scaled down to using 1 mg/ml Ig by reducing the antibody and enzyme concentrations by a factor of 10. The time allowed for the coupling should then be increased to 24 h

10.4
Enzyme-Linked Immunosorbent Assays (ELISA)

One of the solid-phase immunoassays most frequently used by parasitologists is the ELISA. This technique was developed in the 1970s, and has become a valuable approach to the immunological analyses of parasite molecules. Obvious uses include the assay of postinfection sera for the presence of natural antibodies to specific parasite antigens in an immunodiagnostic approach. In addition, the specific analysis of antibody class and subclass responses occurring during infection or exposure to a parasite can give valuable information on factors such as worm burden (Pritchard et al. 1995) or resistance and susceptibility to parasite infection (Hagan et al. 1991; Stewart et al. 1995). The technique not only has applications for the assay of antibodies associated with infection, but also for the detection of antibodies produced by experimental immunization. The development of polyclonal and monoclonal antibodies, as detailed above, has been made easier with the advent of the ELISA.

The basic indirect ELISA technique involves the coating of a polystyrene or polyvinyl chloride microtitre plate with the antigen of choice, and then incubating this with the test sera or solutions containing antibody. Binding of the test antibody to the antigen is detected by the addition of a second enzyme-labelled antibody conjugate which is directed against the first immunoglobulin (e.g. alkaline phosphatase-labelled goat anti-human IgG is used to detect human IgG, recognizing an antigen bound to the plate). The addition of the enzyme substrate will then allow a colorimetric assessment of the amount of antibody binding. The assay can also be used in a direct fashion, for example to make a comparative analysis of total IgG in serum samples. In this case, diluted serum would be bound to the plate and probed directly with the second antibody conjugate.

Binding of the antigen to the plastic plate is not fully understood, but is thought to involve hydrophobic forces (Venkatesan and Wakelin 1993). Binding can be affected by a number of factors, including the nature of the antigenic mixture, temperature and the pH of the coating buffer, and each antigenic preparation may have different binding characteristics. Protein antigens, for example, bind best at a pH of 9.6 whilst carbohydrate antigens may bind better at pH 3 (Venkatesan and Wakelin 1993). Where plates are coated and left at 4 °C overnight, different degrees of antigenic binding may occur on the outermost wells of the plate (edge effect). In addition, the nature of the plate material may affect binding, and commercial suppliers, such as Immulon, market a range of different plates with different binding capacities.

After antigen has been bound to the plate and excess removed by washing, the remaining binding sites on the plate must be blocked to prevent non-specific adsorption of subsequent reagents to occur. In many cases, an additional protein solution such as 5% skimmed milk or 1% bovine serum albumen (BSA) are used, but these must be used with care as the test serum may have antibodies to BSA. An alternative approach is to use a detergent such as Tween 20 which minimizes non-specific reactions. Again, care must be taken with this approach, as the detergent may destroy lipid-containing antigenic epitopes. In performing a set of ELISA experiments it is, therefore, important to make a preliminary assessment of the optimal conditions for running that particular assay. This will include the binding and blocking conditions but, more importantly, will involve using the correct reagent dilutions and this will be described in the next section.

10.4.1
The Basic Technique

Equipment
- 96-Well microtitre plate
- 10–100 μl micropipette
- 100 μl 8-channel multipipette
- Microplate optical density reader

Reagents
- Carbonate-bicarbonate coating buffer, pH 9.6
- Washing buffer (PBS/0.1% Tween 20)
- Blocking buffer (PBS/0.3% Tween 20 or PBS/5% skimmed milk)
- Test serum/antibody containing solution
- Enzyme conjugated second antibody against immunoglobulins of the test serum
- Enzyme substrate (see Table 2)
- Antigenic extract

Procedure, ELISA

1. Dilute antigen to optimum concentration in coating buffer (total protein concentrations in the region of 1–5 μg/ml are most frequently used, but should be checked for each antigen).

2. Coat the plate by adding 100 μl of antigen to each well. Check which well the plate reader will take a blank reading on, and leave this empty of all reagents until the final substrate solution. Leave additional wells uncoated to act as controls (see Sect. 10.5.2).

3. Cover with cling film and incubate at 4 °C overnight. This is frequently done for convenience in timing the assay. Alternatively plates may be incubated at 37 °C for 1 h, but it is important to be consistent about binding procedures when comparing assays.

4. Antigen should be removed from the plates, which are then washed 3–5 × 3 min in PBS/0.1% Tween 20. The plate should be slapped, upside down, onto tissue paper in between washes to remove droplets.

5. Add 150 μl of blocking buffer to each well (including controls) and incubate for 1 h at room temperature.

6. Wash as in 5.

7. The test antibody (and any controls) should be diluted to the appropriate working dilution in blocking buffer and 100 μl is added to the respective wells on the plate. Incubate for 1.5 h at room temperature.

8. Wash as in **5**.

9. Dilute the second antibody conjugate to the appropriate working dilution in blocking buffer and add 100 μl to each well. Incubate for 1 h at room temperature.

10. Wash as in **5**.

11. Add 100 μl substrate solution (Table 1) to each well including the blanking well and read on a microplate reader, at the appropriate wavelength, when the colour develops (usually 10–30 min).

All of the above incubations can be carried out at 37 °C if day-to-day variation is a problem.

10.4.2
Optimizing the Assay

The optimal dilutions of antigen, test antibody and antibody conjugate to be used will vary for each set of assays or batch of reagents used. It is important to standardize the conditions before interpreting results.

The enzyme antibody conjugate

Antibody conjugates bought from commercial suppliers usually have a recommended working dilution supplied with them (usually in the range of 1/1000–1/20 000). This should be taken as a guideline, and it may be necessary to use one or two dilutions less for particular assays. For reagents which have been conjugated in the laboratory, the optimum dilution should be identified by titrating the antibody against a 1/1000 dilution of normal serum, of the specificity required, coated directly on to the microtitre plate (e.g. coat plate with 1/1000 normal human serum to titrate a rabbit anti-human IgG conjugate). The optical density should be read after about 10 min. The results should be plotted on a graph and the dilution nearest the top of the curve, just before it starts to drop off, should be used.

Table 1. Substrate solutions for the most frequently employed enzyme-antibody conjugates

Conjugated Enzyme	Substrate	Buffering solution	Results
Alkaline phosphatase	P-nitrophenyl phosphate (PNPP)	10 mg PNPP in 10 mM diethanolamine (pH 9.5) containing 0.5 mM $MgCl_2$	Positive yellow colour develops after 2–30 min. Read at 405 nm
Horseradish peroxidase	Tetramethyl benzidine (TMB)	1 mg TMB in 10 ml 0.05 M phosphate-citrate buffer (pH 5) plus 0.03% sodium perborate or 10 μl H_2O_2 (30% soln.)	Positive blue colour develops after 5–30 min and can be read at 370 or 650 nm. Reaction can be stopped using 2 M H_2SO_4 and yellow colour can be read at 450 nm
Horseradish peroxidase	2,2′-Azino-bis (3-ethylbenz-thiazoline-6-sulphonic acid (ABTS)	10 mg ABTS in 0.05 M phosphate-citrate buffer (pH 5) plus 0.03% sodium perborate or 10 μl H_2O_2 (30% soln.)	Positive green colour develops in 5–30 min and can be read at 405 nm. Reaction can be stopped with 1% SDS
Horseradish peroxidase	5-amino -sialicylic acid (5ASA)	10 mg dissolved in 50 ml 0.1 M phosphate buffer, pH 6 containing 1 mM EDTA plus 30 μl H_2O_2 (30% soln.)	Positive brown colour develops after 10–60 min and can be read at 450 nm. Reaction may be stopped using 3 N NaOH and read at 550 nm

Note: To improve sensitivity, biotinylated antibodies can be used. After initial probing with these reagents, an additional 30–60-min incubation in avidin/peroxidase or streptavidin/peroxidase conjugates is required. The final development is done using the above peroxidase substrates. Substrates and their appropriate buffers are available in tablet form from Sigma Chemical Co.

The determination of the optimum dilutions of antigen and test antibody to use will depend on the availability of the test serum. If reasonable amounts of a positive serum are available (e.g. a human serum from a patient with advanced infection of the parasite under study), then this should be used. If this is not available, then a hyperimmune serum against the desired antigen should be used to assess the optimum antigen dilution. It is important to remember that this hyperimmune serum may contain a large proportion of high affinity antibody, which may not necessarily be the case with the test antibody. In both cases, results should be compared with a similar negative serum sample.

Antigen and test antibody dilution

The optimization assay should be set up in a "chequer board" form where antigen is titrated in one direction and test antibody in the opposite direction.

1. Determine which well is the blanking well for the plate reader (in many cases this can be set at well A1).

2. To wells B1–H1 add 200 μl of antigen solution at a concentration of approximately 100 μg/ml in coating buffer.

3. To the remainder of the wells on the plate add 100 μl of coating buffer.

4. Using a multipipette, transfer 100 μl from the wells in column 1 to those in column 2, mix and transfer 100 μl to column 3 and so on until column 11. Remove 100 μl from these wells and discard (do not put into column 12).

5. Incubate, block and wash as in Section 10.5.1.

6. To wells in row B add 200 μl of positive antibody solution diluted 1/50 in blocking buffer.

7. Transfer 100 μl from the wells in row B to those of row C and repeat down to row G. Discard the final 100 μl from row G (do not add to row H).

8. Incubate and complete the ELISA as in Section 10.5.1 using optimal dilutions of enzyme antibody conjugate.

9. Read the plate at 10-min intervals for 30 min.

10. Pick the antigen and positive serum dilutions which give the highest optical densities when compared with the negative serum. Since there may be several combinations

which give equally good results, pick the one which will be most economic in terms of reagents. As a guideline, many ELISAs involving human serum IgG are used with an antigen concentration of 1–5 μg/ml and a serum dilution of 1/100–1/400. Analyses of other antibody classes such as IgE, however, may require serum dilutions of 1/25 and overnight incubations.

11. Repeat the ELISA using the negative serum.

Controls Once the optimal conditions for running the ELISA have been established, the testing of unknown samples can begin, and all samples should be assayed at least in duplicate. In all assays, however, it is important to include a number of controls. Since it is possible for both the test antibody and enzyme antibody conjugate to bind non-specifically to the plate, the following control wells should be set up:

- Two wells where antigen has been omitted (test serum and conjugate are included)

- Two wells where antigen and test serum have been omitted (conjugate included)

- Two wells where antigen is included but test antibody is omitted

Any significant reaction in any of these wells will indicate non-specific reactions. In the first case, a colour reaction indicates that the test serum is binding non-specifically to the plate and the blocking step should be improved. In the second case, a reaction indicates that the conjugate is binding non-specifically to the plate and again requires improved blocking. In the third case, a positive reaction indicates a cross-reactivity between the conjugate and the antigen.

Each assay of unknown samples should also include a definite positive control standard to indicate that the assay has worked and to act as a reference for the other results (see below).

Results The assays described above are single-dilution assays where one dilution only of test serum is used. This type of assay is easiest to perform, but can suffer from day-to-day variations. If large numbers of samples are being assayed, for example in an immunodiagnostic screening survey, then optical density readings on one plate on one day may not directly relate to others on a different day. One way of getting around this is to include the same positive control standard on each plate on each day. The end of the reaction can be taken as the point where the positive control standard gives a preset optical density value (e.g. 0.8). The time taken to reach this value may vary from day to day, but in each case the optical densities of the test samples should be comparable.

Another approach is to give the positive standard a comparative value of 1 and express all test values as a ratio of this by dividing the optical density of the test sample by that of the positive control. Thus, for example, if, in one assay, the positive control standard gives an optical density of 0.8 and that of a test sample is 0.3, then the ratio of the optical densities is 0.3/0.8 = 0.375. If, on another day, the same two samples are run and the positive control standard gives an OD of 0.6 and the test sample an OD of 0.225, this will still give a ratio of 0.225/0.6 = 0.375. Day-to-day variation can also be reduced by calculating the end point of an antibody titration (titre) relevant to a negative control sample. This, however, is less economical in terms of time and reagents.

10.4.3
Capture ELISA

Antigen capture assays can be used to detect and quantify antigens, and depend on a sandwich of two antibodies recognizing different sites on the antigen molecule. This method has been most frequently employed to detect parasite antigens in circulation or in host urine or faeces. The basic assay depends on binding either an antigen-specific polyclonal or monoclonal antibody to a microtitre plate, adding a solution which contains an antigen; and quantifying the amount of antigen binding to the plate using either the same polyclonal or a monoclonal antibody, recognizing a different antigenic determinant, which has been conjugated to an enzyme. In most cases, it is beneficial to titrate the antigen containing solution across the plate.

Prior to the assay, it is necessary to purify both antibodies and to conjugate one to an enzyme. The optimum working dilutions of each should also be defined using a standard **The Basic Protocol, ELISA**

ELISA as in Section 10.4.2, with the desired antigen bound directly to the plate. For the capture assay:

1. Dilute the unlabelled antibody to its optimum working dilution in carbonate/bicarbonate coating buffer (pH 9.6) and add 100 μl to each well. If the optimum dilution is not known, the antibody should be coated in excess at a concentration of around 20 μg/ml. Cover plate and leave overnight at 4 °C.

2. Wash plates with PBS 0.1% Tween.

3. Block the remaining active sites on the plate with PBS/ 0.3% Tween or PBS/0.5% skimmed milk, for 1 h at room temperature.

4. Wash plates three times in PBS/0.1% Tween.

5. Add 100 μl of antigen containing solution and incubate for 1 h at 37 °C. For most assays, it is best to titrate any solutions containing an unknown concentration of antigen across the plate. The initial dilution will depend on having an approximate idea of how much antigen may be there. For example, if serum is being assayed for circulating antigen, then it is best to start at a dilution of 1 in 2 in PBS and titrate using doubling dilutions. An assay of parasite culture medium for secreted antigen may require a more dilute starting point of around 1 in 20. In any case, it is useful to run a titration of a known antigen concentration to act as a calibration.

6. Wash three times in PBS/0.1%Tween.

7. Add 100 μl/well of the second conjugated antibody at an optimum dilution and incubate for 1 h at 37 °C.

8. Wash three times in PBS/0.1% Tween

9. Add the enzyme substrate and read at the appropriate wavelength when the colour develops.

It is important to include appropriate controls in the layout of the plate. These should include wells where no capture antibody is present but antigen and antibody conjugate are; and also wells where no antigen is present but both antibodies are.

- The medium in which the antigen is to be detected can affect the assay. In serum, host antibodies can affect the assay, and therefore the correct dilution is important. It may also be necessary to remove some of the host antibody by passing the serum through a Protein A column before loading the sample onto the plate. Other media, such as faeces, contain proteolytic enzymes which may damage or remove the coated antibodies, and some workers have found that the addition of a non-specific protein source, such as foetal calf serum, can reduce this effect (Allan et al. 1992).

- Capture assays of this type can also be used to detect host components produced in small amounts in response to parasitic infections, e.g. cytokines.

10.5
Immunoblotting

Blotting is the term used for transferring molecules to a matrix on which they are immobilized. It was originally described by Southern (1975) for DNA. Subsequently, Towbin et al. (1979) devised a protocol for the electrophoretic transfer of proteins from polyacrylamide gels to nitrocellulose sheets and this is referred to as Western blotting or immunoblotting.

The technique is dependent on denatured parasite polypeptides from SDS-PAGE gels (see Chap. 2) being immobilized on nitrocellulose paper. Antigens can then be probed by antibody. The development of mini-gel electrophoresis systems (Hoeffer Scientific Instruments. Biorad) has meant that reagents can be used more sparingly.

The immunoblotting technique is particularly useful for identifying diagnostic antigens when parasite extracts are probed with sera from infected individuals. This method is currently employed to provide diagnostic confirmation of several parasitic infections including cysticercosis caused by *Taenia solium* (Tsang et al. 1989) and hydatid disease caused by *Echinococcus granulosus* (Maddison et al. 1989). Information on the existence of stage-specific parasite antigens can also be obtained in this way. Additionally, immunoblotting can provide a screening mechanism for monoclonal antibody production and allow the selection of antibodies to specific antigenic bands. Specific antibodies to selected bands on an immunoblot can also be eluted from the blot and re-used to screen cDNA libraries. This method has been used in the production of the first commercial recombinant vaccine to a helminth parasite, *Taenia ovis* (Johnson et al. 1989).

10.5.1
The Basic Technique

Equipment
- Blotting tank
- Transformer/power pack
- Gel cassette
- Scotchbrite scouring pads
- Whatman No. 3 filter paper
- Nitrocellulose sheets (0.1–0.4 μM pore size)
- Buffer: (25 mM Tris/192 mM glycine, pH 8.3. 20% v/v methanol improves the transfer and binding of proteins < 60 kDa but hinders large proteins > 100 kDa. Transfer of these is improved by adding 0.1% SDS).
- SDS/PAGE gel previously run as in Chapter 2.

Preparation of the sandwich

1. Soak the SDS-PAGE gel in transfer buffer for 20 min. (It is advisable to trim one corner for orientation purposes).

2. Cut a piece of nitrocellulose paper, slightly bigger than the gel, and slowly lower it on to the surface of some transfer buffer. Let the paper soak for 15 min. (Always wear gloves when handling the nitrocellulose to prevent contamination with proteins on the fingers).

3. Assemble the blotting sandwich using two pieces of filter paper and two Scotchbrite pads, all of which have been soaked in transfer buffer, as follows. Open the carrier for the blot apparatus and lay one Scotchbrite pad on one side of the cassette (if the cassette has two different-coloured sides, place the pad on the white or clear side).

Lay a piece of filter paper on top of the pad and place the nitrocellulose paper on top of the filter paper. Make sure that there are no air bubbles trapped between the layers.

Lay the gel carefully on top of the nitrocellulose again ensuring that no bubbles are trapped.
Lay the second piece of filter paper on top of the gel and finally cover with the second Scotchbrite pad.
Close the outer (black) half of the cassette, lock and place in the blotting tank with the black side of the cassette towards the cathode (–ve).

Transfer conditions will vary with the size of the protein, with high molecular weight proteins transferring more slowly than lower ones. Transfer can be performed rapidly in 1–2 h with a current of 250–500 mA. It is often convenient to transfer overnight, and in this case a lower voltage and current should be used. Transfer for 16 h at 30–60 V with a current of about 100–150 mA. It is preferable to have the blotting tank in a cold room or to use a water-cooling system, particularly with rapid transfer with high voltages. The apparatus can also be placed on a magnetic stirrer to circulate the buffer. **Transfer conditions**

Transfer of proteins when mini-gel systems have been used can be done in approximately 2 h at 200–250 mA.

Once the blot has been removed, it is important to label the orientation. The blot can then be air-dried, placed in a plastic bag and stored at –20 °C for several weeks.

Prior to probing, it is essential to mark the position of the molecular weight markers. If prestained markers (Novex MultiMark) have been used, they will automatically appear as different-coloured bands on the nitrocellulose. If unstained markers are used, they must be visualized by staining with a dye; 0.5% Ponceau S in 10% acetic acid is suitable and will stain the markers in 1 min. Excess stain is washed off with distilled water. The markers should preferably be cut off before staining, although staining of the entire blot in this way does not appear to alter antigenicity significantly.

10.5.2
Probing the Blot with Antibody

- Blotting tray 10×5-ml trough for large blots; 8×2-ml trough (Accurtran^{TM}) for mini-blots; or multi-blot clamp (Biometra)   **Equipment**
- Rocking platform
- Incubation buffer, PBS/0.1%(v/v)Tween 20

- Blocking solution , 5%(w/v) skimmed milk powder in PBS
- Primary antibody
- Secondary antibody probe against primary antibody
- Detection system for secondary antibody probe

Choice of antibody probe

In performing immunoblotting, the choice of which antibody probe to use will be dependent on the origin of the primary antibody and the class of the particular immunoglobulin that is of interest. Commercially produced antibody probes are available for a wide variety of human and animal immunoglobulins and may be directed against a particular immunoglobulin class (e.g total IgG, IgM, IgA, IgE), subclass (e.g. IgG4) or antibody fragment (e.g. Fab, Fc, whole molecule). It is important to remember that parasite extracts may contain host molecules including immunoglobulins.

For less common experimental species, availability may be limited to anti-IgG antibodies only. For animals such as gerbils, cotton rats, voles etc. where no commercially produced probes are available, the best option is to purify the immunoglobulin in question (Hudson and Hay) and to raise an antibody probe against it in a species such as rabbit. The specific rabbit anti-immunoglobulin can then be purified and conjugated to an enzyme (see Sect. 3.3). It may, however, be worth checking cross-reactivity of commercial probes in detecting the particular antibody in question (i.e does anti-rat IgG bind to gerbil IgG?). Protein A and Protein G probes should also be considered.

A number of detection systems are also available for immunoblotting. These mainly involve enzyme or radio-labelled reagents. The choice of probe is again based on personal choice.

The use of enzyme-coupled probes is now more common then radiolabelled probes such as ^{125}IODINE because of safety.

The main enzyme systems available and their associated substrates are shown in Table 2.

Probing procedure

All protein-binding sites on the nitrocellulose paper must be blocked to prevent non-specific binding. Bovine serum albumen (BSA) or skimmed milk are suitable blocking agents, although PBS/0.3% Tween 20 is also effective. All the incubations below concern enzyme-conjugated detection systems and should be performed at room temperature on a rocking platform.

1. Block the nitrocellulose using 5% skimmed milk (w/v) in PBS for 1 h

2. Wash 3×10 min in PBS/Tween.

3. Cut the nitrocellulose into individual strips and place into incubation troughs or clamp into a multi-trough incubation unit.

4. Add the primary antibody at the required dilution in PBS/Tween containing 5% (w/v) skimmed milk, for 1.5 h.

5. Wash 3×10 min in PBS/Tween

6. Incubate in the second conjugated antibody (Table 2) at the appropriate dilution in PBS/Tween for 1 h.

7. Wash 3×10 minutes in PBS/Tween.

8. Give a final 5-min wash in substrate buffer without Tween (Table 2).

9. Incubate the nitrocellulose in the appropriate enzyme substrate solution (see Table 1) until coloured bands appear (5–60 min).

10. Stop the reaction and remove the nitrocellulose to a piece of filter paper and allow to air-dry.

- Nitrocellulose paper from various suppliers can vary in protein binding. It is best to try samples from different companies before embarking on large scale blotting (Sartorius and Biotrace papers are generally acceptable. The pore size should be no greater than 4 μM).   **Note**

- To save on reagents, low-volume blotting chambers are available to enable screening of large numbers of samples. Such systems enable the screening of up to 30 samples from one mini-gel blot and are particularly useful for monoclonal antibody screening.

- The optimum dilutions of primary and secondary antibodies used should be determined prior to large-scale blotting experiments. As a rough guide, primary antiserum dilutions of 1/100–1/400 tend to give acceptable

Table 2. Most commonly used antibody conjugates/substrate systems for immunoblotting

Antibody/enzyme conjugate	Substrate buffer	Substrate solution	Comments
Horseradish peroxidase (HRP)	PBS or TBS	30 mg 4-chloro-1-napthol dissolved in 10 ml ethanol and made up to 50 ml with PBS plus 30 μl of hydrogen peroxidase (30%)	Gives a blue-black colour. The reaction can be stopped by washing in distilled water
Horseradish peroxidase (HRP)	PBS or TBS	10 mg 3,3'-diaminobenzidine (DAB) dissolved in 10 ml PBS plus 10 μl of hydrogen peroxidase (30%)	Gives brown coloration. Reaction can be stopped by washing in distilled water Possibly carcinogenic
Alkaline phosphatase	100 mM NaCl, 5 mM MgCl$_2$, 100 mM diethanolamine (pH 9.5)	33 μl bromochloroindoyl phosphate (BCIP; 0.5 g dissolved in 100% dimethylformamide) plus 66 μl nitroblue Tetrazolium (NBT; 0.5 g dissolved in 70% dimethylformamide) added to 10 ml of substrate buffer	Gives a blue-black colour. Reaction may be stopped by washing in PBS containing 20 mM EDTA Slightly more sensitive than peroxidase methods

Note: To improve sensitivity biotinylated antibodies can be used. After initial probing with these reagents an additional 30–60 min incubation in avidin/peroxidase or streptavidin/peroxidase conjugates is required. The final development is done using the above peroxidase substrates. Substrates and their appropriate buffers are available in tablet form from Sigma Chemical Co.

results. Purified primary antibodies may require a higher dilution, whilst supernatants from monoclonal-antibody producing cells are best used undiluted.

- Incubation times can be reduced slightly by incubating at 37 °C. Likewise, if necessary, primary antibody incubations can be carried out overnight at 4 °C. There is, however, a possibility that some leeching of antigen from the blot may occur during prolonged incubations in the presence of Tween.

10.6
Affinity Chromatography

10.6.1
Antibody Purification

Once either polyclonal or monoclonal antibodies to significant antigens have been experimentally produced, it may be necessary to not only purify the IgG fraction as mentioned above, but also to isolate the immunoglubulins that are specific for the desired antigen. The production of isolated antigen-specific immunoglobulins is particularly important if the antibodies are to be used as specific probes with minimal cross-reactivity. Such antibody preparations can be prepared by binding the antigen in question to Sepharose beads in a chromatography column and passing the antibody solution through the column. Specific immunoglobulins will stick to the column and can be subsequently eluted by altering the column conditions. This technique can also be used to remove undesired immunoglobulinsfrom polyclonal preparations. Such is the case where host proteins may have been present in the parasite immunization preparation. Antibodies against the host proteins can be removed from the antiserum by passing it through a column containing bound host protein.

Reagents

- Cyanogen bromide (CNBr)-activated Sepharose 4B (Pharmacia)
- Coupling buffer (0.1 M NaHCO$_3$, pH 8.3, containing 0.5 M NaCl)

- Blocking buffer (1 M ethanolamine or 0.2 M glycine, pH 8)
- 0.1 M acetate buffer, pH 4, containing 0.5 M NaCl
- Protein antigen to be coupled to gel at 5–10 mg protein/ml gel
- 2 ml chromatography column or syringe barrel and a three-way tap
- Spectrophotometer for reading O.D. at 280 nm or Uvcord monitor and chart recorder

Coupling an antigen to the gel

1. Weigh out required amount dried CNBr-activated Sepharose 4B. (1 *g* dried material gives about 3.5 ml final gel volume.)

2. Swell gel and wash on a sintered glass filter with 1 mM HCl (200 ml/g).

3. Dissolve protein in coupling buffer to a concentration of 5–10 mg/ml gel.

4. Mix protein solution with gel suspension at a ratio of 1:2 by rotating end over end for 2 h at room temperature or overnight at 4 °C.

5. Let the gel settle and transfer to blocking buffer for 2 h at room temperature to block the remaining active groups.

6. Wash away excess protein by washing with alternating coupling buffer and acetate buffer (in two cycles).

7. Store gel at 4 °C in coupling buffer.

Running the column

1. Plug the bottom of the syringe with glass wool and attach a three-way tap to the outlet.

2. Resuspend the gel and carefully pour it into the syringe with the tap closed. Allow the gel to settle and adjust the volume of buffer, using the tap until all of the gel has been transferred to the column.

3. Using a peristaltic pump or buffer reservoir, run 10–20 column volumes of 0.05 M Tris-HCl, 0.5 M NaCl, pH 8, through the column and obtain a baseline trace if using a UVicord and chart recorder.

4. Dilute 1 ml of the serum to be purified 1:1 in 0.05 M Tris-HCl, 0.5 M NaCl, pH 8 and load on to the column at a

flow rate of around 1 ml/min. For monoclonal antibodies run up to 10 ml of culture medium.

5. Start collection of 2-ml fractions as the antibody runs through the column until the OD 280 drops back to baseline levels. Keep these fractions if trying to remove unwanted antibodies from the sample.

6. Add eluting buffer (100 mM glycine-HCl, 0.5 M NaCl, pH 2.8) and collect fractions corresponding the second protein peak. This will contain antibodies to the antigen immobilized on the column.

7. Regenerate the column by running a further 10–20 vol of 0.05 M Tris-HCl, 0.5 M NaCl, pH 8.

8. Pool the elution fractions and raise the pH to around 8 by adding a small amount of solid Tris directly to the tube.

9. Dialyze the eluted antibody against several changes of PBS overnight at 4 °C.

10. Store at 4 °C or at –20 °C with 1% BSA. N.B. External protein sources such as BSA should not be added if the antibody is to be labelled.

11. The column can be reused if stored on the presence of 0.02% sodium azide at 4 °C.

It may be necessary to pass a sample through the column **Note** more than once either to remove unwanted antibodies or to extract the maximum specific antibody from the column.

10.6.2
Antigen Purification

Polyclonal or monoclonal antibodies against significant antigens can provide a useful tool for purifying the antigens themselves. The binding of the antibody to an affinity column means that a crude antigenic mixture can be passed through to isolate the antibody-reactive components. In most cases, it is better to use monoclonal antibodies for this procedure, as the immunoglobulins will bind to identical antigenic epitopes. The use of polyclonal antibodies means that there will be binding to different sites on the antigen molecule, resulting in a higher avidity. The harsh conditions needed to elute an antigen when it is bound by several antibodies may damage the column and denature the antigen. Strong binding can be reduced by loading saturating amounts of antigen onto the column. Polyclonal antibodies can be suitable if they have been raised against synthetic peptides or defined regions of the antigen.

An additional use for this technique is to remove cross-reacting antigens from an antigenic mixture. This may be particularly useful where human or animal postinfection serum or antiserum is used. Antigens which cross-react with heterologous sera can be removed to produce a more specific diagnostic antigen preparation. An example of this was shown by Gottstein (1985), where an immunoglobulin fraction prepared from an antiserum against *Echinococcus granulosus* antigens was coupled on to a column. An antigenic extract from *Echinococcus multilocularis* was then passed through the column to remove any cross- reacting antigens and leave a purified specific preparation which was only recognized by people with alveolar hydatidosis.

Antibody coupling technique An antibody preparation can be coupled to cyanogen bromide, as indicated in Section 10.5.1, but this will mean that the orientation of the antibody molecules on the column will be random. This may result in some of the antigen-binding sites being obscured, leading to a lower binding capacity. It is usually better to couple the antibody to the beads by the Fc region of the molecule. This can be done using Sepharose beads already coupled to Protein A. The antibody is permanently linked to Protein A using dimethylpimelimidate.

Reagents
- Protein A Sepharose CL-4B (Pharmacia)
- 0.1 M phosphate buffer, pH 8
- 0.2 M sodium borate, pH 9
- Dimethylpimelimidate
- 0.2 M ethanolamine

Purification procedure
1. Swell the Protein A Sepharose CL-4B beads in 0.1 M phosphate buffer (pH 8).

2. Let the Sepharose settle and remove most of the supernatant.

3. Add 2 mg of antibody in 1 ml of 0.1 M phosphate buffer (pH 8) per ml of swollen beads.

4. Mix by gentle rocking for 1 h at room temperature.

5. Wash the beads twice in 10 vol of 0.2 M sodium borate (pH 9) by centrifugation at 3000 g for 5 min.

6. Resuspend in 10 vol of 0.2 M sodium borate (pH 9) and add enough solid dimethylpimelimidate to bring the final concentration to 20 mM.

7. Mix for 30 min at room temperature on a rocking platform.

8. Stop the reaction by washing the beads once in 0.2 M ethanolamine (pH 8), then incubating for 2 h at room temperature in 0.2 M ethanolamine with gentle mixing.

9. Pour the Sepharose into the column and store at 4 °C under 20% ethanol for up to 1 year.

10. The column should be run as in Section 10.5.1, using 0.1 M glycine/HCl (pH 2.8) for elution.

Note. Remember to be certain whether it is the column run-through or the eluate that is required.

Notes

• The coupling efficiency can be checked by taking a sample of beads at steps **5** and **7**, boiling in reducing sample buffer and running on an SDS/PAGE gel. The sample from step **5** should contain a band corresponding to immunoglobulin heavy chain (about 55 kDa) which is absent from the step **7** sample.

• This method is only suitable for antibody classes and subclasses which have a high affinity for Protein A. For other immunoglobulins it is possible to add an additional coupling step involving an anti-immunoglobulin second antibody. For example, Protein A Sepharose coupled to anti-mouse IgG (Fc), followed by specific monoclonal mouse IgG1.

References

Allan AC, Craig PS et al. (1992) Coproantigen detection for immunodiagnosis of echinococcosis and taeniasis in dogs and humans. Parasitology 104:347–355

Akerstrom B, Bjorck L (1986) A physicochemical study of Protein G, a protein with unique immunoglobulin G-binding properties. J Biol Chem 261:10240–10247

Avrameas S (1969) Coupling enzymes to proteins using glutaraldehyde. Use of conjugates for detection of antigens and antibodies. Immunochemistry 6:43–52

Boulanger D, Reid GDF, Sturrock RF et al. (1991) Immunization of mice and baboons with the recombinant Sm28GST affects both worm viability and fecundity after experimental infection with *Schistosoma mansoni*. Parasite Immunol 13:473–479

Butterworth AE (1984) Cell-mediated killing of helminths. Adv Parasitol 23:143–235

Capron M, Capron A (1986) Rats, mice and men – models for immune effector mechanisms against schistosomiasis. Parasitol Today 2:69–75

Craig PS (1986) Detection of specific circulating antigen, immune complexes and antibodies in human hydatidosis from Turkana (Kenya) and Great Britain, by enzyme-immunoassay. Parasite Immunol 8:171–188

Garside P, Behnke JM (1989) *Ancylostoma celanicum* in the hamster – observations on the host-parasite relationship. Parasitology 98:283–289

Gillespie SH, Hawkey PM (1994) Medical parasitology – a practical approach. IRL Press, Oxford University Press, Oxford

Gottstein B (1985) Purification and characterization of a specific antigen from *Echinococcus multilocularis*. Parasite Immunol 7:201–212

Hagan P, Blumenthal UJ, Dunne D, Simpson AJG, Wilkins HA (1991) IgE, IgG4 and resistance to reinfection with *Schistosoma haematobium*. Nature (Lond) 349:243–246

Harlow E, Lane D (1988) Antibodies. A laboratory manual. Cold Spring Harbour Laboratory, Cold Spring Harbour

Johnson KS, Harrison GBL, Lightowiers HW et al. (1989) Vaccination against ovine cycticercosis using a defined recombinant antigen. Nature (Lond) 338:585–587

Khalife J, Dunne DW, Capron M et al. (1986) Immunity in human schistosomiasis: regulation of protective immune mechanisms by IgM blocking antibodies. J Immunol 142:4422–4426

Maddison SE, Slemenda SB, Schantz PM et al. (1989) A specific diagnostic antigen of *Echinococcus granulosus* with an apparent molecular weight of 8 kDa. Am J Trop Med Hyg 40:377–383

Maizels RM, Blaxter ML, Robertson BD, Selkirk ME (1991) Parasite antigens-parasite genes. A laboratory manual for molecular parasitology. Cambridge University Press, Cambridge

McManus DP, Barrett NJ (1985) Isolation, fractionation and partial characterization of the tegumental surface from protoscoleces of the hydatid organism, *Echinococcus granulosus*. Parasitology 90:111–129

Nakane PK, Kawaoi A (1974) Peroxidase labelled antibody: a new method of conjugation. J Histochem Cytochem 22:1084–1091

Pritchard DI, McKean PG, Rogan MT (1988) Cuticle preparations from *Necator americanus* and their immunogenicity in the infected host. Mol Biochem Parasitol 28:275–284

Pritchard DI, Leggett KV, Rogan MT, McKean PG, Brown A (1991) *Necator americanus* secretory acetylcholinesterase and its purification from excretory-secretory products by affinity chromatography. Parasite Immunol 13:187–199

Pritchard DI, Quinnell RJ, Walsh EA (1995) Immunity in humans to *Necator americanus*: IgE, parasite weight and fecundity. Parasite Immunol 17:71–75

Richman DD, Cleveland PH, Oxman MN, Johnson KM (1982) The binding of staphylococcal Protein A by the sera of different animal species. J Immunol 128:2300–2305

Rogan MT, Craig PS (1992) In vitro killing of taeniid oncospheres, mediated by human sera from hydatid endemic areas. Acta Trop 51:291–296

Rogan MT, Morris DM, Pritchard DI, Perkins AC (1990) *Echinococcus granulosus*: the potential use of specific radiolabelled antibodies in diagnosis by immunoscintigraphy. Clin Exp Immunol 80:225–231

Sanchez F, Garcia J, March F et al. (1991) Immunological localization of major hydatid fluid antigens in protoscoleces and cysts of *Echinococcus granulosus* from human origin. Parasite Immunol 13:583–589

Southern EM (1975) Detection of specific sequences among DNA fragments separated by gel electrophoresis. J Mol Biol 98:503–517

Stewart GS, Elson L, Araugo E et al. (1995) Isotype specific characterization of antibody responses to *Onchocerca volvulus* in putatively immune individuals. Parasite Immunol 17:371–380

Towbin H, Staehelin T, Gordon J (1979) Electrophoretic transfer of proteins from polyacrylamide gels to nitrocellulose sheets: procedure and some applications. Proc Natl Acad Sci USA 76:4350–4354

Tsang VCW, Brand JA, Boyer AE (1989) An enzyme-linked immunoelectro-transfer blot assay and glycoprotein antigens for diagnosing human cysticercosis (*Taenia solium*). J Infect Dis 159:50–59

Venkatesan P, Wakelin D (1993) ELISAs for parasitologists: or lies, damned lies and ELISAs. Parasitol Today 9:228–232

Voller A, Bidwell DE, Bartlett A (1976) Microplate enzyme immunoassays for the immunodiagnosis of virus infections. Manual of Clinical Immunology, pp 506–512

Subject Index

Acanthamoeba spp.   262
acetone powder   4
acetylcholine esterase   85, 91–99
  purification   95–97
  quantitation   92
actin   287, 288, 289
ADP, AMP, ATP   22, 24, 27
affinity chromatography   326, 328,
    352–357
agglutination
  direct   305, 313
  non-specific   307, 312, 313, 314
  parasite   307, 313, 314, 317
  scoring system   307, 313, 314, 316
  trypanosomatids   304, 305, 308
alkaline phosphatase   337, 342, 352
amidomethylcoumarins   149–151, 162, 163
amidotrifluoromethylcoumarins   149
amino-9-ethylcarbazole   153
amino acryl nitrophenol   153
ammonium sulphate   326
Ancylostoma spp.   97, 100, 107, 121, 145,
    176
antibody   188–191
  heterologous   190
  penetration   193
  production, polyclonal, monoclonal
      323–336
  region specific   190
antigen   324, 329, 330, 338, 340, 343
  carbohydrate   339
  circulating   321
  copro-   321
  protein   339
  storage   322
  surface   254, 270, 292
Artioposthia triangulata   202, 206
Ascaris suum   196, 201, 203, 204, 219

autofluorescence   252
azocasein   143

Bdelloura candida   202
bile   297
biotinylated antibody   342, 352
blood coagulation   100–110
  clotting pathways   102

Caenorhabditis elegans   92, 134, 202, 204
calcium ions   293, 294
carbohydrates   229, 235, 241
carbonate bicarbonate buffer   340, 346
catalase   10, 116
cell
  cloning   334–335
  freezing   335–336
  fusion   332–333
  suspension   234
ceramide   280, 281, 282, 290, 291
chemical leaching   235
chemotherapy   263
cholesterol   280, 281
confocal scanning laser microscopy (CSLM)
    197
critical point drying   244
cryofixation   244
cuticle   293
cyanogen bromide (CnBr)   353, 354
cytoskeleton   287

detergent   322
  solubilisation   33, 35
Diazo-2AM   293
Diclodophora merlangi   196, 198
diethylaminoethyl (DEAE) cellulose   327,
    328
dimethylpimelimidate   356, 357
Diphyllobothrium dendriticum   195

Dirofilaria immitis   115, 176
DNA
  purification 67
  quantification 69
  probes
    32P labelling 73
    biotin labelling 75
domains 286
Dugesia tigrina 202, 203

Echinococcus spp. 177, 263, 347, 356
edrophonium sepharose 96
  affinity chromatography 96
electroelution 55, 298
electrophoresis 32–65, 95, 97, 98, 112,
    124, 146–147, 161–165
ELISA 325, 333, 336, 338–347
  controls 344
  standards 341–343
Ellman assay 94
embedding media 237
  acrylic resins 256, 260, 261
    Lowicryl 258–261
    LR-Gold 258, 260, 261
    LR-White 256–258, 260, 261
  epoxy resins 237
    Araldite 237
    Epon 237
    Spurr 237
endocytic vesicles 291
Entamoeba histolytica 139, 153, 169,
    172, 177
enzyme
  activity 7
  assay 6, 9–15
  coupled assay 10–11
  factors affecting activity 11–13
  glycosidases 311
  glycoside hydrolases 312
  isotopic assays 11
  preparation 1–6
  spectrophotometric assays 7, 10
  stains 242
  statistics 3–15
epicuticle 275, 292
extinction coefficient 10
extracellular volume 17–18
extraction
  buffers 3–5, 204
  peptide 201, 205

Fasciola hepatica 87, 170, 174, 194, 195
fatty acid binding protein 297–298
fatty acids 280, 281
feeder cells 331–332
FITC detection
  flow cytometry 304, 308
  UV fluorescence 308–311
fixation 229, 233, 234, 235, 241, 247, 253,
    255, 257, 262, 308, 310, 312, 314–316
  immersion 230
  microwave 230
  perfusion 230
FRAP 284–287
freeze clamping 16
freeze fracture 228
freeze substitution 259–260

gels
  gradient 47
  IEF 57, 63
  NEPHAGE 58, 59
  preparitive 55
  SDS-PAGE 36
  tricine 48
Globodera spp. 145, 174, 282
glutathione-S-transferase 110–115
  cloning 114–115
  purification 113
glycoprotein analysis 51
  endoglycosidase 51
Golgi 290

haemoglobin 143
Haemonchus contortus 87, 203
Haplometra spp. 170
Heligmosomoides polygyrus 87, 96, 97,
    114, 118, 121
hide powder azure 144
horsraddish peroxidase 337, 342, 352
HPLC 26–27, 147
Hymenolepis spp. 5
hypoxanthine guanine phosphoribosyl
  transferase (HGPRT) 328

IgG separation 326–328
in situ hybridization 241
in vitro culture 84–90
immunization 324, 325, 329–331
  foot-pad 330–331
immunocytochemistry 188, 189

cryostat section   192, 194
    fixation   192
    SEM   246
    wholemount   192, 195
immunogold labelling   198
    double   200
    post-embedding   198, 251, 256–262
    preembedding   251, 254–256
inhibitors
    use in determining pathways   27–29
insect tissues   305, 308, 309, 310, 312,
        314
ion-exchange chromatography   327

keyhole limpet haemocyanin   118–120

Leishmania spp.   137, 147, 154, 167, 169,
        171, 173, 178, 262, 263
leishmanolysin   154, 163, 173
low temperature embedding   258
luminometry   24

metabolic pathways
    criteria for demonstration   1
    use of inhibitors   27–29
metabolites
    enzyme cycling assays   23
    extraction   16–17
    glycolytic   21–22
    measurement   15–27
    tricarboxylic acid cycle   21–22
metal stains   228, 239
    lead citrate   240
    uranyl acetate   236, 240, 253
micelles   301
microscopy   197, 198, 227–268, 304, 311,
        316, 317
microwave fixation   230
mitochondria   291
molar extinction   299
Monezia expansa   196, 200, 202, 204, 206,
        207, 210, 211, 215
Moniliformis moniliformis   282
monolayers   230, 234, 237
muscle cells
    dispersed   220–222
    patch clamp recording   223
    visualisation of   222
muscle physiology   215, 217
    action of transmitters   219

microelectrodes   219, 220
    muscle strips   216
    photo-optic devices   218
myeloma cells   328, 332–333

NAD/NADH   17
NADP/NADPH   17, 23
Necator americanus   83, 88, 91, 96, 97,
        100, 106, 107, 108, 109, 125
neuraminidase   311, 312
neuropeptides   188, 201
    characterisation   200
    FMRF amide-related peptides
        (FaRPs)   188, 190, 198, 202, 204,
        206, 211
    GNFFRF amide   198, 200, 204, 206, 211
    neuropeptide F (NPF)   188, 190, 196,
        200, 204, 214, 215
    purification of   200, 207, 210, 212
neutrophil inhibition   120–123
Nippostrongylus brasiliensis   88, 92, 96, 97
NITR-5   270
nitroanilides   151–152
NMR   24–26, 27
northern blotting   73

oelic acid   300
Onchocerca spp.   115, 177

Panagrellus redivivus   202, 204, 206
parasite extracts   322
PCR   78
    RT-PCR   79
Peldri II   244
peptide mapping   53
    V8 protease   54
periodic acid-silver technique   241
pH
    effect on enzyme activity   12
phalloidin   287–290
phosphate buffered saline   326
plasmepsin   154
Plasmodium spp.   262, 291
platelet aggregation   83, 106–110
Ponceau S   349
Porocephalus spp.   176
praziquantel   286
Protein A   246, 247, 248, 326, 327, 328,
        347, 350, 356
    affinity of   327

Protein A (*Contd.*)
  binding capacity   328
protein determination   7–9
Protein G   247, 248
proteinase   53, 99, 133–186, 213–215, 311, 316, 322–347
    inhibitor   34, 134, 137, 156, 157, 158–161, 322
prothrombin   105

radioimmunoassay (RIA)   201, 205, 206
raman fluorescence   301
recalcification time   103–104
Russel viper venom   106
ruthenium red   235, 241

Schistosoma spp.   88, 89, 116, 137, 145, 152, 153, 174, 196, 216, 220, 221, 262, 275, 276, 277, 279, 280, 282, 283, 286, 289, 320
shadowing techniques   228
silver enhancement   246, 247, 250
Southern blotting   72
Spirometra spp.   174
sputter coating   244
stypven   105
subcellular fractionation   4–6
superoxide dismutase   6, 10, 115–117

Taenia spp.   115, 347, 348
tannic acid   235
tegument   281
temperature
    effect on enzyme activity   12
thin layer chromatography   148, 282
thromboplastin   104
tissue blocks   230, 236, 237
Toluidine blue   239
Trichinella spiralis   89, 115, 282, 283, 287, 292, 293
trichloroacetic acid   142
Trichomonas spp.   167
Trichostrongylus colubriformis   89, 91
Trichuris spp.   96, 176
Trypanosoma spp.   162, 168, 169, 171, 172, 173, 262
trypsin   286, 311, 312

ultramicrotome   238

variable surface glycoproteins   311

wet cleaving technique   229
whole mount preparations   192, 195, 228, 251, 252–253
Wuchereria bancrofti   90

Springer-Verlag
and the Environment

We at Springer-Verlag firmly believe that an international science publisher has a special obligation to the environment, and our corporate policies consistently reflect this conviction.

We also expect our business partners – paper mills, printers, packaging manufacturers, etc. – to commit themselves to using environmentally friendly materials and production processes.

The paper in this book is made from low- or no-chlorine pulp and is acid free, in conformance with international standards for paper permanency.

MIX
Papier aus verantwortungsvollen Quellen
Paper from responsible sources
FSC® C105338

If you have any concerns about our products,
you can contact us on
ProductSafety@springernature.com

In case Publisher is established outside the EU,
the EU authorized representative is:
Springer Nature Customer Service Center GmbH
Europaplatz 3, 69115 Heidelberg, Germany

Printed by Libri Plureos GmbH
in Hamburg, Germany